Communications and Control Engineering Series

Editors: A. Fettweis · J. L. Massey · M. Thoma

Scientific Fundamentals of Robotics 6

M. Vukobratović
V. Potkonjak

Applied Dynamics and CAD of Manipulation Robots

With 187 Figures

Springer-Verlag
Berlin Heidelberg New York Tokyo

D. Sc., Ph. D. MIOMIR VUKOBRATOVIĆ, corr. member of
Serbian Academy of Sciences and Arts
Institute »Mihailo Pupin«, Belgrade,
Volgina 15, POB 15, Yugoslavia

Ph. D. VELJKO POTKONJAK, assistent professor,
Electrical Engineering Faculty of Belgrade University
Belgrade, Yugoslavia

Library of Congress Cataloging in Publication Data.
Vukobratović, Miomir:
Applied dynamic and CAD of manipulation robots
M. Vukobratović; V. Potkonjak.
Berlin; Heidelberg, New York; Tokyo: Springer, 1985.
(Scientific fundamentals of robotics; 6)
(Communications and control enginieering series)

ISBN 3-540-13074-8 Springer-Verlag Berlin Heidelberg New York Tokyo
ISBN 0-387-13074-8 Springer-Verlag New York Heidelberg Berlin Tokyo

Printed in Germany

Offsetprinting: Mercedes-Druck, Berlin
Bookbinding: Lüderitz & Bauer, Berlin
2061/3020 5 4 3 2 1 0

Preface

This book is a logical continuation of Volume 1 of the series entitled "Scientific Fundamentals of Robotics" which presents all of the basic methods for computerized construction of dynamics of manipulation robots as well as the essential concepts of computer-aided design of their mechanics. Vol. 1 of the Series also contains the main practical results from the elastodynamics of manipulation robots, having in mind a need for forming a computer procedure which allows efficient checks of elastic deformations of a manipulator tip or some other of its characteristic points.

Wishing to add a highly applications-oriented dimension to the dynamic aspect of studies of manipulation robots, the authors have made a kind of a topic-based selection by leaving unconsidered some aspects of studies of robots, such as elasticity, and discussing others, more important in their opinion, to such an extent as suffices to make them practically applicable.

The authors have decided not to treat in detail the problem of flexible manipulation robots for two reasons. The first results from the attitude that the permissible (desired) robot elasticity may, satisfactorily well, be tested using the method described in Vol. 1 of the Series. Although somewhat conservative, this method very efficiently, within the dynamic analysis of robots, provides the information about whether or not the constraint imposed on the elastic displacement of, e.g., manipulator tip is satisfied. The second reason stems from a serious authors' dilemma: is there any practical sense in "deliberately" designing elastic robots, knowing that this would require more precise methods for calculating the elastodynamic effects of robot structure to be able to synthesize the control for such elastic manipulation robots?

In contrast to this, omitted, aspect of robot dynamics studies, in the present Volume the authors have made great efforts towards solving the other, more important, problems to such an extent as allows their practical application in industrial robotics.

We will mention some of them only.

First of all, the book presents the complete results of dynamic analysis starting from a wide variety of tasks defined in the external coordinates which are the most suitable for robot technology users.

Then, keeping in mind just the actual requirements of industrial robotics, the authors have tried to approach the delicate problem of dynamics of closed manipulation chains from their functional aspect. However, this has resulted in the lack of generality in considering complex kinematic chains of closed configurations.

Finally, this book presents an attempt to promote computer-aided design of robot mechanisms and an adequate choice of their actuators based on the complete models of motion dynamics and the proposed power consumption criterion.

This procedure has resulted in a software package for dimensioning and selecting the actuators of manipulation robots. The package incorporates their complete dynamic analysis.

The book is primarily intended for researchess engaged in developing methods for c.a.d. robot design as well as for robot designers who could, using sufficiently exact information about manipulation robot dynamics, implement systems with no exess conservativism in calculating the structure and selecting the actuators of robots. Of course, the book is also intended for students at postgraduate courses. We hope that it will also be usable at undergraduate courses in practical dynamics of robots and their computer-aided design.

Chapter 1 presents a general discussion on robotic systems. The basic definitions and classifications of robotic structures are discussed. The Chapter also considers the general problems of computer-aided design of machines.

Chapter 2 deals with the dynamic analysis of manipulation robot structures. We derive the general scheme of the algorithm for dynamic analysis and then elaborate each block of this scheme. The method for the formation of dynamic model is discussed first. Jacobian form allows us to formulate a convenient methodology for manipulation task definition. There are several dynamic characteristics which can be calculated and tested: driving forces and torques, power requirements, torque-speed diagrams, stresses in segments, elastic deformations, etc. This Chapter also discusses the synthesis of nominal dynamics. The synthesis uses the complete dynamic model including actuator devices. Four examples are presented.

Chapter 3 deals with the closed chain robot structures. The first case of closed chains is the mechanism with a kinematic parallelogram. Another case of closed chain in robotics appears when an open chain robot performs a manipulation task which imposes some constraints upon robot gripper motion. For instance, we can consider a task of writing on a surface, or a peg-in-hole task. General methodology for dynamic modelling of such closed chains is derived and the most interesting cases elaborated in detail. Impact problems are also discussed. The Chapter contains two examples.

Chapter 4. is devoted to computer-aided design of manipulation robots. The optimality criteria and the constraints are defined. The algorithms and characteristics explained in Ch. 2 and Ch. 3 allow us to define the optimization procedures. For the choice of optimal dimensions, energy consumption criterion is suggested. The systematic procedure for the choice of optimal actuators and reducers is based on power-dynamic power characteristic. At the end, a program package for CAD of manipulation robots is considered.

The authors express their gratitude to Mr D. Katić, who contributed in the procedure for optimal choice of hidraulic actuators.

The authors also are grateful to Miss G. Aleksić for her help in preparing english version of this book. Our thanks also go to Miss V.Ćosić for her careful and excellent typing of the whole text.

Belgrade, Yugoslavia, December 1984 The Authors

Contents

Chapter 1:
General About Manipulation Robots and Computer-Aided Design of Machines

1.1. General About Manipulation Robots

1.1.1. Introduction

A manipulation robot, regardless of its function performs the task of positioning and orientation of an object, called the terminal device. This can be a gripper, holding some working object, a spray-painting gun, or some other type of tool. The position of the terminal device, in the general case arbitrary and variable, is defined by the position of one of its points and by its orientation with respect to that point.

A manipulation robot defines its general structure by:

- a mechanical structure, which supports the terminal device which should be positioned,

- actuators, which act on the mentioned structure with the task to change its position and, thereby, the position of the terminal device,

- various sensors, which are necessary for the control, among which the proprioceptive sensors can be distinguished, which provide for the knowledge of the mechanical state of the manipulation robot, and the exteroceptive sensors, which indicate the state of the environment around the manipulation robot,

- a control system, commanding the actuators of the manipulation robot, starting from the definition of the movement to be performed, obtained from the decision system and the information from the proprioceptive sensors,

- a decision system which ensures functioning of logic and elaborates the motion of the manipulation robot, starting from the definition of the task to be performed, which has been feed - in by the operator by means of the communication system. The basic functions of the decision system are the interpretation and understanding of operator's commands, solving the imposed tasks, generation of plans and messages to the operator, as well as the preparation of the data base necessary for this system,

- a communication system which processes the messages from the decision system to the operator via an alphanumeric or graphic displey, or a system of voice analysis or synthesis, or a "teaching box", etc.

A manipulation robot can adapt itself to every change of the environment acting upon it. Most today's industrial manipulation robots do not possess this capability of adaptation, because of the absence of the decision system.

1.1.2. Definition of position of an object in space

In order to define the position of an object in space, it is possible to fix the positions of three non-colinear points of that object, or nine parameters, which are not independent, because the coordinates of these points are connected by three relations which express the invariance of the distances between these three points.

Consequently, definition of the position of a free object requires in the general case, the knowledge of six independent parameters:

- three independent parameters which define the position of one point of the object (in Cartesian, cylindric, spheric, or other coordinates),

- three independent parameters which define the orientation of the object with respect to the stated point (Euler's angles, Euler's parameters, or similar).

By definition, this object possesses six degrees of freedom.

1.1.3. Structure of an industrial manipulation robot

The mechanical structure of an industrial manipulation robot is an assembly of theoretically rigid bodies linked by means of connections.

class of task	nr. of links	nr. of d.o.f.	TYPES OF PAIRS: I			II			III		
I	1	5	nr, of movem.	rot.	lin.						
			allowed	3	2						
			restricted	0	1						
			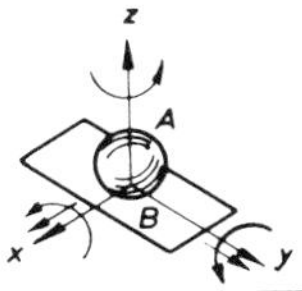								
II	2	4	nr. of movem.	rot.	lin.	nr. of movem.	rot	lin.			
			allowed	3	1	allowed	2	2			
			restricted	0	2	restricted	1	1			
			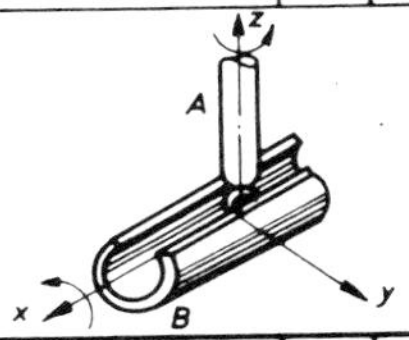			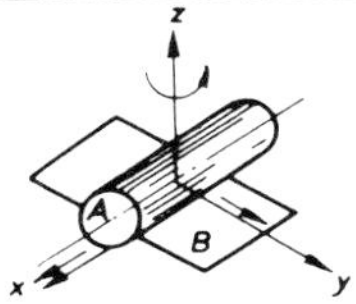					
III	3	3	nr. of movem.	rot.	lin.	nr. of movem.	rot.	lin.	nr. of movem.	rot.	lin.
			allowed	3	0	allowed	2	1	allowed	1	2
			restricted	0	3	restricted	1	2	restricted	2	1
			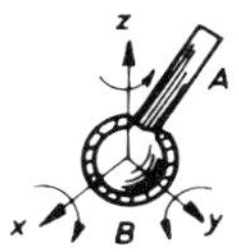			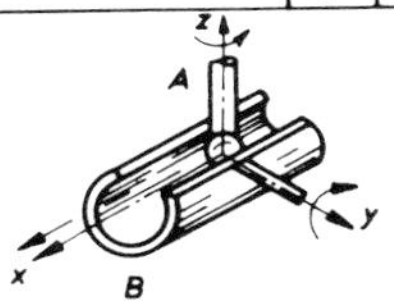			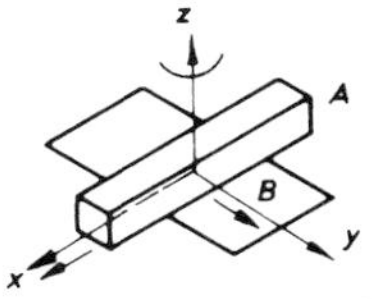		
IV	4	2	nr. of movem.	rot.	lin.	nr. of movem.	rot.	lin.			
			allowed	2	0	allowed	1	1			
			restricted	1	3	restricted	2	2			
			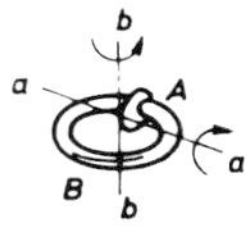			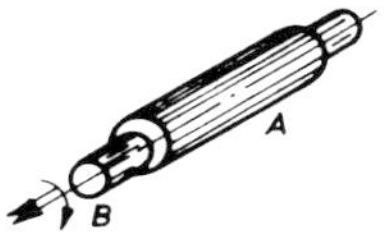					
V	5	1	nr. of movem.	rot.	lin.	nr. of movem.	rot.	lin.	nr. of movem.		
			allowed	1	0	allowed	0	1	allowed	0	0
			restricted	2	3	restricted	3	2	restricted	2	2
						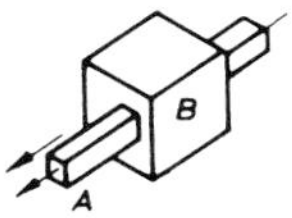			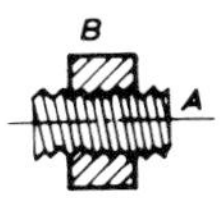		

Fig. 1.1. Table of kinematic pairs

A connection exists between two bodies and permits relative motion between them (Fig. 1.1)

The number of degrees of freedom of some connection equals the minimal number of parameters which determine the position of some body B_2 in its relative motion to a body B_1 (B_1 - body No 1, B_2 - body No 2).

Class of a connection is the supplement of the number of degrees of freedom to six.

All sorts of connections are not used in industrial manipulation robots, keeping in mind the actual state of the development of actuators which perform the relative motion of B_1 and B_2.

Thus it is sufficient to consider furtheron:

- the connection enabling relative rotational motion,
- the connection enabling relative translatory motion.

These connections are of class 5.

In order to standardize the terms with those in international literature, we adopt (R) for rotational connections and (P) for prismatic - translational ones (the term "linear" is used too).

1.1.4. Disposition of segments and their connections

Bodies B_i and connections C_i, which form a manipulation robot can be arranged starting from a reference body B_o, fixed or mobile, in the following manner:

- arrangement in the form of a simple chain (Fig. 1.2),
- arrangement in the form of a branched chain (Fig. 1.3),
- arrangement in the form of a complex chain (Fig. 1.4).

The last type of structure is characterized by the presence of "mechanical loops", which means various paths enabling the terminal device to be reached, starting from one body, over a series of bodies and connections.

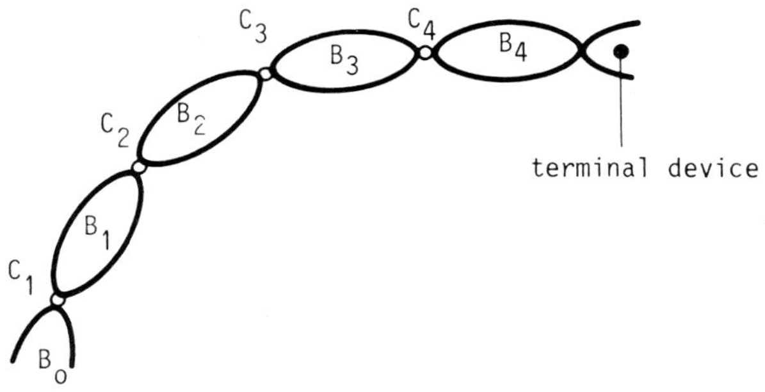

Fig. 1.2. Manipulation robot with a simple chain structure

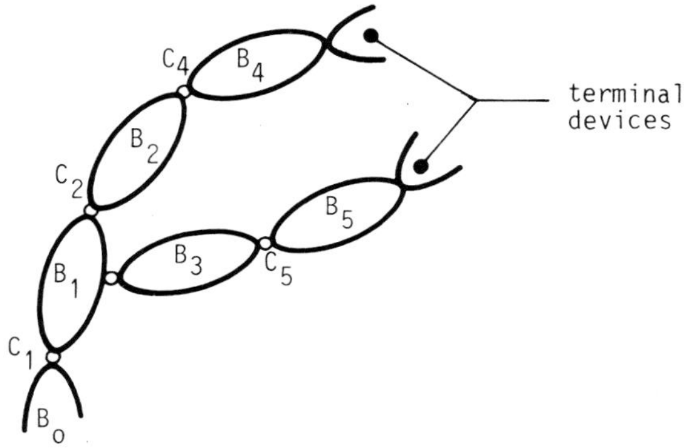

Fig. 1.3. Manipulation robot with a branched chain structure

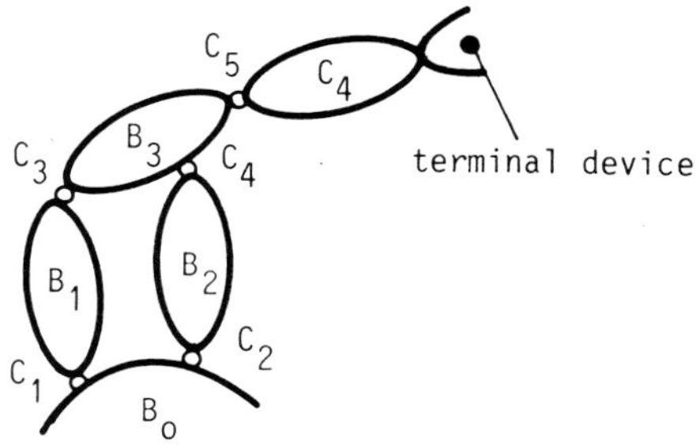

Fig. 1.4. Manipulation robot with a complex chain structure

In the following text of this chapter, only manipulation robots with simple chain structure will be considered.

1.1.5. Simple chain structure types

Keeping in mind what was said before, the manipulation robots have a simple structure, characterized by a set of n letters R or P which define the sequence and type of connections, progressing from the reference body towards the terminal device.

Table T.1 presents a few examples of structure.

Manipulation robot	Structure
Cincinati Milacron T3	R R R R R R
Unimate 4000	R P P R R
Unimate PUMA 560	R R R R R R
ASEA IRb 60	R R R R R R
IMP*) UMS-3	R P P R R R

Table T.1. Some examples of structures of industrial manipulation robots

1.1.6. Mobility index and degrees of freedom of a manipulation robot

Except for the mobile robots, the reference body B_o is fixed and called the base, while bodies B_1 to B_n are moving. The positions of these n bodies are defined by means of 6n parameters.

A class m of connection between two bodies defines m of these parameters. Therefrom if follows that, if the manipulation robot possesses N_m connections of class m

$$M = 6n - \sum_{m=1}^{5} m N_m$$

which represents the number of variable parameters, determining its configuration.

M is the Mobility Index of a manipulation robot.

Although the manipulation robots with complex chain configurations have some advantages as compared with those with a simple or branched chain structure, e.g. greater rigidity, almost all industrial manipulation robots have a simple chain structure. Under these conditions, if all connections are rotational and/or prismatic, the general relation for the mobility index of these manipulation robots reduces to:

$$M = n$$

*) IMP - Mihailo Pupin Institute, Belgrade, Yugoslavia.

Thus, as six parameters are necessary for the definition of the position of the terminal device in space, a manipulation robot of a simple chain structure possesses at least six rotational and/or prismatic connections.

The number of degrees of freedom of a manipulation robot (considered as a mechanical system) equals the number of independent parameters defining the position of the chain. The number of d.o.f. of the terminal device is less then or equal to M.

Example 1

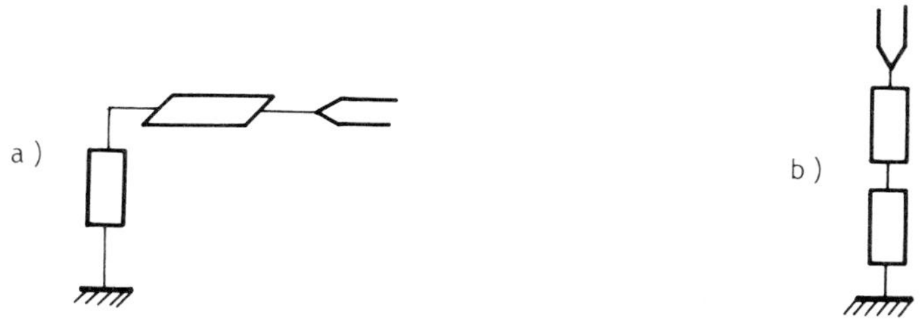

Fig. 1.5. Mobility index and degrees of freedom for two manipulation robots with two prismatic connections

The manipulation robot consists of 2 moving bodies and 2 prismatic connections:

with perpendicular axes	with parallel axes
d.o.f. of the terminal device	
d.o.f. = 2	d.o.f. = 1
because the terminal device is	
parallel to a plane	parallel to a straight line
keeping a constant orientation	

Example 2

a)

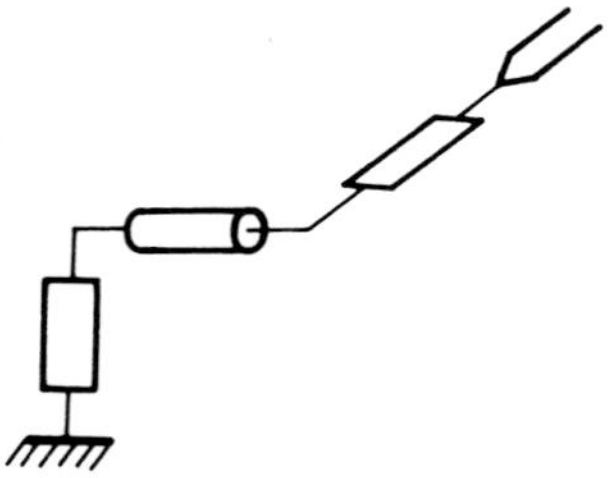

b)

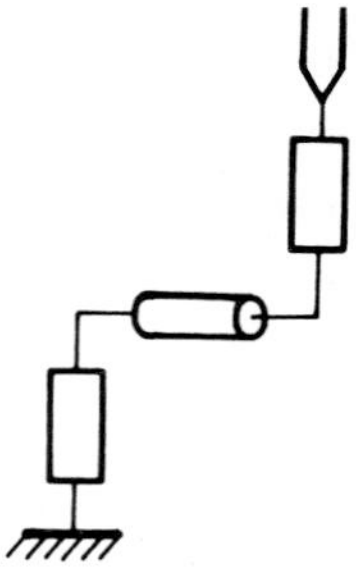

Fig. 1.6. Mobility index M and d.o.f. for a manipulation robot with three connections

The manipulation robot consists of 3 moving bodies, 1 rotational and 2 translational connections:

with perpendicular axes	with parallel axes
mobility index M=3	
degrees of freedom of the terminal device	
d.o.f. = 2	d.o.f. = 1
because the terminal device moves locally	
parallel to a plane with some arbitrary orientation	parallel to a plane with some imposed orientation

1.1.7. Redundancy and singularity

If number of d.o.f. of the terminal device equals M for any position of manipulation robot it is non-redundant.

This is the case of the manipulation robot from example 1(a).

In the opposite case, i.e. when d.o.f. <M, two situations are possible:

- this inequality is satisfied for all positions which the manipulation robot can take, and it is now defined as redundant (case of example example 1(b)),

- this inequality is satisfied for some configurations which the manipulation robot can take, and it is now defined as locally redundant.

The corresponding configurations are called singular. This is the case of the manipulation robots from example 2(b).

1.1.8. Degrees of freedom of a task: (d.o.f.t)

Number of degrees of freedom of a task (d.o.f.t.) equals the number of independent parameters enabling all the desired positions to be attained by the terminal device.

1.1.9. Compatibility

When d.o.f. = d.o.f.t., which is a necessary, but not sufficient condition so that a manipulation robot can perform some given task, the notion of compatibility expresses the possibility to find the configuration of the manipulation robot which enables it to attain the desired state of the terminal device.

1.1.10. Decoupling the orientation from the position of the terminal device

Consider the manipulation robot, the structure of which is presented in Fig. 1.7.

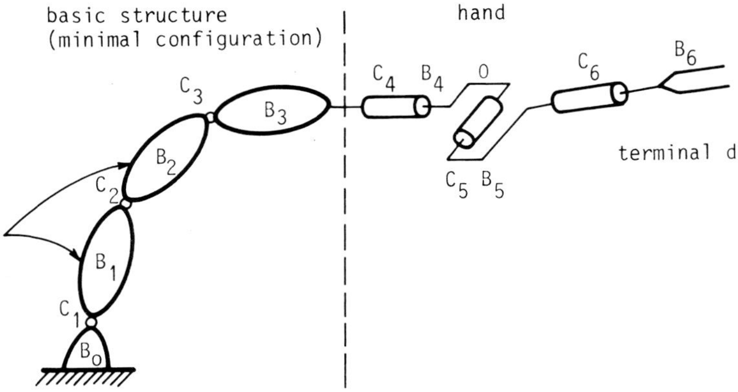

Fig. 1.7. Structure of a decoupled manipulation robot

It possesses six connections of class 5 and enables the following general states:

- the first three connections are arbitrary,
- the last three connections are rotational with axes intersecting at point 0 and mutually perpendicular,
- position of point 0 depends only on the position of bodies B_1, B_2 and B_3,
- position of bodies B_4, B_5 and B_6 determines the orientation of the terminal device (last segment - gripper in Chapter 2) with respect to point 0.

This decoupling is intended to reduce the problem of determining the six parameters of the manipulation robot configuration to two independent problems, each having three parameters only.

However, point 0 is rarely the point of the terminal device to be positioned. Each change of its position causes thus a change of its orientation and vice versa. Thus, the decoupling has not been solved quite generally.

1.1.11. Different minimal configurations

They are derived from different reference systems of a point in space.

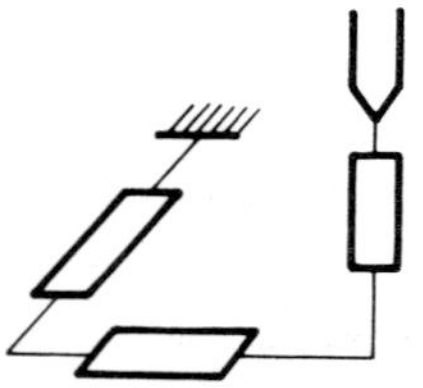

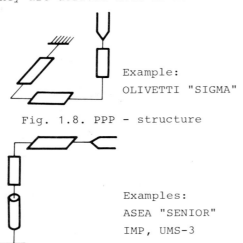

Example:
OLIVETTI "SIGMA"

Fig. 1.8. PPP - structure

This type of structure, used with approx. 14% of industrial manipulation robots, is well suited for determining the point 0 in Cartesian coordinates.

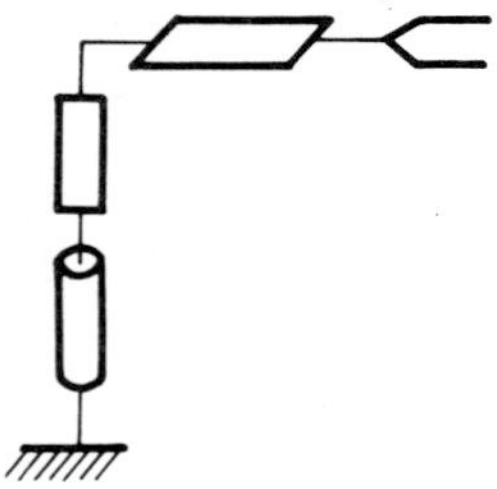

Examples:
ASEA "SENIOR"
IMP, UMS-3

Fig. 1.9. RPP - structure

This type of structure, used with approx. 45% of industrial manipulation robots, is well suited for determining the point 0 in cylindrical coordinates.

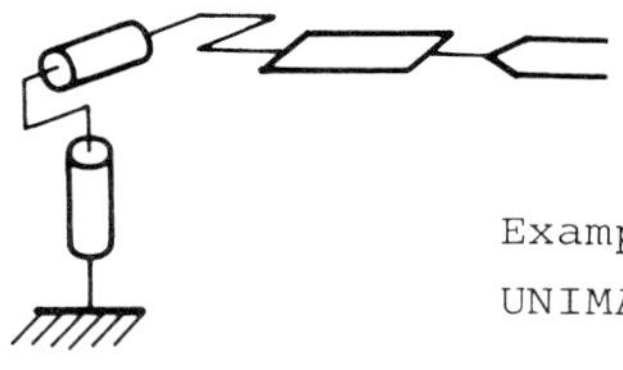

Example:
UNIMATE 4000

This type of structure, used with approx. 15% of industrial manipulation robots is well suited for determining the point 0 in spherical coordinates.

Fig. 1.10. RRP - structure

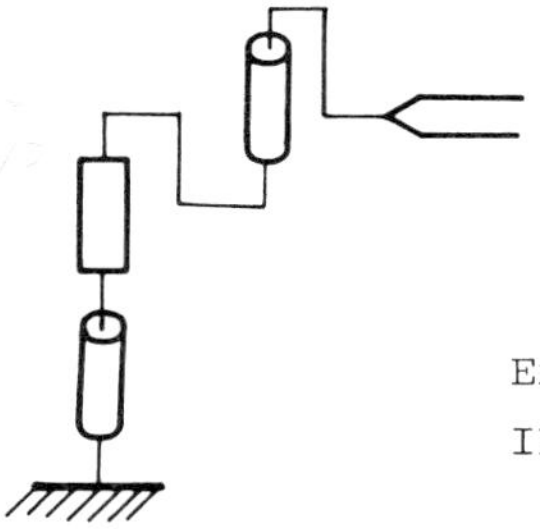

Example:
IBM 7535

This type of structure, used with approx. 1% of industrial manipulation robots, is well suited for determining the point 0 in thorical coordinates.

Fig. 1.11. RPR - structure

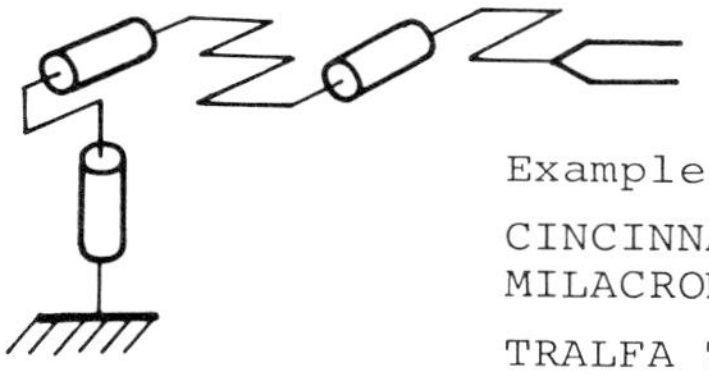

Examples:
CINCINNATI - MILACRON T3
TRALFA TR3000

This type of structure, used with approx. 25% of industrial manipulation robots, is called arthropoide structure.

Fig. 1.12. RRR - structure

1.1.12. Workspace

Workspace of a manipulation robot is the space, physically swept by a point of the terminal device during motion of the robot configuration.

Let us note that the choice of the point mentioned above, is arbitrary. Some producers assume it as the center 0 of the wrist and thus obtain the nominal workspace. Others take the point on the tip of the terminal device.

In addition, the orientation of the terminal device does not appear in this workspace. But, although this is only an approximative characteristic of the manipulation robot performance, it anyhow permits comparison of different basic structures (minimal robotic configurations).

1.1.13. Comparison of the workspaces of different minimal configurations

To this end we will make the following assumptions:

- permissible rotation of each rotational connection is 360°,
- translation of each prismatic connection equals L,
- the "principal" (greatest) dimension of each segment of the manipulation robot equals L.

The workspaces illustrated suppose some arbitrary wrist, the center of which, 0, is the reference point.

(a) PPP structure

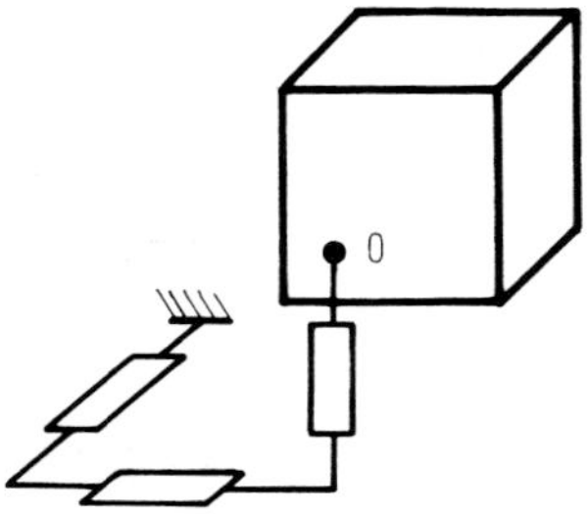

Workspace is a cube of side L

$$V = L^3$$

Fig. 1.13. Workspace of PPP - structure

(b) RPP - structure (or PRP)

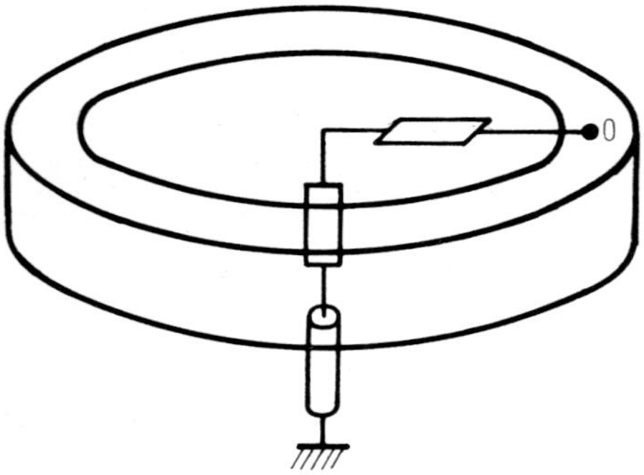

Workspace is a thorus of square cross-section of mean radius L and external radius 2L

$$V = 3\pi L^3 \approx 9L^3$$

Fig. 1.14. Workspace of RPP - structure

(c) RRP - structure

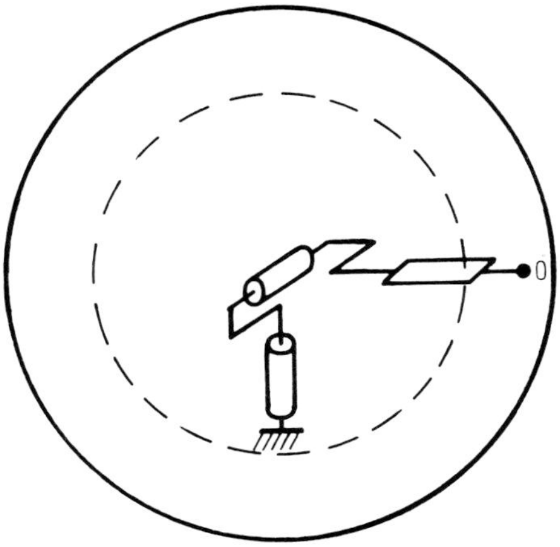

Workspace is a hollow sphere of interior radius L and external radius 2L

$$V = \frac{28}{3}\pi L^3 \approx 29L^3$$

Fig. 1.15. Workspace of RRP - structure

(d) RPR - structure

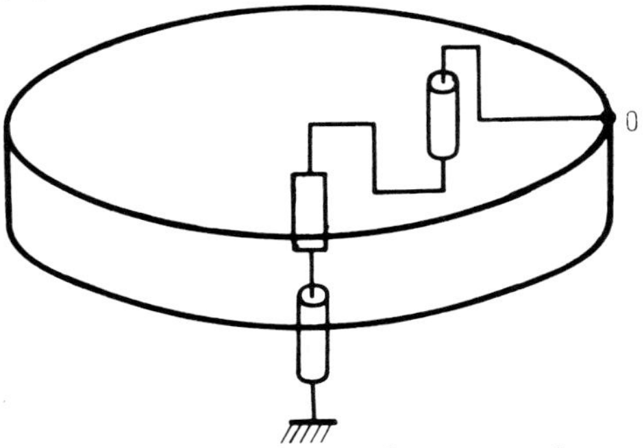

Workspace is a cylinder of radius 2L and height L

$$V = 4\pi L^3 \approx 13L^3$$

Fig. 1.16. Workspace of RPR - structure

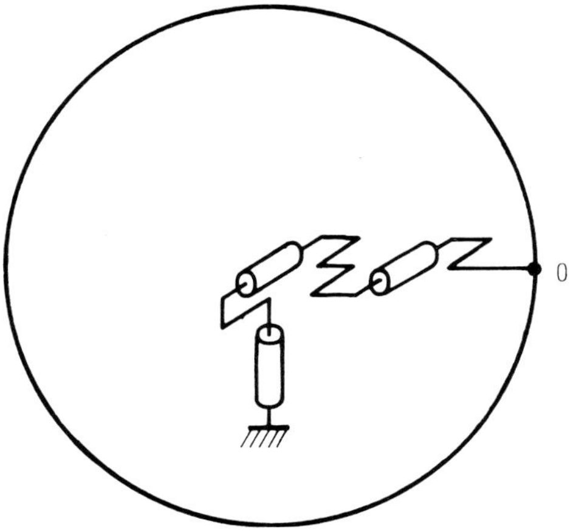

Workspace is a sphere of radius 2L

$$V = \frac{32}{3}\pi L^3 \approx 34L^3$$

Fig. 1.17. Workspace of RRR - structure

This comparison demonstrates the evident superiority of the RRP and RRR structures, possessing a workspace approximatelly 30 times greater than the PPP - structure. The RPP and RPR - structures, with workspaces approximatelly 10 times greater, thus offer medium sized workspaces.

1.2. General Remarks on Up-To-Date Methods for Design of Machines

1.2.1. Task specification and starting data

We witness today a considerable growth of the complexity of problems that should be solved in the process of designing constructions and machines. Realization of machines of a qualitatively new level assumes the use of important achievements of fundamental sciences, design and technology, protection of servicing personnel against vibration, noise and injuries. The task of improving the quality of machines should be solved in the stage of design when it is necessary and possible to thoroughly consider a construction, i.e., take into account a large number of, frequently, contradictory requirements, such as a minimum mass providing a sufficient rigidity and a sufficient reliability, high-speed operation with a lower dynamic load, a low price and a long lifecycle, etc. In the design of machines and mechanisms it is necessary to achieve an optimal choice of their parameters (structural, kinematic, dynamic, exploitative) which are best suited to the imposed, often numerous, requirements. In the present design practice this task is solved by studying a number of alternative variants and performing appropriate calculations. Elaboration of a large number of alternative variants, based on conventional approaches, connot, in principle, give to a designer the idea about machine capabilities. To illustrate this, let us say that, e.g., if ten different values are assigned to each of ten parameters, it follows that 10^{10} variants-tasks should be solved, and this exceeds even the performances of contemporary computers.

The costs associated with solving such tasks by classical methods constantly increase, and the negative effects of accepting nonoptimal solutions become more and more serious. A further aggravating circumstance is the fact that these are multicriteria tasks with conflicting objective functions. It is therefore difficult for a designer to select a compromise solution applying the classical methods for finding the extremum, and most new optimization procedures are predetermined

Many efforts remain to be made towards creating a dialogue methodology and accepting design solutions in man-machine systems intended for design automation. The need for dialogue in integral design automation systems results from a poor formalism of many design operations and from the impossibility to give their unifold computerized models. A solution to dialogue-related problems is attempted by creating appropriate logical procedures over the models describing the task.

The integral design automation systems should incorporate all design stages. They should include computers and automation devices much more powerful than are those used today. Their implementation requires the solution complex problems regarding the theory of optimal, logical solutions, heuristic procedures, etc. as well as a further development of mathematical models of all parts of design process to provide a basis for creating specialized mathematical support. Such systems should be modular, with high-level hierarchical design languages and data design languages and data banks, with applications-oriented programs capable of synthesizing the design process.

It is thus possible to create an integral design automation system as a man-machine system which incorporates:

- technical means providing input, processing and output of information about project results in an acceptable form and at a particular time;
- data bases and data base management systems permitting a direct access for all participants in design;
- a system for accepting design solutions and a library of design solutions modules for all levels of design process;
- operating systems to control the processes of improving and renewing object modules and data bases in design process.*)

---

*) A part of the text presented in this chapter is based on the text given in Ch. 2 of the book: Modèles des robots manipulateurs-application à leur commande, by B.Gorla and M.Renaud, Cepadues-editions, Toulouse, 1984.

Chapter 2: Dynamic Analysis of Manipulator Motion

2.1. Introduction

General ideas about the dynamic analysis of manipulator motion are explained in this chapter. We derive a computer algorithm for such an analysis. First, it is necessary to explain what it is meant by the notion of dynamic analysis. By this notion we mean the calculation of all dynamic characteristics which can be useful for a designer in the process of manipulator design or synthesis of its control algorithm. Let us be more precise. We prescribe some manipulation task, start the algorithm, and obtain, as output, dynamic variables such as driving forces and torques in manipulator joints, some other characteristics of actuators, stresses in manipulator segments, the value of elastic deformations, etc. All these pieces of information help the designers and the engineers in the application of the device. Such a calculation procedure is sometimes called the simulation algorithm. The notion of the simulation of dynamics usually involves only the calculation of motion for the prescribed driving forces and torques, i.e., the integration of the differential equations of motion. But, it we use the term simulation somewhat more liberally, then the calculation of all dynamic variables for the prescribed motion can also be called simulation.

Formerly, all this work had to be done by hand. The dynamic equations were written and solved by hand. Such a task is extremely difficult for any real robot configuration. We mention only two reasons for that. The problem to be kept in mind is the always present risk of making numerous errors when handling such a complex task. Further, even if we perform the whole task without mistakes, the equations obtained are so complex that such dynamic model can hardly be used. In addition, being hardly adaptable to different configurations, such a model does not make much sense.

This analytic approach to dynamic analysis was the reason that theoretic investigations in robotics could not help the engineers in practice. So, the theory and the practice in robotics have grown almost independently.

The appearance of computer-aided methods*) for setting and solving dynamic models**) of active mechanisms in robotics [1 - 22] represents significant progress in the theory of robots and manipulators. All the work about the formation of dynamic model is transferred to the computer. The methods are general enough to cover all relevant robot configurations, i.e., the configuration represents input data. In addition, the model obtained has a suitable and compact matrix form. Such methods bring the theory closer to applications. The first two computer-aided methods (c.-a. methods in the subsequent text) have been developed indepedently in Yugoslavia and USSR [1 - 6] and later on several methods have appeared [7 - 22]. These c.-a. methods play now the central role in the algorithms for the dynamic analysis.

In this chapter we describe a computer algorithm for the complete dynamic analysis of manipulator motion. We first explain the general block - scheme of the algorithm and then elaborate each block. The c.-a. method for the formation of dynamic model is presented first (paragraph 2.3).

A special problem in this dynamic analysis is the definition of the manipulation task in a form which is suitable for the computer and simple for the user, but still general enough. So, special attention is paid to this question (paragraph 2.4). We derive the so-called general algorithm for dynamic analysis which has the manipulation task input in the form of acceleration and angular acceleration of manipulator gripper. The adjective "general" indicates that this algorithm serves as a basis for the development of a new algorithm which covers some typical classes of manipulation tasks, and which is, thus, more convenient for users. In 2.5. we discuss dynamic characteristics which are of interest for the application and which have thus to be computed. A problem of special interest is the calculation of elastic deformations and Para. 2.5.4. is devoted to the solution of this problem. Since the algorithm is dedicated to computer-aided design of manipulators it contains some testings of relevant dynamic characteristics (2.6). In 2.7 we demonstrate the practical realization of the algorithm. Finally, the algorithm is expanded by introducing the mathematical models of actua-

---

*) Computer oriented methods, automatic methods.

**) Mathematical model, dynamic equations, differential equations of motion.

tor units in order to synthesize the programmed control of functional movements of the manipulator in nominal dynamics (2.9).

2.2. Block-Scheme of the Algorithm for Dynamic Analysis

In the previous paragraph it has been said that the computer-aided (c.-a.) method plays the central role in the algorithm for dynamic analysis of manipulator motion. The method is general and operates for arbitrary manipulator configuration. The manipulator is considered as a mechanical system with n degrees of freedom (d.o.f. in the sequel). The c.-a. method sets its dynamic model in the matrix form

$$W\ddot{q} = P + U \tag{2.2.1a}$$

or

$$P = W\ddot{q} - U \tag{2.2.1b}$$

where q is an n-dimensional vector of generalized coordinates ($\ddot{q}$ is the generalized accelerations vector) and P is an n-dimensional vector of driving forces and torques in manipulator joints. W is an $n\times n$ inertial matrix depending on the generalized coordinates q. U is an $n\times 1$ matrix depending on generalized coordinates q and generalized velocities $\dot{q}$.

Sometimes a more general form of the model is used [22]:

$$P = g(q, \dot{q}, \ddot{q}) \tag{2.2.2}$$

This form is used because there are some c.-a. methods which do not form the model (2.2.1) but directly compute the drives P. The function g in (2.2.2) represents such an algorithm. The c.-a. method used in this book forms the model (2.2.1) and computes the matrices W, U.

We now explain the procedure for solving the dynamic model (2.2.1) for a finite time interval T. From the standpoint of dynamics two problems may be considered. One problem is to compute the motion of the dynamic system if the driving forces and torques (the drives in the sequel) are prescribed and the other is to compute the driving forces and torques which produce the prescribed motion. Here, we are interested in the latter problem.

Let us discretize the time interval T into small subintervals Δt_k by introducing a sequence of time instants $t_o, t_1, \ldots, t_{end}$. It may be said that a c.-a. method operates and forms the model (2.2.1) for one time instant in which the state $(q, \dot{q})$ is known. This means that it computes numerically the system matrices W and U starting with the known state $(q, \dot{q})$ in that time instant. Hence, the system (2.2.1) turns to a system of algebraic equations with respect to generalized accelerations $\ddot{q}$.

If we wish to compute the motion for known drives P, then we solve these algebraic equations (2.2.1) to obtain $\ddot{q}$. Further, with known q, $\dot{q}$ and $\ddot{q}$, integration is performed over the subinterval Δt to obtain the state in the next time instant. Then the whole procedure is repeated for that new time instant.

We are here interested in the other problem i.e. the computation of drives P(t) for the prescribed motion. In principle, it is sufficient to carry out the model forming c.-a. procedure at each time instant. In this way one obtains the function P(t) in a discrete form. If one wishes to use this idea, then one has to know the state q, $\dot{q}$ in order to compute the system matrices W, U and $\ddot{q}$ also in order to compute the drives P from (2.2.1). These variables have to be known at each time instant. But we notice that only $\ddot{q}$ has to be the input value because q and $\dot{q}$ at some time instant can always be obtained by integration from the previous one. If the subinterval Δt is small enough, then $\ddot{q}$ may be considered as constant and the integration has the simple form

$$\begin{aligned} q(t+\Delta t) &= \frac{1}{2}\ddot{q}(t)\Delta t^2 + \dot{q}(t)\Delta t + q(t) \\ \dot{q}(t+\Delta t) &= \ddot{q}(t)\Delta t + \dot{q}(t) \end{aligned} \qquad (2.2.3)$$

So, the input is the value of $\ddot{q}$ at each time instant and only the initial state $q(t_o)$, $\dot{q}(t_o)$. If the manipulation task is defined in generalized coordinates then the problem of computation of P(t) is solved (Fig. 2.1).

The generalized coordinates for mechanical systems which appear in manipulator theory usually represent the internal coordinates defining the relative position of two segments connected by a joint. In real manipulation tasks it is not easy to determine these internal coordinates. We usually want the manipulator to perform some task which is defined in terms of some external coordinates (e.g.: the law of manipulator tip motion and gripper orientation in the working space). Then, the algorithm has to perform the transfer from the external coordinates to internal ones. In order to prescribe the task correctly, it is necessary that the external coordinates (vector X) and the internal coordinates (vector q) depend, and depend only, on each other. If it is not the case, for instance if a surplus of the manipulator d.o.f. is present, then a special problem arises which will not be treated here. We only mention that the surplus of the d.o.f. can be compensated by introducing some additional requests into the manipulation task (e.g.: the request that the manipulator bypasses some obstacle in the working space). Here, we assume that the vectors X and q are of the same dimensions.

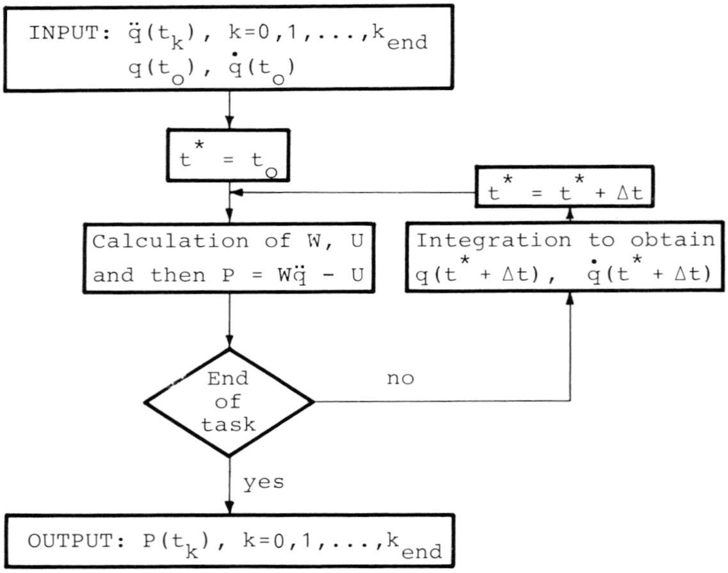

Fig. 2.1. The scheme for the calculation of drives

Let us designate by η the function which transforms the generalized coordinates q into external ones X,

$$X = \eta(q) \tag{2.2.3}$$

where q and X are n-dimensional vectors. The function η is one-place and can always be determined (not explicitly but as a computational algorithm). The problem lies in the difficulty of calculating q from such a system of equations (2.2.3) resulting from the impossibility to either express q explicitly or even approximate it numerically because of the complexity of the system which has to be solved.

But remember the following: to compute the drives P by using the model forming c.-a. method, it is necessary to know q, $\dot{q}$, $\ddot{q}$. However, only $\ddot{q}$ appears as input since q, $\dot{q}$ are obtained by integration from the previous time instant. So, in order to realize the coordinates transfer, it is enough to develop the procedure for calculating $\ddot{q}$ from the known state q, $\dot{q}$ and known external coordinates $X(t)$.

By double differentation,

$$\dot{X} = \frac{\partial \eta}{\partial q}\dot{q} \tag{2.2.4}$$

$$\ddot{X} = \frac{\partial \eta}{\partial q}\ddot{q} + \frac{\partial^2 \eta}{\partial q^2}\dot{q}^2 \tag{2.2.5}$$

Let us introduce the notation $X^a = \ddot{X}$, $J = \frac{\partial \eta}{\partial q}$, $A = \frac{\partial^2 \eta}{\partial q^2}\dot{q}^2$. Equation (2.2.5) then becomes

$$X^a = J\ddot{q} + A \tag{2.2.6}$$

J is an $n\times n$ Jacobian matrix depending on q, and A is an $n\times 1$ adjoint matrix depending on q, $\dot{q}$. Hence, it is necessary to prescribe X^a in a series of time instants, or calculate it starting from the manipulation task prescribed in some way. Then methods must be found for numerically calculating the matrices J, A for a known state q, $\dot{q}$. Then from system (2.2.6), with the assumption that the Jacobian matrix J is nonsingular, one obtains the required generalized accelerations $\ddot{q}$:

$$\ddot{q} = J^{-1}(X^a - A) \tag{2.2.7}$$

Now it is possible to use the c.-a. method to form and solve the dynamic model. After calculating q, $\dot{q}$, $\ddot{q}$ and P one can find other dynamic characteristics. The complete time-iterative procedure for dynamic analysis of manipulator motion can mainly be presented by the block-scheme in Fig. 2.2.

Each block from this scheme will be elaborated in the sequel.

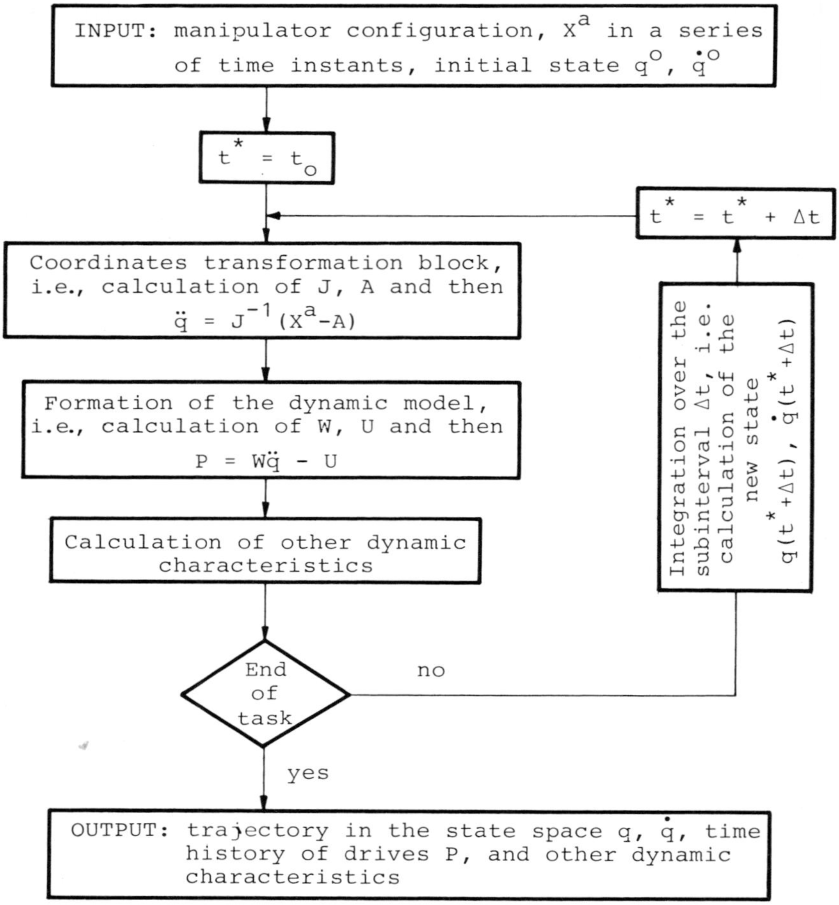

Fig. 2.2. Block scheme of the dynamic analysis algorithm

2.3. Computer-Aided Method for the Formation of Manipulator Dynamic Model

Some basic ideas of computer-aided formation of dynamic model have been discussed in 2.2. We now derive the complete c.-a. model forming procedure. The method is based on Appel's equations [16, 22] (see Appendix).

A mechanism with n degrees of freedom is considered so we introduce an n-dimensional vector of generalized coordinates to define its position:

$$q = [q_1 q_2 \cdots q_n]^T \tag{2.3.1}$$

The dynamics of such a mechanical system can be described by a system of differential equations in matrix form:

$$W\ddot{q} = P + U \tag{2.3.2}$$

where P is the n-dimensional vector of driving forces and torques in mechanism joints. As previously stated the n×n inertial matrix W depends on q and the n×1 matrix U depends on the state q, $\dot{q}$. By the term formation of dynamic model we mean the calculation of the matrices W and U. Hence, we now derive the procedure for numerical computation of W, U for the known mechanism state q, $\dot{q}$ and the configuration considered.

Mechanism configuration. This method considers the mechanism of open chain type consisting of n arbitrary rigid bodies (Fig. 2.3). Also, there is no branching in the mechanism.

The joints connecting the mechanism segments have one d.o.f. each. That d.o.f. may be rotational or linear. A rotational joint S_i (Fig. 2.4) allows a relative rotation around an axis determined by a unit vector $\vec{e}_i$. A linear joint S_j (Fig. 2.5) allows a relative translation along an axis determined by a unit vector $\vec{e}_j$.

C_i, C_j and quadrats are used to mark the centers of gravity (c.o.g. in the sequel) of each segment in the figures in the text. s_i, s_j are indicators determining the type of joints:

$$s_k = \begin{cases} 0, & \text{if } S_k \text{ is a rotational joint} \\ 1, & \text{if } S_k \text{ is a linear joint.} \end{cases} \tag{2.3.3}$$

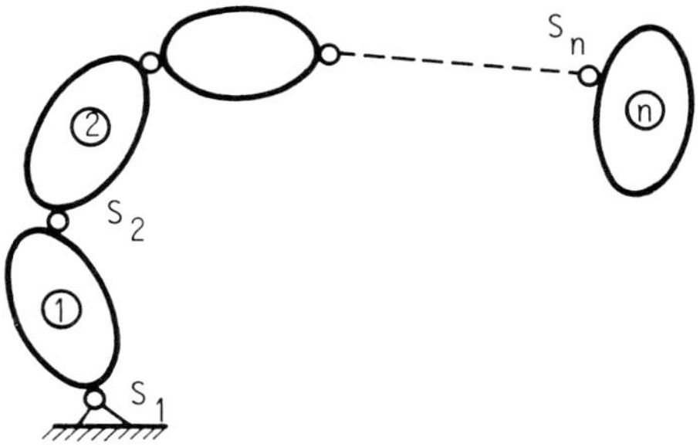

Fig. 2.3. Open kinematic chain without branching

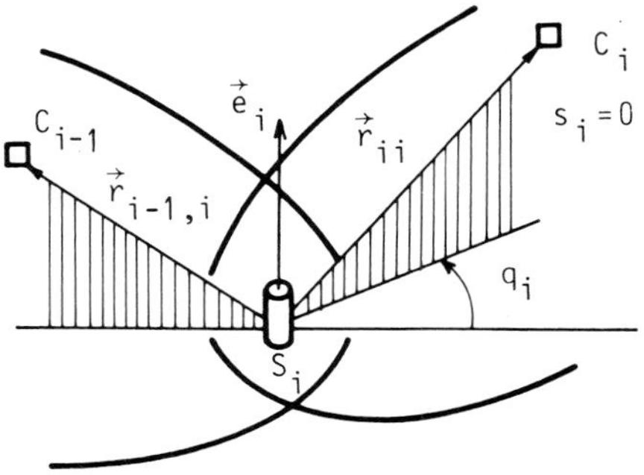

Fig. 2.4. Rotational joint

The prescription of the configuration will be discussed later.

Driving forces and torques. There is a driving motor in each mechanism joint. So, there is a driving torque $\vec{P}_i$ acting in the rotational joint S_i:

$$\vec{P}_i = P_i^M \vec{e}_i \tag{2.3.4a}$$

and a driving force $\vec{P}_j$ acting in the linear joint S_j:

$$\vec{P}_j = P_j^F \vec{e}_j \tag{2.3.4b}$$

Now, the vector of the drives is

$$P = [P_1 \cdots P_n]^T \tag{2.3.4c}$$

In the expression (2.3.4c) the upper indices M, F are omitted because the indicators s_k are used to determininе the type of each joint and each drive.

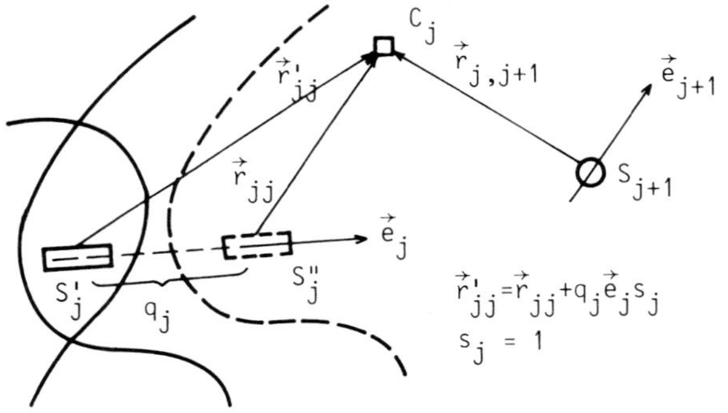

Fig. 2.5. Linear joint

Generalized coordinates. A set of n generalized coordinates $q_1,\ldots,q_n$ is used to determine the mechanism position. Each generalized coordinate corresponds to one d.o.f., i.e., to one joint.

For a rotational joint S_i the corresponding generalized coordinate q_i is defined as an angle of rotation in the joint around the axis $\vec{e}_i$. That angle may be regarded as the angle between the projections of the vectors $-\vec{r}_{i-1,i}$ and $\vec{r}_{ii}$ onto the plane perpendicular to the joint axis $\vec{e}_i$ (Fig. 2.4).

A particular case occurs when $\vec{r}_{ii} \| \vec{e}_i$ or $\vec{r}_{i-1,i} \| \vec{e}_i$. Then, the angle of rotation may not be considered in the previous way. If $\vec{r}_{i-1,i} \| \vec{e}_i$ we call it the "specificity" of the (i-1)-th segment on the upper end. Then we introduce a unit vector $\vec{r}^{\,*}_{i-1,i}$ perpendicular to $\vec{e}_i$ ($\vec{r}^{\,*}_{i-1,i} \perp \vec{e}_i$) (Fig. 2.6a). Further, the vector $\vec{r}^{\,*}_{i-1,i}$ is used instead of $\vec{r}_{i-1,i}$ for determining the generalized coordinate q_i. If $\vec{r}_{ii} \| \vec{e}_i$ we call it the "specificity" of the i-th segment on the lower end. Then we introduce a unit vector $\vec{r}^{\,*}_{ii}$ perpendicular to $\vec{e}_i$ ($\vec{r}^{\,*}_{ii} \perp \vec{e}_i$) (Fig. 2.6b) and use it instead of $\vec{r}_{ii}$.

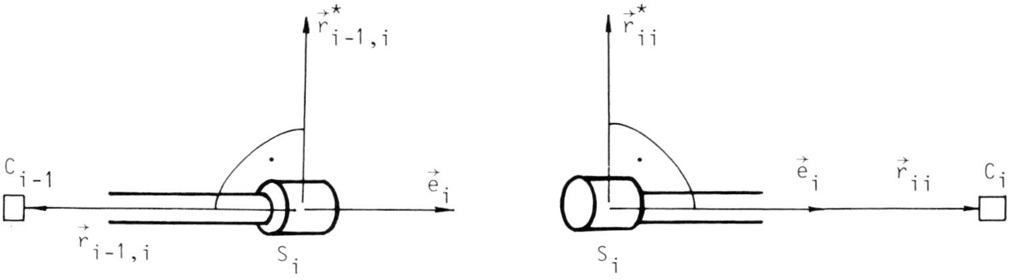

Fig. 2.6.(a) "Specificity" of (i-1)-th segment on the upper end (b) "Specificity" of i-th segment on the lower end

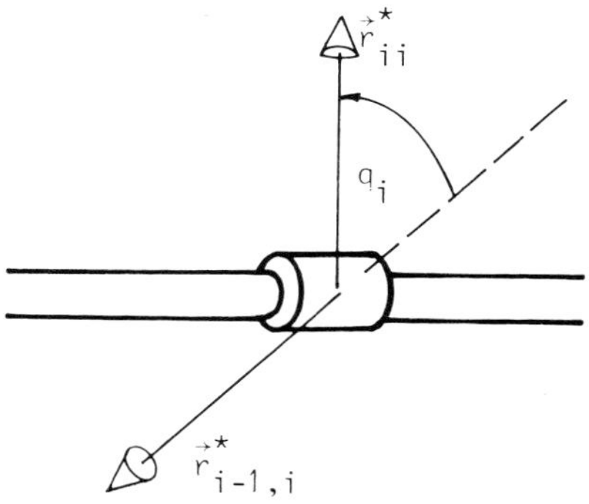

Fig. 2.7. Definition of the generalized coordinate in the case of "specificity"

The definition of generalized coordinate in the case of "specificity" is shown in Fig. 2.7.

The existence of "specificity" has to be given to the algorithm via special indicators.

If S_j is a linear joint, then the corresponding generalized coordinate

q_j is defined as a relative linear displacement along the joint axis $\vec{e}_j$ i.e. $q_j = |\overline{S_j' S_j''}|$ (Fig. 2.5).

<u>Coordinate systems and transition matrices</u>. Let us introduce coordinate systems. First, there is an external immobile Cartesian coordinate system Oxyz. A vertical z-axis is suitable but is not obligatory. Further, for each segment "i", a body-fixed (b.-f. in the sequel) Cartesian coordinate system $O_i x_i y_i z_i$ is defined. The origin O_i of such a system coincides with the c.o.g. C_i of the segment and the axes are oriented along the inertial principal axes. Such orientation of b.-f. axes is also suitable but is not obligatory.

We introduce the notation $\vec{a}_i$ to designate a vector corresponding to the i-th segment or i-th joint and which is expressed via three projections onto the axes of the external coordinate system. $\vec{\tilde{a}}_i$ designates the same vector but expressed by projections onto axes of the i-th body-fixed system. $\underset{\sim}{\vec{a}}_i$ denotes the same vector but expressed with respect to the (i-1)-th b.-f. system.

Now, the transition matrix from the i-th b.-f. system to the external system (matrix A_i) is defined as follows:

$$\vec{a}_i = A_i \vec{\tilde{a}}_i \tag{2.3.5a}$$

There is also a relative transition matrix $A_{i-1,i}$ from the i-th to the (i-1)-th b.-f. system:

$$\underset{\sim}{\vec{a}}_i = A_{i-1,i} \vec{\tilde{a}}_i \tag{2.3.5b}$$

or inversely:

$$\vec{\tilde{a}}_i = A_{i-1,i}^{-1} \underset{\sim}{\vec{a}}_i = A_{i,i-1} \underset{\sim}{\vec{a}}_i \tag{2.3.5c}$$

A few things should be pointed out. The vectors $\vec{r}_{ii}$ and $\vec{r}_{i,i+1}$ (Figs. 2.4, 2.5) which determine the position of joints relative to the segment c.o.g. are proper to each segment. So they are constant vectors if expressed by projections onto the axes of i-th b.-f. system. That is, $\vec{\tilde{r}}_{ii}$ and $\vec{\tilde{r}}_{i,i+1}$ are constants. Further, the axis $\vec{e}_i$ of the joint S_i has a constant position with respect to the i-th and (i-1)-th system. So the axis vector is constant if expressed via projections onto the i-th or (i-1)-th b.-f. system. That is $\vec{\tilde{e}}_i$ and $\underset{\sim}{\vec{e}}_i$ are constants. Such vectors $\vec{\tilde{r}}_{ii}$, $\vec{\tilde{r}}_{i,i+1}$, $\vec{\tilde{e}}_i$, $\underset{\sim}{\vec{e}}_i$ which determine the geometry of the i-th segment and the i-th joint have to be prescribed for each segment and joint.

The transition matrices are obtianed recursively. In each iteration a new segment is added to the chain and the corresponding relative transition matrix is computed. So, when adding the i-th segment we compute $A_{i-1,i}$ (and $A_{i,i-1} = A_{i-1,i}^{-1}$). If the absolute transition matrix is needed then

$$A_i = A_{i-1}A_{i-1,i} \tag{2.3.6}$$

We now show the calculation of relative matrix $A_{i-1,i}$.

Let us consider a joint S_i and suppose that it is rotational. The process of calculation of a transition matrix is divided into two phases: the phase of "assembling" a joint and the phase of "rotation". The following vectors should be computed:

$$\underset{\sim}{\vec{a}}_i = \frac{-\underset{\sim}{\vec{e}}_i \times (\vec{\tilde{r}}_{i-1,i} \times \underset{\sim}{\vec{e}}_i)}{|\underset{\sim}{\vec{e}}_i \times (\vec{\tilde{r}}_{i-1,i} \times \underset{\sim}{\vec{e}}_i)|} \quad ; \quad \vec{\tilde{a}}_i = \frac{\vec{\tilde{e}}_i \times (\vec{\tilde{r}}_{ii} \times \vec{\tilde{e}}_i)}{|\vec{\tilde{e}}_i \times (\vec{\tilde{r}}_{ii} \times \vec{\tilde{e}}_i)|} \tag{2.3.7}$$

(a) (b)

The vectors $\underset{\sim}{\vec{a}}_i$ and $\vec{\tilde{a}}_i$ are perpendicular to $\underset{\sim}{\vec{e}}_i$ and $\vec{\tilde{e}}_i$ respectively. $\vec{a}_i$ is the unit vector of the axis "a" and (2.3.7b) holds for $q_i=0$ (Fig. 2.8).

Introducing $\underset{\sim}{\vec{b}}_i = \underset{\sim}{\vec{e}}_i \times \underset{\sim}{\vec{a}}_i$, the three linearly independent vectors $\{\underset{\sim}{\vec{e}}_i, \underset{\sim}{\vec{a}}_i, \underset{\sim}{\vec{b}}_i\}$ are obtained (on the (i-1)-th segment). Introducing $\vec{\tilde{b}}_i = \vec{\tilde{e}}_i \times \vec{\tilde{a}}_i$, we also obtain the three linearly independent vectors $\{\vec{\tilde{e}}_i, \vec{\tilde{a}}_i, \vec{\tilde{b}}_i\}$ (on the i-th segment).

Let $A_{i-1,i}^{o}$ be the transition matrix corresponding to $q_i = 0$. Then (2.3.7b) holds and so:

$$\underset{\sim}{\vec{e}}_i = A_{i-1,i}^{o}\vec{\tilde{e}}_i, \quad \underset{\sim}{\vec{a}}_i = A_{i-1,i}^{o}\vec{\tilde{a}}_i, \quad \underset{\sim}{\vec{b}}_i = A_{i-1,i}^{o}\vec{\tilde{b}}_i \tag{2.3.8}$$

Now matrix notation will be introduced. Let e_i be the 3×1 matrix corresponding to the vector $\vec{e}_i$. Analogous matrix notation will be used for all other vectors in the text. In this way each 3×1 matrix a_i has its corresponding vector $\vec{a}_i$. We always use that notation which is more suitable for the expession considered. Now, expressions (2.3.8) can be written together in matrix form

$$[\underset{\sim}{e}_i \; \underset{\sim}{a}_i \; \underset{\sim}{b}_i] = A^o_{i-1,i}[\tilde{e}_i \; \tilde{a}_i \; \tilde{b}_i] \tag{2.3.9}$$

It follows that

$$A^o_{i-1,i} = [\underset{\sim}{e}_i \; \underset{\sim}{a}_i \; \underset{\sim}{b}_i][\tilde{e}_i \; \tilde{a}_i \; \tilde{b}_i]^{-1} \tag{2.3.10}$$

By computing the matrix $A^o_{i-1,i}$, the process of "assembling" the joint is completed.

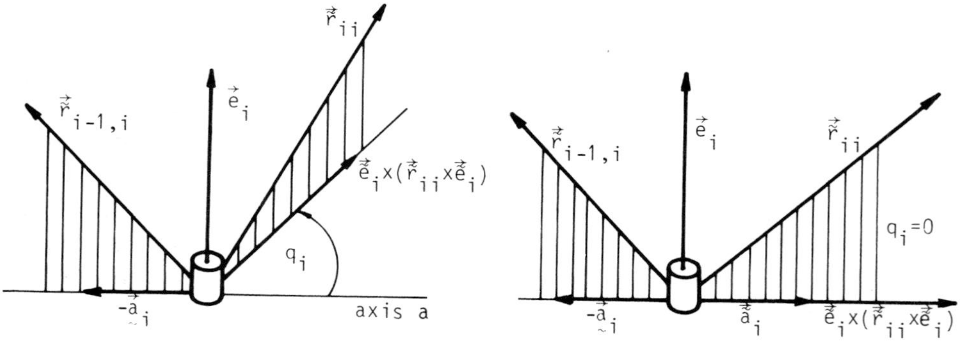

Fig. 2.8. Determination of the transition matrix

The columns of the matrix $A^o_{i-1,i}$ represent the unit vectors of the i-th b.-f. system expressed via projections onto the axes of the (i-1)--th system, but for $q_i=0$. So, rotation should be performed. The finite rotation formula is used to rotate each unit vector around the axis $\vec{e}_i$ for the angle q_i.

Let v_{i1}, v_{i2}, v_{i3} denote the columns of the matrix $A^o_{i-1,i}$, i.e., the unit vectors of the b.-f. system:

$$A^o_{i-1,i} = |v_{i1} \; v_{i2} \; v_{i3}| \tag{2.3.11}$$

Now, by rotation:

$$\vec{V}_{ij} = \vec{v}_{ij}\cos q_i + (1-\cos q_i)(\vec{\underset{\sim}{e}}_i \times \vec{v}_{ij}) \cdot \underset{\sim}{e}_i + \vec{\underset{\sim}{e}}_i \times \vec{v}_{ij}\sin q_i \quad j=1,2,3 \tag{2.3.12}$$

where V_{ij} is the j-th column (i.e. the unit vector) after rotation. So, the transition matrix corresponding to the angle q_i will be:

$$A_{i-1,i} = [V_{i1}\ V_{i2}\ V_{i3}] \tag{2.3.13}$$

In the case of "specificity" of the (i-1)-th segment on the upper end the vector $\vec{\tilde{r}}^*_{i-1,i}$ is used instead of $\vec{\tilde{r}}_{i-1,i}$. If there is a "specificity" of the i-th segment on the lower end, the vector $\vec{\tilde{r}}^*_{ii}$ is used instead of $\vec{\tilde{r}}_{ii}$.

If S_i is a linear joint, then there is only the "assembling" plase.

Thus, we have divided the computation of a transition matrix into two phases: "assembling" and "rotation". It should be emphasized that the "assembling" phase is performed only in the first iteration (for t_o) and in all later time-iterations only "rotation" is performed.

<u>Input data for the algorithm</u>. Let us return to the definition and prescription of the mechanism configuration. By the term "configuration" we mean the structure and the parameters. By the structure we mean the number of segments and the number and type of joints i.e. the kinematic scheme. By the parameters we mean all the information about the segments (dimensions, inertial properties etc.). So here is a list of the input data defining the configuration:

n = number of d.o.f. (=number of segments = number of joints),

s_i, $i=1,\ldots,n$, determine the types of joints,

$\vec{\tilde{e}}_i$, $\vec{e}_{\sim i}$, $i=1,\ldots,n$, determine the orientation of joint axis relative to the connected segments,

$\vec{\tilde{r}}_{ii}$, $\vec{\tilde{r}}_{i,i+1}$, $i=1,\ldots,n$ and $\vec{r}_{o1}$, determine the position of joints relative to segment c.o.g.,

$\vec{\tilde{r}}^*_{ii}$ or $\vec{\tilde{r}}^*_{i,i+1}$, in the case of "specificity" of i-th segment on the lower or upper end,

m_i, $\tilde{J}_i$, $i=1,\ldots,n$, m_i is the mass of i-th segment and $\tilde{J}_i$ is the inertia tensor of the same segment with respect to the corresponding b.-f. system.

The mechanism state q, $\dot{q}$ is obtained by integration from the previous time instant and $\ddot{q}$ is calculated from the external acceleration $\ddot{x}^a$.

<u>Kinematic relations</u>. Let $\vec{v}_i$ be the velocity and $\vec{w}_i$ the acceleration of

i-th segment c.o.g. Further, let $\vec{\omega}_i$ be the angular velocity and $\vec{\varepsilon}_i$ the angular acceleration of the same segment. In the kinematic chain considered the recursive expressions for velocities and accelerations may be found. It is more suitable to write these expressions in the b.-f. coordinate systems. Thus, the velocities are

$$\vec{\omega}_i = A_{i,i-1}\vec{\omega}_{i-1} + \dot{q}_i(1-s_i)\vec{e}_i, \tag{2.3.14}$$

$$\vec{v}_i = A_{i,i-1}(\vec{v}_{i-1}-\vec{\omega}_{i-1}\times\vec{r}_{i-1,i})+\vec{\omega}_i\times\vec{r}\,'_{ii}+\dot{q}_i s_i\vec{e}_i,$$

$$\vec{r}\,'_{ii} = \vec{r}_{ii} + q_i\vec{e}_i s_i, \tag{2.3.15}$$

and the accelerations are

$$\vec{\varepsilon}_i = A_{i,i-1}\vec{\varepsilon}_{i-1} + \left[\ddot{q}_i\vec{e}_i+\dot{q}_i(\vec{\omega}_i\times\vec{e}_i)\right](1-s_i), \tag{2.3.16}$$

$$\vec{w}_i = A_{i,i-1}\left[\vec{w}_{i-1}-\vec{\varepsilon}_{i-1}\times\vec{r}_{i-1,i}-\vec{\omega}_{i-1}\times(\vec{\omega}_{i-1}\times\vec{r}_{i-1,i})\right] +$$

$$+ \vec{\varepsilon}_i\times\vec{r}\,'_{ii}+\vec{\omega}_i\times(\vec{\omega}_i\times\vec{r}\,'_{ii}) + \left[\ddot{q}_i\vec{e}_i+2\dot{q}_i(\vec{\omega}_i\times\vec{e}_i)\right]s_i, \tag{2.3.17}$$

with initial conditions

$$\vec{v}_o = 0, \quad \vec{\omega}_o = 0, \quad \vec{w}_o = 0, \quad \vec{\varepsilon}_o = 0. \tag{2.3.18}$$

Dynamic approach. As a dynamic approach the Appel's equations are chosen. We explain the general ideas. For a dynamic system representing a set of material points we form the function

$$G = \sum_{\nu} \frac{1}{2} m_\nu \ddot{\vec{r}}_\nu^{\,2} \tag{2.3.19}$$

which is called the energy of acceleration. $\vec{r}_\nu$ is the position vector of the point ν and m_ν is its mass. $\sum_\nu$ designates the sum over all material points. If the generalized coordinates are used to describe such a system then the function G becomes a quadratic form with respect to generalized accelerations. If there are n generalized coordinates then the dynamics of the system may be described by a system of equations (Appel's equations):

$$\frac{\partial G}{\partial \ddot{q}_i} = Q_i, \qquad i=1,\ldots,n \tag{2.3.20}$$

Q_i is a generalized force corresponding to the coordinate q_i.

Let us now apply this approach to the manipulation mechanism.

<u>Forming the system of equations</u>. Now we consider the kinematic chain and form the Appel's equations. These equations (2.8.20), can be written in matrix form

$$\frac{\partial Q}{\partial \ddot{q}} = Q \tag{2.3.21}$$

where

$$Q = [Q_1 \; \cdots \; Q_n]^T \tag{2.3.22}$$

is the vector of generalized forces.

The function of the "acceleration energy" G for the whole chain under consideration represents the sum of the corresponding functions for each segment, i.e.,

$$G = \sum_{i=1}^{n} G_i, \tag{2.3.23}$$

and for the individual segment, the function G_i is given by means of a known expression [25]:

$$G_i = \frac{1}{2} m_i \vec{w}_i^2 + \frac{1}{2} \vec{\varepsilon}_i \tilde{J}_i \vec{\varepsilon}_i - \left[2(\tilde{J}_i \vec{\omega}_i) \times \vec{\omega}_i \right] \vec{\varepsilon}_i . \tag{2.3.24}$$

In order to form the iterative matrix computer algorithm (in each iteration the next segment is considered), let us define the following matrices:

Ω is a $3\times n$ matrix, the columns of which are the coefficients of the generalized accelerations in the expression for $\vec{w}_i$ (in the i-th iteration), Θ is a 3×1 matrix, containing the free term of the same expression.

Now the acceleration $\vec{w}_i$ can be written in the matrix form

$$\tilde{w}_i = \Omega \ddot{q} + \Theta . \tag{2.3.25}$$

Let us introduce notation for the columns:

$$\left. \begin{aligned} \Omega &= \left[\beta_1^i \; \cdots \; \beta_i^i \; 0 \; \cdots \; 0 \right] \\ \Theta &= [\delta^i] \end{aligned} \right\} \tag{2.3.26}$$

Further, let us define the following matrices:

Γ is a 3×n matrix, the columns of which are, in the i-th iteration, the coefficients of the generalized accelerations in the expression for $\vec{\varepsilon}_i$, Φ is a 3×1 matrix, containing the free term of the same expression.

Now, we write the angular acceleration in the form

$$\tilde{\varepsilon}_i = \Gamma\ddot{q} + \Phi \tag{2.3.27}$$

and let us write for the columns

$$\left.\begin{aligned} \Gamma &= \begin{bmatrix}\alpha_1^i & \cdots & \alpha_i^i & 0 & \cdots & 0\end{bmatrix}, \\ \Phi &= [\gamma^i]. \end{aligned}\right\} \tag{2.3.28}$$

In each iteration, a new segment is added to the shain, and modifications of and supplements to the matrices Ω, Θ, Γ, Φ are obtained so that they correspond to the new segment. The expressions for modifications and supplements are derived from the recursive expressions (2.3.14) - (2.3.17). Thus, for the i-th iteration,

$$\left.\begin{aligned} \vec{\alpha}_j^i &= A_{i,i-1}\vec{\alpha}_j^{i-1}; \qquad j=1,\ldots,i-1, \\ \vec{\alpha}_i^i &= \vec{e}_i(1-s_i), \end{aligned}\right\} \tag{2.3.29}$$

$$\vec{\gamma}_i = A_{i,i-1}\vec{\gamma}^{i-1} + \dot{q}_i(1-s_i)(\vec{\omega}_i\times\vec{e}_i), \tag{2.3.30}$$

$$\left.\begin{aligned} \vec{\beta}_j^i &= A_{i,i-1}\vec{\beta}_j^{i-1} - A_{i,i-1}(\vec{\alpha}_j^{i-1}\times\vec{r}_{i-1,i}) + \vec{\alpha}_j^i\times\vec{r}\,'_{ii}; \quad j=1,\ldots,i-1 \\ \vec{\beta}_i^i &= \vec{\alpha}_i^i\times\vec{r}\,'_{ii} + \vec{e}_i s_i, \end{aligned}\right\} \tag{2.3.31}$$

$$\begin{aligned} \vec{\delta}^i &= A_{i,i-1}\vec{\delta}^{i-1} - A_{i,i-1}(\vec{\gamma}^{i-1}\times\vec{r}_{i-1,i}) + \vec{\gamma}^i\times\vec{r}\,'_{ii} + \vec{h}, \\ \vec{h} &= -A_{i,i-1}\left[\vec{\omega}_{i-1}\times(\vec{\omega}_{i-1}\times\vec{r}_{i-1,i})\right] + \\ &\quad + \vec{\omega}_i\times(\vec{\omega}_i\times\vec{r}\,'_{ii}) + 2\dot{q}_i(\vec{\omega}_i\times\vec{e}_i)s_i. \end{aligned} \tag{2.3.32}$$

Substituting (2.3.25) and (2.3.27) into (2.3.24), the function G_i acquires the form

$$G_i = \frac{1}{2}\ddot{q}^T W_i \ddot{q} + V_i \ddot{q} + D_i, \qquad (2.3.33)$$

where

$$\begin{aligned} W_i &= m_i \Omega^T \Omega + \Gamma^T \tilde{J}_i \Gamma, \\ V_i &= m_i \Theta^T \Omega + \Phi^T \tilde{J}_i \Gamma - 2\tilde{u}^T \Gamma, \\ D_i &= \frac{1}{2} m_i \Theta^T \Theta + \frac{1}{2} \Phi^T \tilde{J}_i \Phi - 2\tilde{u}^T \Phi, \\ \vec{u} &= (\tilde{J}_i \vec{\omega}_i) \times \vec{\omega}_i . \end{aligned} \qquad (2.3.34)$$

Taking care about (2.3.23) and (2.3.33), it is clear that the "acceleration energy" function G for the whole chain will have the form

$$G = \frac{1}{2}\ddot{q}^T W \ddot{q} + V\ddot{q} + D. \qquad (2.3.35)$$

Let us substitute (2.3.35) into Appel's equations (2.3.21):

$$\frac{\partial G}{\partial \ddot{q}} = W\ddot{q} + V^T = Q. \qquad (2.3.36)$$

The calculation of the generalized forces will be treated in more detail somewhat later and it will be shown that they can be calculated in the form

$$Q = P + Y, \qquad (2.3.37)$$

where P is the vector of drives and Y can be calculated independently of P. Thus, (2.3.36) acquires the form

$$W\ddot{q} = P + Y - V^T \qquad (2.3.38)$$

and by introducing

$$U = Y - V^T. \qquad (2.3.39)$$

the form (2.3.2) is obtained, or

$$W\ddot{q} = P + U. \qquad (2.3.40)$$

Let us now consider the calculation of the matrices W and V. Due to (2.3.23) and (2.3.33) it is evident that

$$W = \sum_{i=1}^{n} W_i, \qquad V = \sum_{i=1}^{n} V_i, \tag{2.3.41}$$

which yields the possibility of recursively calculating the matrices W and V. In the i-th iteration this is

$$W^{(i)} = W^{(i-1)} + W_i,$$
$$V^{(i)} = V^{(i-1)} + V_i, \tag{2.3.42}$$

the upper index in parenthesis indicating that a matrix with no index in the algorithm is involved, i.e., its value is calculated iteratively and the upper index (i) designates the value of that matrix in the i-th iteration.

The problem of formulating the left-hand side of Appel's equations (2.3.21) has been solved. On the right-hand side of the equations the generalized forces appear.

Calculation of the generalized forces. Let us designate by Q_i the generalized force corresponding to the coordinate q_i. The expression for the generalized force will be derived by means of virtual displacement method.

Let us first consider a rotational joint S_i. Let us allow the coordinate q_i to have some virtual displacement δq_i (Fig. 2.9), kepping all the other coordinates constant. Let us now find the expression for the virtual work of active forces over that displacement. In this mechanical system the active forces are the drives and the gravity forces.

Over the virtual displacement δq_i the work is performed by the driving torque P_i^M and the gravity forces of segments i, i+1,...,n (Fig. 2.9).

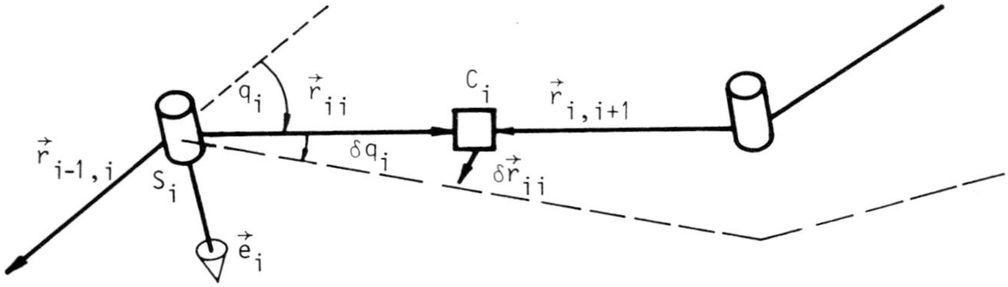

Fig. 2.9. Rotational virtual displacement

Work of the driving torque is

$$\delta A^{P} = P_i^M \cdot \delta q_i \qquad (2.3.43)$$

Work of the gravity forces is

$$\delta A^{g} = \delta A^{i} + \delta A^{i+1} + \dots + \delta A^{n} \qquad (2.3.44)$$

where i,i+1,... are not exponents but the upper indices. δA^k designates the work of the gravity force of segment k. For this segment, using notation from Fig. 2.10, the work done by the gravity force has the form

$$\delta A^{k} = m_k \vec{g} \cdot \delta \vec{r}_k^{(i)} \qquad (2.3.45)$$

where $\vec{g}$ is the gravitational acceleration $\vec{g} = \{0, 0, -9.81\}$.

The vector $\vec{r}_k^{(i)}$ is determined as

$$\vec{r}_k^{(i)} = \overrightarrow{S_i C_k} = \sum_{p=i}^{k-1} (\vec{r}\,'_{pp} - \vec{r}_{p,p+1}) + \vec{r}\,'_{kk} \qquad (2.3.46)$$

Then

$$\delta \vec{r}_k^{(i)} = \vec{e}_i \times \vec{r}_k^{(i)} \delta q_i \qquad (2.3.47)$$

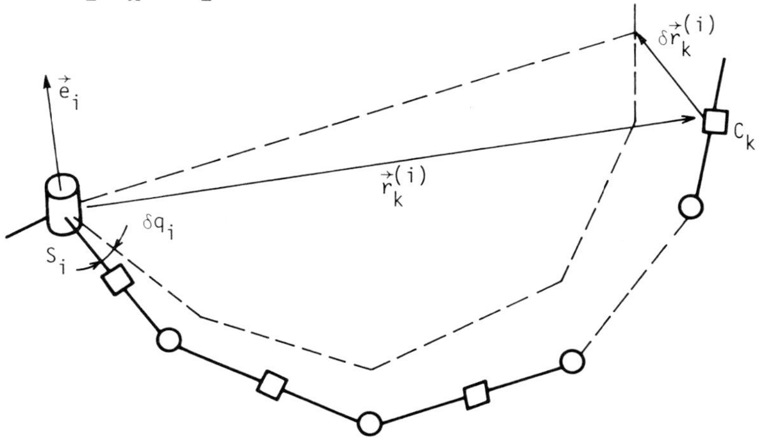

Fig. 2. 10. Determining the work of gravity forces

Substituting (2.3.47) into (2.3.45) we find

$$\delta A^{k} = m_k \left[\vec{g},\ \vec{e}_i,\ \vec{r}_k^{(i)} \right] \delta q_i \qquad (2.3.48)$$

where the expression in square brackets represents the vector box product.

Now the total work of gravity forces, according to (2.3.44), (2.3.48), is equal to

$$\delta A^g = \sum_{k=i}^{n} m_k \left[\vec{g}, \vec{e}_i, \vec{r}_k^{(i)} \right] \delta q_i \tag{2.3.49}$$

and the total work of active forces, according to (2.3.43), (2.3.49), is

$$\delta A = \delta A^P + \delta A^g = \left\{ P_i^M + \sum_{k=i}^{n} m_k \left[\vec{g}, \vec{e}_i, \vec{r}_k^{(i)} \right] \right\} \delta q_i \tag{2.3.50}$$

Now, by definition, the generalized force is

$$Q_i^{(rot)} = P_i^M + \sum_{k=i}^{n} m_k \left[\vec{g}, \vec{e}_i, \vec{r}_k^{(i)} \right] \tag{2.3.51}$$

where $\vec{r}_k^{(i)}$ is given by (2.3.46).

Let us now consider a linear joint S_j, and let us allow the linear coordinate q_j to have a virtual displacement δq_j, keeping all other coordinates constant (Fig. 2.11). Over that displacement the work will be done by the driving force P_j^F and the gravity forces of segments $j, j+1, \ldots, n$.

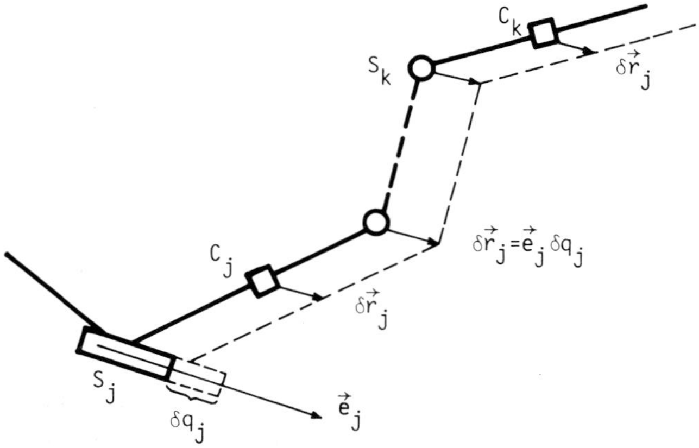

Fig. 2.11. Linear virtual displacement

For driving force

$$\delta A^P = P_j^F \cdot \delta q_j \tag{2.3.52}$$

Note that with the linear virtual displacement the whole part of chain (from S_j to the free end) is moved linearly by $\delta\vec{r}_j = \vec{e}_j\delta q_j$ (Fig. 2.11). Thus, the work of gravity is

$$\delta A^g = \sum_{k=j}^{n} m_k\vec{g}\cdot\delta\vec{r}_j = \vec{e}_j\vec{g}\left(\sum_{k=j}^{n} m_k\right)\cdot\delta q_j \tag{2.3.53}$$

By (2.3.52) and (2.3.53) the total work is

$$\delta A = \delta A^P + \delta A^g = \left\{P_j^F + \vec{e}_j\vec{g}\sum_{k=j}^{n} m_k\right\}\delta q_j \tag{2.3.54}$$

By definition the generalized force is

$$Q_j^{(transl)} = P_j^F + \vec{e}_j\vec{g}\sum_{k=j}^{n} m_k \tag{2.3.55}$$

By considering expressions (2.3.51) and (2.3.55), it may be noted that the expressions for generalized forces may be written in the form

$$Q = P + Y \tag{2.3.56}$$

where P is the drives vector given by (2.3.4c) and Y is a vector given as

$$Y = \begin{bmatrix} y_1 & \cdots & y_n \end{bmatrix}^T \tag{2.3.57a}$$

$$y_i = \begin{cases} \sum_{k=i}^{n} m_k\left[\vec{g},\ \vec{e}_i,\ \vec{r}_k^{(i)}\right], & s_i = 0 \\ \sum_{k=i}^{n} \vec{e}_i\vec{g}m_k, & s_i = 1 \end{cases} \tag{2.3.57b}$$

and that Y is calculated independently of P.

The generalized forces are calculated recursively in the algorithm. This calculation can be done in the external or in the b.-f. system.

Globally, the algorithm of the c.-a. model forming i.e. for the computation of matrices W, U may be represented by the block scheme in Fig. 2.12.

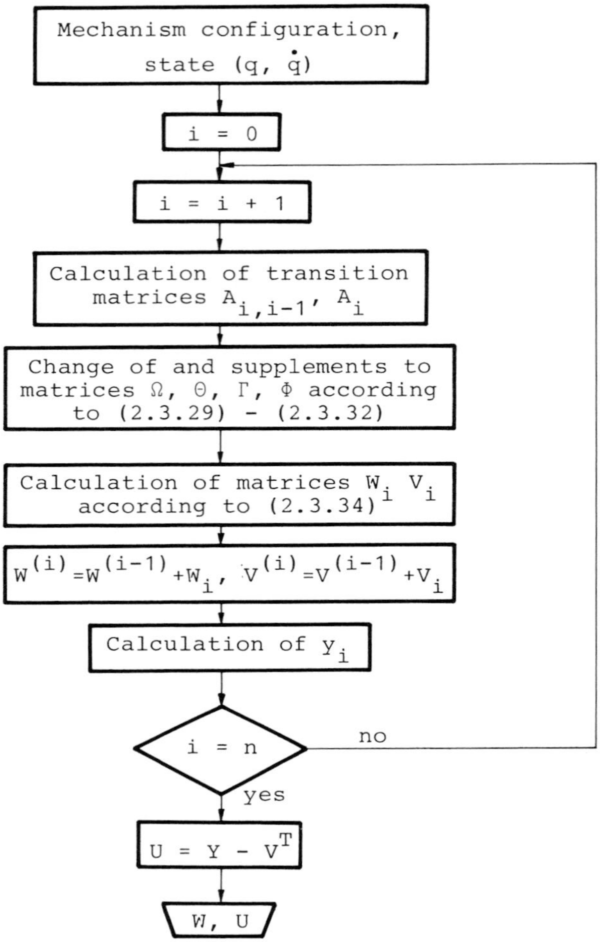

Fig. 2.12. Blok-scheme of c.-a. model forming procedure

2.4. Definition of Manipulation Task

Definition of the manipulation task is a rather complex question. First, it should satisfy two basic requirements: to be general enough to cover all manipulation tasks in practice and to be simple enough so that a user can easily operate with the algorithm. Second, the manipulators which appear in practice have different numbers of degrees of freedom since they are dedicated to different classes of manipulation task. In order to derive a suitable way of task prescription it is necessary to analyze the need for a certain number of degrees of freedom of the manipulator and their use in various categories of tasks.

Something more should be said about the number of manipulator d.o.f. and about the number of d.o.f. necessary for performing a particular manipulation task. First, we note the difference between the number of d.o.f. of the whole manipulator (considered as a dynamic system), and the number of d.o.f. of the gripper (considered as the last rigid body in the chain). These two numbers need not be equal. The number of manipulator d.o.f. depends of the number of joints and the number of d.o.f. in each joint. For instance, if a manipulator represents an open chain without branching and consists of n segments and n one - d.o.f. joints, then it has n degrees of freedom. On the other hand, the gripper, considered as a rigid body, cannot have more than six d.o.f. Even if n is less than six, those two numbers of d.o.f. need not be equal. For instance, it may happen that the manipulator has five d.o.f. and the gripper (i.e. the last segment) has four d.o.f. This means that one d.o.f. is lost. An extensive discussion on kinematics is necessary in order to explain the loss of some d.o.f. Such a loss can occur in some special cases of relative positions of joints axes. Such special cases are called kinematic singularities. We do not give here the whole kinematic discussion on singularities but only demonstrate it through some examples. The first example is a theoretic one. Let us consider a mechanism as shown in Fig. 2.13. It has six d.o.f. But the last body (number 2) in this chain has five d.o.f. and so it cannot move and rotate in all directions. The loss of one d.o.f. happens along the axis "a" connecting S_1 and S_2. The essential conclusion is that in the case of kinematic singularities the last body in the chain (with manipulators it is the gripper) really cannot move and rotate in all directions. Thus, such singularities are called the true singularities. The mechanism of Fig. 2.13. is always singular.

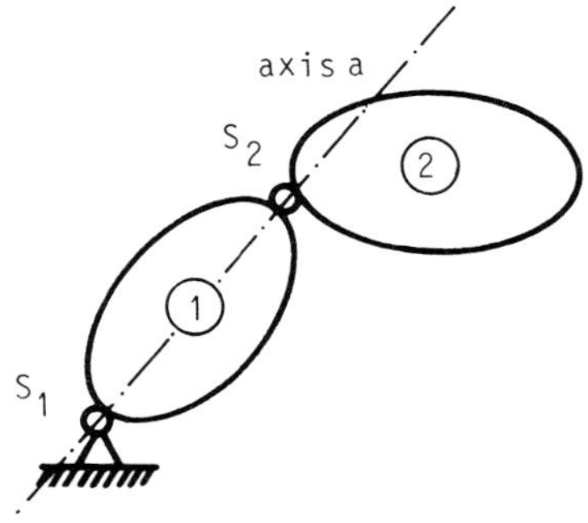

Fig. 2.13. Two-body mechanism

A more practical example is given in Fig. 2.14. Fig. 2.14a shows a manipulator with 6 d.o.f. Its gripper also has 6 d.o.f. and so it can move and rotate in all directions. But if the same manipulator comes into position of Fig. 2.14b then it still has 6 d.o.f. but its gripper has 5 d.o.f. which means that one d.o.f. is lost. The gripper really cannot rotate around the axis "c".

So this manipulator has a true singularity in position (b).

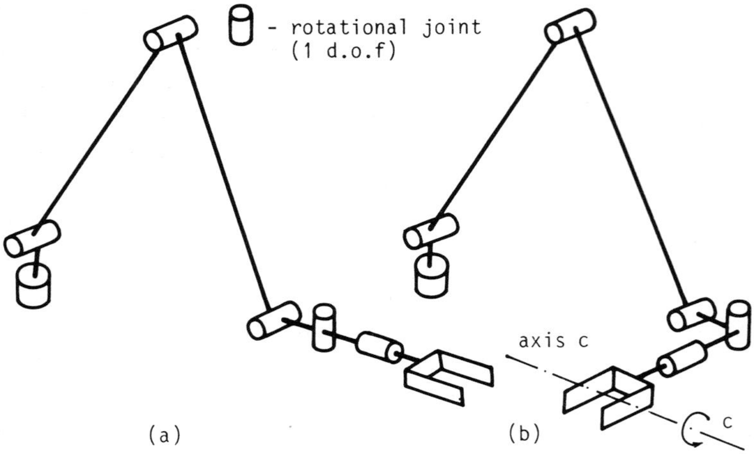

Fig. 2.14. Manipulator with 6 d.o.f.

It should be explained why it is important to notice the difference betwen those two numbers of d.o.f. (of the whole manipulator and of its gripper). If a manipulator with n d.o.f. ($n \leqslant 6$) has to perform some prescribed manipulation task, it is essential that the gripper has enough d.o.f. to solve the task geometrically. This is due to the fact that for such manipulators the tasks are usually imposed on the gripper. But, for solving the dynamics of that task, it is necessary to solve dynamic equations for the whole manipulator, and such equations include all n d.o.f. If we consider a manipulator with $n>6$ (a redundant manipulator), then the discussion is rather complicated. Six d.o.f. of the gripper are enough for all manipulation tasks. So the tasks for manipulators with $n>6$ usually contain some additional requirements which are not imposed on the gripper but on the manipulator as a whole (for instance, that a manipulator bypasses some obstacle in the working space). Hence, for solving the task geometry, not only the gripper d.o.f. but the d.o.f. of the manipulator as a whole are also important. An extensive discussion on kinematics with exact answers about the conditions for the disappearance of some d.o.f. will not be given here. So we shall restrict out consideration to the cases when there are no singularities. Let us be more precise. For a manipulator with $n \leqslant 6$ d.o.f. we shall assume that its gripper also has n d.o.f.; for a manipulator with $n>6$ d.o.f. we shall assume that its gripper has six d.o.f., i.e., the maximal possible number. It is an important fact that this assumption holds for almost all practical problems. For instance, when imposing a task on a certain manipulator we always take care about avoiding the singularity points, i.e., we choose the trajectories which can be performed. Keeping in mind the above assumption, we shall simply talk about degrees of freedom, regardless of whether the whole manipulator, or its gripper only, is concerned. The following consideration will be restricted to mechanisms having one d.o.f. joints only. This means that number of segments equals the number of d.o.f.

We now derive the so-called general algorithm for dynamic analysis. It is a basis for the development of a new algorithm which covers some typical classes of manipulation tasks and is more suitable for users.

2.4.1. General algorithm for dynamic analysis

Let us consider a manipulator as an open chain of rigid bodies, without branches, as shown in Fig. 2.15. Let the manipulator have six d.o.f.

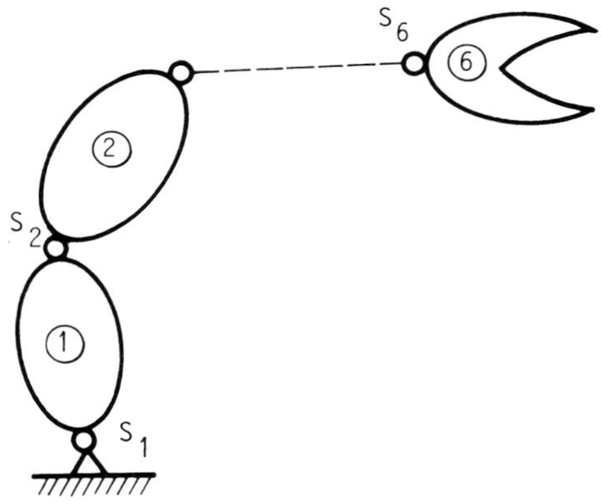

Fig. 2.15. Manipulator as a kinematic chain

The last body (segment) of the chain represents the manipulator gripper, i.e., in the phase of transferring some work object, the last segment is the gripper and object combined. Thus, the manipulation task can usually be considered as a prescribed motion of a rigid body (the last segment) in space. In the development of the general simulation algorithm, one started from the fact that a rigid body motion can, in the most general way, be prescribed by means of a known initial state, and a known time function of the center of gravity acceleration (or of some other point on the segment) $\vec{w}(t)$, and a known time-function of the angular acceleration $\vec{\varepsilon}(t)$ of the body. Such an approach is justified because for many manipulation tasks these values can easily be prescribed. For instance, gripper center of gravity motion is usually prescribed quite easily by means of the trajectory and the velocity profile. Thus, we will now describe the algorithm (considering the manipulation task) in terms of $\vec{w}(t)$ and $\vec{\varepsilon}(t)$, i.e., these are the input values [23, 24]. So, in matrix notation

$$X^a = \begin{bmatrix} w \\ \varepsilon \end{bmatrix}, \tag{2.4.1}$$

where w designates a 3×1 matrix corresponding to vector $\vec{w}$, likewise for ε and all other vecotrs in the sequel. Now, it is clear why we introduced the notation $X^a = \ddot{X}$ (see (2.2.5), (2.2.6)). It is due to the fact that ε is not the second derivative of any coordinates.

In Para. 2.3. when the c.-a. method for model formation was described, it was shown how the matrices Ω, Θ, Γ, Φ were derived and calculated, so that it holds:

$$\tilde{w} = \Omega\ddot{q} + \Theta, \tag{2.4.2a}$$

$$\tilde{\varepsilon} = \Gamma\ddot{q} + \Phi. \tag{2.4.2b}$$

The matrices are calculated from the recursive expressions for velocities and accelerations of segments. There are no indices in the equations (2.4.2a,b). $\tilde{w}$ is the acceleration of gripper c.o.g. and $\tilde{\varepsilon}$ is the angular acceleration of the gripper. The matrices Ω, Θ, Γ, Φ are calculated recursively during the formation of dynamic model. In each iteration they correspond to the new segment. At the end of model forming procedure these matrices correspond to the gripper. This is the reason that no indices are used with these matrices.

The vectors $\tilde{w}$ and $\tilde{\varepsilon}$ hold for the gripper and are expressed in the corresponding body-fixed coordinate system. In the external coordinate system it holds

$$w = A_6\Omega\,\ddot{q} + A_6\Theta\,, \tag{2.4.3a}$$

$$\varepsilon = A_6\Gamma\,\ddot{q} + A_6\Phi \tag{2.4.3b}$$

where A_6 is the transition matrix of the gripper system. These two expressions can be combined

$$\begin{bmatrix} w \\ \varepsilon \end{bmatrix} = \underbrace{\begin{bmatrix} A_6\,\Omega \\ A_6\,\Gamma \end{bmatrix}}_{J}\ddot{q} + \underbrace{\begin{bmatrix} A_6\,\Theta \\ A_6\,\Phi \end{bmatrix}}_{A} \tag{2.4.4}$$

to give

$$X^a = \begin{bmatrix} w \\ \varepsilon \end{bmatrix}_{(6\times 1)} = J\ddot{q} + A \tag{2.4.5}$$

where the Jacobian and the adjoint matrix are

$$J = \begin{bmatrix} A_6\,\Omega \\ A_6\,\Gamma \end{bmatrix}_{(6\times 6)}, \quad A = \begin{bmatrix} A_6\,\Theta \\ A_6\,\Phi \end{bmatrix}_{(6\times 1)} \tag{2.4.6}$$

The internal (generalized) accelerations are now $\ddot{q} = J^{-1}(X^a - A)$.

It should be exphasized that this approach to the Jacobian is very suitable. When we carry out the c.-a. model forming procedure then we have the matrices Ω, Θ, Γ, Φ calculated. So, we immediately have the Jacobian and the adjoint matrix with no additional calculation.

Equations (2.4.2) - (2.4.6) hold for the gripper c.o.g. If motion of some other gripper point is prescribed instead of its c.o.g. then the Jacobian matrix is changed. Let us consider a point A determined by a vector $\vec{p} = \overrightarrow{C_6A}$ (Fig. 2.16). Then, for the acceleration it holds

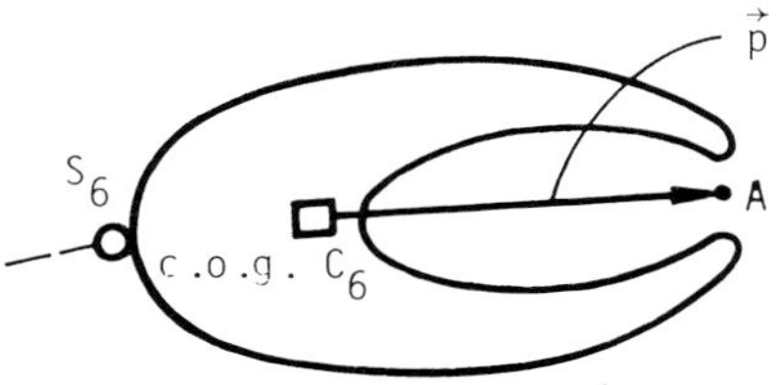

Fig. 2.16. Positioning of a gripper tip

$$\vec{w}_A = \vec{w}_{c_g} - \vec{p}\times\vec{\varepsilon}_g + \vec{\omega}_g\times(\vec{\omega}_g\times\vec{p}) \quad (2.4.7)$$

In order to obtain a more general form the index "g" has been used for gripper in (2.4.7) instead of the index "6". Let $\vec{k} = \vec{\omega}_g\times(\vec{\omega}_g\times\vec{p})$ and let us use matrix notation. Then:

$$\tilde{w}_A = \tilde{w}_{c_g} - \underline{\underline{\tilde{p}}}\tilde{\varepsilon}_g + \tilde{k} \quad (2.4.8)$$

where $\underline{\underline{\tilde{p}}}$ designates the matrix

$$\underline{\underline{\tilde{p}}} = \begin{bmatrix} 0 & -\tilde{p}_3 & \tilde{p}_2 \\ \tilde{p}_3 & 0 & -\tilde{p}_1 \\ -\tilde{p}_2 & \tilde{p}_1 & 0 \end{bmatrix} \quad (2.4.9)$$

which corresponds to the vector $\vec{p} = \left\{\tilde{p}_1,\ \tilde{p}_2,\ \tilde{p}_3\right\}$ and is used to form the vector product by matrix calculus.

If we introduce (2.4.2) into (2.4.8) instead of $\tilde{w}_{c_g}$ and $\tilde{\varepsilon}_g$, we obtain

$$\tilde{w}_A = (\Omega-\underline{\underline{\tilde{p}}}\Gamma)\ddot{q} + \Theta - \underline{\underline{\tilde{p}}}\Phi + \tilde{k} \quad (2.4.10)$$

or

$$\tilde{w}_A = \Omega^*\ddot{q} + \Theta^* \quad (2.4.11)$$

where

$$\Omega^* = \Omega - \underline{\underline{\tilde{p}}}\Gamma, \qquad \Theta^* = \Theta - \underline{\underline{\tilde{p}}}\Phi + \tilde{k} \quad (2.4.12)$$

After transition to the external coordinate system and combination with (2.4.3b) it follows

$$X^a = \begin{bmatrix} {}^{w}A \\ \\ \varepsilon \end{bmatrix} = \begin{bmatrix} A_g\ \Omega^* \\ \cdot \\ A_g\ \Gamma \end{bmatrix} \ddot{q} + \begin{bmatrix} A_g\ \Theta^* \\ \\ A_g\ \Phi \end{bmatrix} \tag{2.4.13}$$

Hence the new Jacobian and the new adjoint matrix are

$$J = \begin{bmatrix} A_g\ \Omega^* \\ \\ A_g\ \Gamma \end{bmatrix} = \begin{bmatrix} A_g(\Omega - \underline{\underline{\tilde{p}}}\Gamma) \\ \\ A_g\ \Gamma \end{bmatrix}, \quad A_g(\Omega - \underline{\underline{\tilde{p}}}\Gamma) = \Omega'$$

$$A = \begin{bmatrix} A_g\ \Theta^* \\ \\ A_g\ \Phi \end{bmatrix} = \begin{bmatrix} A_g(\Theta - \underline{\underline{\tilde{p}}}\Phi + \tilde{k}) \\ \\ A_g\ \Phi \end{bmatrix}, \quad A_g(\Theta - \underline{\underline{\tilde{p}}}\Phi + \tilde{k}) = \Theta' \tag{2.4.14}$$

It was said that this algorithm had the manipulation task input in terms of accelerations. There are some algorithms which use the task input in terms of velocities. Then

$$\dot{X} = J\dot{q} \Rightarrow \dot{q} = J^{-1}\dot{X}$$

The Jacobian is the same as in (2.4.4), (2.4.14). It may seem to someone that the prescription of external velocity ($\dot{X}$) is more suitable than the prescription of acceleration ($X^a = \ddot{X}$). As we have chosen the acceleration input, here are some reasons, justifying such an approach. If the velocity input is used, then we obtain $\dot{q}$ and it is necessary to perform numerical differentiation in order to compute $\ddot{q}$ which is needed for the calculation of drives P. The numerical differentiation is still an undesirable procedure. On the other hand, the manipulation task is very often given by a trajectory and a velocity profile along it. The velocity profile in fact means the acceleration. Special conveniences appear in some practical cases. The most common velocity profile is the triangular and the trapezoidal one (Fig. 2.17). It can be seen that the corresponding accelerations are constant and thus very easy for prescription. This acceleration approach will be worked out in the sequel.

This general algorithm is sometimes extremely convenient. Suppose a manipulation task in which a 6 d.o.f. manipulator should move an object from the point A to B (Fig. 2.18) along the straight line keeping all the time the initial orientation of the object. The velocity profile is assumed to be triangular. Let the execution time be T. First, we ha-

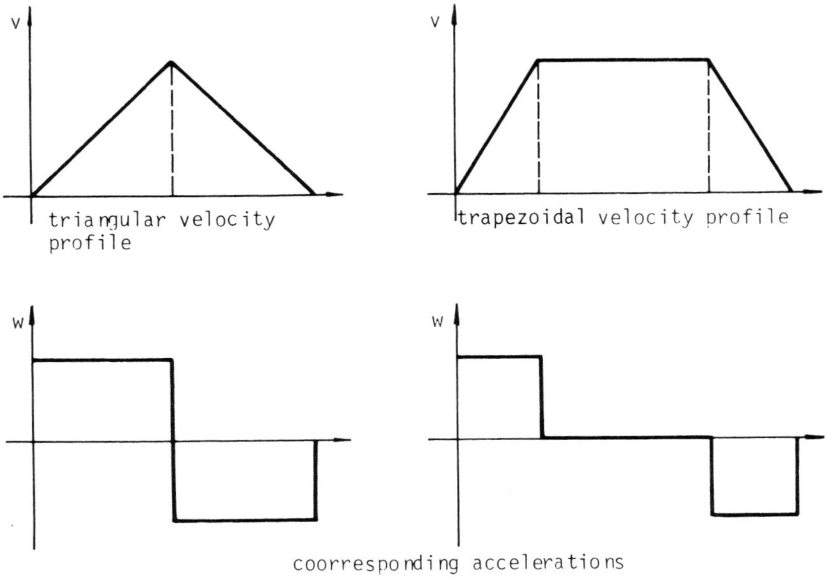

Fig. 2.17. Velocity profiles and the corresponding accelerations

ve to prescribe the initial state $q(t_o) = q^o$, $\dot{q}(t_o) = 0$. In order to prescribe the linear motion of the object in terms of acceleration it is enough to give:

$$w = \begin{bmatrix} w_x & w_y & w_z \end{bmatrix}^T$$

$$w_x(t) = \begin{cases} 4\ell_x/T^2, & 0 < t < T/2 \\ -4\ell_x/T^2, & T/2 < t < T \end{cases}, \quad w_y(t),\ w_z(t) \text{ analogously}$$

In order to prescribe the constant orientation we use

$$\varepsilon = \begin{bmatrix} \varepsilon_x & \varepsilon_y & \varepsilon_z \end{bmatrix}^T = 0$$

There is another convenience of the general algorithm. With this algorithm the Jacobian matrix is singular if and only if the manipulator has a true singularity in the position considered. With some other algorithms the Jacobian singularity may happen even if there is no true singularity of the manipulator.

We have seen that the general algorithm has its advantages but the problems which arise with its use still result from the input $[w\ \varepsilon]^T$. In some manipulation tasks the changes in object orientation are not suitable for prescription in terms of $\vec{\varepsilon}$. Hence the input $[w\ \varepsilon]^T$ needs some further development.

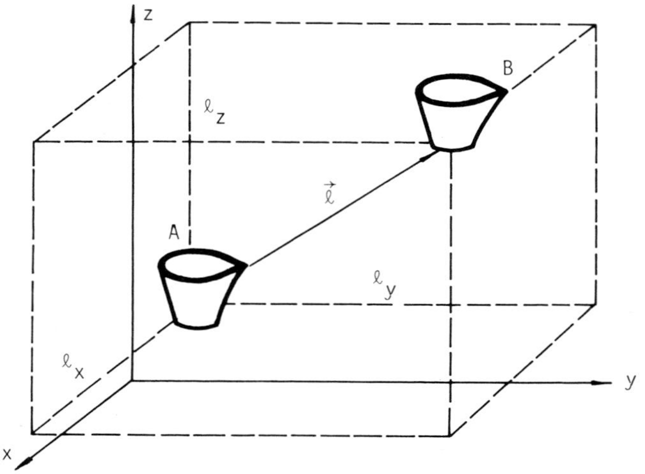

Fig. 2.18. Manipulation task scheme

2.4.2. Practical approach to manipulation task definition

It was said that the general algorithm described above may sometimes be unsuitable because of the input $\vec{w}$, $\vec{\varepsilon}$. For a manipulator motion from one position to another it is necessary to define $\vec{w}(t)$ and $\vec{\varepsilon}(t)$. While $\vec{w}(t)$ can easily be defined, definition of orientation change in terms of $\vec{\varepsilon}$ may be difficult. Since our intention is to make the algorithm more convenient for the users, we make some further elaboration of the task-input procedure.

We now search for the three parameters which define the orientation of the gripper (or of a rigid body in general) in a convenient way. The most common parameters defining the body orientation are Euler's angles but they are inconvenient for definition. Such orientation definition does not follow from the functional motion of manipulator and these angles can hardly be measured. So, we choose another set of angles (θ, φ, ψ) defining the orientation (Fig. 2.19). Let us notice that this orientation definition understands one direction (b) and the angle of rotation around this direction. Such approach will be shown to be very convenient because it really follows from the functional movements of manipulators in practical operation. However, this angle representation sometimes results in some singularity problems but it is the price we have to pay for simple handling of the algorithm. These problems and the way of avoiding them will be discussed later.

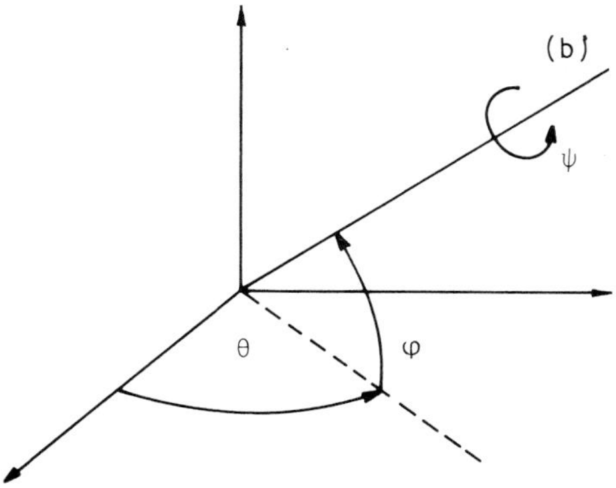

Fig. 2.19. Definition of orientation

In practice, we meet manipulators with different numbers of d.o.f. (most often: 4, 5 or 6 d.o.f.). They are intended for different classes of manipulation tasks. Hence, we now analyze the need for a certain number of d.o.f. and their use in certain classes of tasks [22, 26].

Let us first define more precisely some notions which will be used:

Positioning means moving some given point of the last segment to some desired point in the working space, i.e., the motion of that point of the last segment along a prescribed trajectory according to a prescribed motion law.

Full orientation of a body in space means an exactly determined angular position of the body with respect to the external space. It can also be considered in the following way: some given body axis (or some arbitrary fixed direction on the body) coincides with a prescribed direction in the space and the rotation angle around this direction is also prescribed. The term total orientation is used also.

Partial orientation of a body only means that the given body-fixed direction concides with the prescribed direction in space (which can be changeable according to some law).

The difference between the partial and the total orientation is shown in Fig. 2.20. In Fig. 2.20(a), (b) the task of transferring a container with liquid is shown. In case (a) it is only important that the axis (*) is vertical and so it is the case of partial orientation. In case (b) not only the direction (*) but also the direction (**) is important. For the given direction (*) the direction (**) may be replaced by angle ψ. Hence, the case (b) represents the total orientation. Fig. 2.20(c), (d) shows something analogous but for assembly tasks.

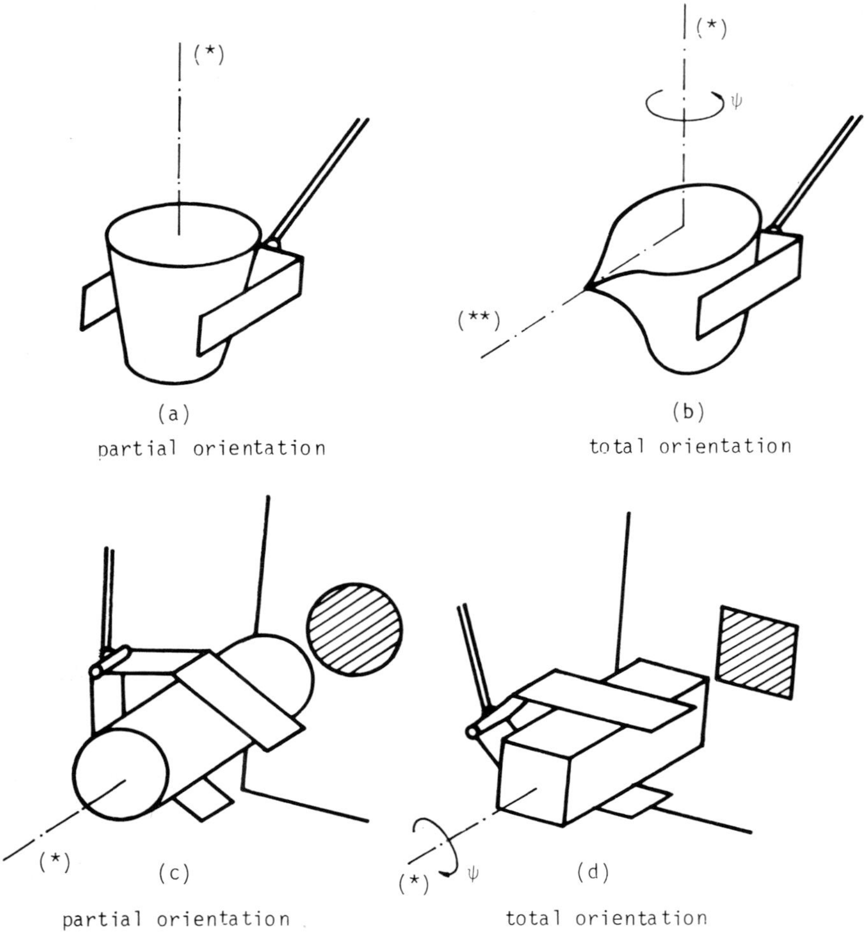

Fig. 2.20. The partial and the total orientation

We now consider two examples which demonstrate that the orientation representation chosen (direction and angle) really follows from practical manipulation tasks. Fig. 2.21. presents a task of spraying powder along a prescribed trajectory. The task reduces to the need to realize the motion of an object (container) along the trajectory (a) (i.e. positioning) and along the trajectory the container should rotate around the direction (b) according to the prescribed law $\psi(t)$. So, the orientation representation in terms of one direction and the rotation angle directly coincide with the functional motion of manipulator. In this case the choice of the direction is different from that in Fig. 2.20(a), (b) because of different manipulation tasks.

The next example represents screwing a bolt in (Fig. 2.22).

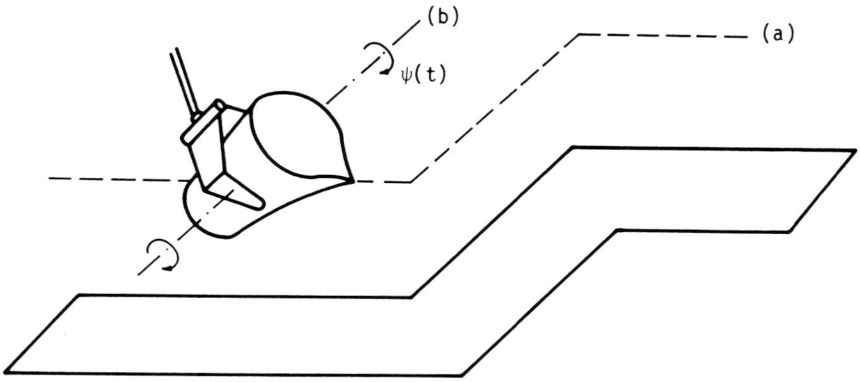

Fig. 2.21. Spraying powder along the prescribed trajectory

Let us now analyze the need for a certain number of d.o.f.:

To solve the positioning task, which is a part of every manipulation task, three d.o.f. are needed.

To solve the positioning task along with the task of partial orientation, five d.o.f. are necessary.

To solve the positioning task along with the task of full orientation, six d.o.f. are necessary.

Manipulators with four, five and six d.o.f. will be considered here. In order to obtain a simple handling algorithm we derive a special adapting block for each manipulator class (4, 5 or 6 d.o.f.). This is due to the fact that these manipulators are dedicated to different classes of manipulation tasks and each class of tasks is suitable for definition in its own way. Even in one class (for instance manipulators with 6 d.o.f.) the task may be defined in a few different ways. Thus, the adapting blocks enable a user to apply the task definition which is most suitable (in his opinion) to his manipulator and to the task which has to be performed. He does it by choosing one of the adapting blocks which is already incorporated in the algorithm or he may even program its own adapting block.

Let us define the generalized position vector X_g. It represents a set of parameters which define the position of a manipulator. This vector is of dimension n (n = number of d.o.f.). It may be chosen to equal the internal coordinates ($X_g = q$) or the external coordinates ($X_g = X$) or it may be mixed (X_g contains some internal and some external coordinates). Each adapting block operates with its special position vector and we choose the most suitable one. This position vector X_g represents some generalization, for it is used instead of external coordinates vector X.

The aim of each adapting block is to compute $\ddot{q}$ which is necessary for the calculation of driving forces and torques. This computation of $\ddot{q}$ is performed by means of Jacobian method or sometimes more directly. Anyhow, $\ddot{q}$ is computed starting from $\ddot{X}_g$ which has to be known.

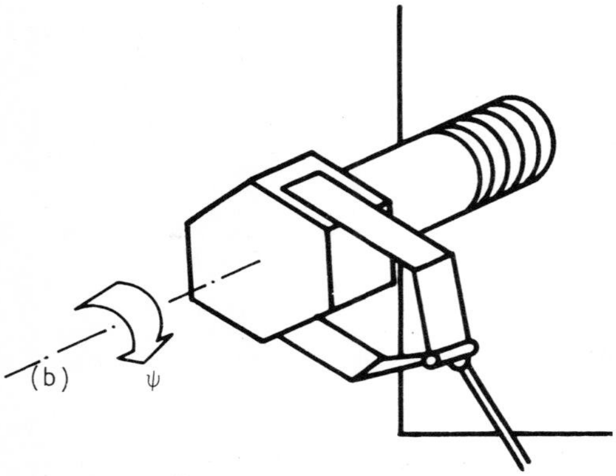

Fig. 2.22. Screwing a both in

We here present these adapting blocks. But before that we have to derive some mathematical transformations which are common for most adapting blocks.

At the beginning of 2.4.2. it was said that the set of angles θ, φ, ψ (Fig. 2.19) was chosen to define the total orientation of a body. In order to make easier the later derivation of adapting blocks we now give some transformations.

Let us introduce a new Cartesian coordinate system which corresponds to the chosen set of angles (θ, φ, ψ). This system is obtained from the external one in the following way: rotation is first made around the z-axis (angle θ), and then around the new y-axis in the negative sense (angle φ); finally, the rotation around the new x-axis represents the angle ψ (Fig. 2.23). This system will be called the orientation system. Let us notice that the x-axis of such system coincides with the previously introduced direction (b).

Let A be the transition matrix of the orientation system. Then

$$A = A_\theta \, A_\varphi \, A_\psi \tag{2.4.15}$$

where

$$A_\theta = \begin{bmatrix} \cos\theta & -\sin\theta & 0 \\ \sin\theta & \cos\theta & 0 \\ 0 & 0 & 1 \end{bmatrix}, \quad A_\varphi = \begin{bmatrix} \cos\varphi & 0 & -\sin\varphi \\ 0 & 1 & 0 \\ \sin\varphi & 0 & \cos\varphi \end{bmatrix}$$

$$A_\psi = \begin{bmatrix} 1 & 0 & 0 \\ 0 & \cos\psi & -\sin\psi \\ 0 & \sin\psi & \cos\psi \end{bmatrix} \tag{2.4.16}$$

Let us find the first and the second derivatives of the transition matrix

$$\dot{A} = \dot{A}_\theta A_\varphi A_\psi + A_\theta \dot{A}_\varphi A_\psi + A_\theta A_\varphi \dot{A}_\psi \tag{2.4.17}$$

$$\begin{aligned} \ddot{A} = {} & \ddot{A}_\theta A_\varphi A_\psi + A_\theta \ddot{A}_\varphi A_\psi + A_\theta A_\varphi \ddot{A}_\psi + \\ & + 2\dot{A}_\theta \dot{A}_\varphi A_\psi + 2\dot{A}_\theta A_\varphi \dot{A}_\psi + 2A_\theta \dot{A}_\varphi \dot{A}_\psi \end{aligned} \tag{2.4.18}$$

where

$$\begin{aligned} \dot{A}_\theta &= \frac{\partial A_\theta}{\partial \theta} \dot{\theta}, \quad & \ddot{A}_\theta &= \frac{\partial A_\theta}{\partial \theta} \ddot{\theta} + \frac{\partial^2 A_\theta}{\partial \theta^2} \dot{\theta}^2 \\ \dot{A}_\varphi &= \frac{\partial A_\varphi}{\partial \varphi} \dot{\varphi}, \quad & \ddot{A}_\varphi &= \frac{\partial A_\varphi}{\partial \varphi} \ddot{\varphi} + \frac{\partial^2 A_\varphi}{\partial \varphi^2} \dot{\varphi}^2 \\ \dot{A}_\psi &= \frac{\partial A_\psi}{\partial \psi} \dot{\psi}, \quad & \ddot{A}_\psi &= \frac{\partial A_\psi}{\partial \psi} \ddot{\psi} + \frac{\partial^2 A_\psi}{\partial \psi^2} \dot{\psi}^2 \end{aligned} \tag{2.4.19}$$

and

$$\frac{\partial A_\theta}{\partial \theta} = \begin{bmatrix} -\sin\theta & -\cos\theta & 0 \\ \cos\theta & -\sin\theta & 0 \\ 0 & 0 & 0 \end{bmatrix}, \quad \frac{\partial^2 A_\theta}{\partial \theta^2} = \begin{bmatrix} -\cos\theta & \sin\theta & 0 \\ -\sin\theta & -\cos\theta & 0 \\ 0 & 0 & 0 \end{bmatrix}$$

$$\frac{\partial A_\varphi}{\partial \varphi} = \begin{bmatrix} -\sin\varphi & 0 & -\cos\varphi \\ 0 & 0 & 0 \\ \cos\varphi & 0 & -\sin\varphi \end{bmatrix}, \quad \frac{\partial^2 A_\varphi}{\partial \varphi^2} = \begin{bmatrix} -\cos\varphi & 0 & \sin\varphi \\ 0 & 0 & 0 \\ -\sin\varphi & 0 & -\cos\varphi \end{bmatrix} \tag{2.4.20}$$

$$\frac{\partial A_\psi}{\partial \psi} = \begin{bmatrix} 0 & 0 & 0 \\ 0 & -\sin\psi & -\cos\psi \\ 0 & \cos\psi & -\sin\psi \end{bmatrix}, \quad \frac{\partial^2 A_\psi}{\partial \psi^2} = \begin{bmatrix} 0 & 0 & 0 \\ 0 & -\cos\psi & \sin\psi \\ 0 & -\sin\psi & -\cos\psi \end{bmatrix}$$

Introducing (2.4.19) and (2.4.20) into (2.4.18) one obtains:

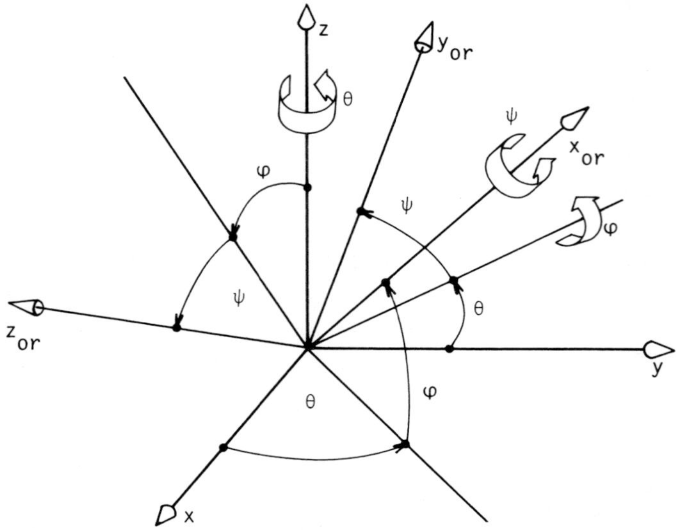

Fig. 2.23. Set of angles determining the total orientation

$$\ddot{A} = \underbrace{\frac{\partial A_\theta}{\partial \theta} A_\varphi A_\psi}_{F_\theta} \ddot{\theta} + \underbrace{A_\theta \frac{\partial A_\varphi}{\partial \varphi} A_\psi}_{F_\varphi} \ddot{\varphi} + \underbrace{A_\theta A_\varphi \frac{\partial A_\psi}{\partial \psi}}_{F_\psi} \ddot{\psi} +$$

$$G \begin{cases} + \dfrac{\partial^2 A_\theta}{\partial \theta^2} A_\varphi A_\psi \dot{\theta}^2 + A_\theta \dfrac{\partial^2 A_\varphi}{\partial \varphi^2} A_\psi \dot{\varphi}^2 + A_\theta A_\varphi \dfrac{\partial^2 A_\psi}{\partial \psi^2} \dot{\psi}^2 + \\ + 2 \dfrac{\partial A_\theta}{\partial \theta} \dfrac{\partial A_\varphi}{\partial \varphi} A_\psi \dot{\theta}\dot{\varphi} + 2 \dfrac{\partial A_\theta}{\partial \theta} A_\varphi \dfrac{\partial A_\psi}{\partial \psi} \dot{\theta}\dot{\psi} + 2A_\theta \dfrac{\partial A_\varphi}{\partial \varphi} \dfrac{\partial A_\psi}{\partial \psi} \dot{\varphi}\dot{\psi} \end{cases} \quad (2.4.21)$$

So (2.4.21) reduces to

$$\ddot{A} = F_\theta \ddot{\theta} + F_\varphi \ddot{\varphi} + F_\psi \ddot{\psi} + G \quad (2.4.22)$$

The matrices F_θ, F_φ, F_ψ are determined by (2.4.21), (2.4.20), (2.4.16). Let it be noticed that the orientation coordinate system has no direct relations with the previously introduced b.-f. system. Let us now derive the adapting blocks. The Jacobian matrices which appear in different adapting blocks are different. But, they will be always marked by the same letter J.

2.4.3. Manipulator with four degrees of freedom

The first block (block 4-1). For a manipulator it is possible to choose a position vector to be equal to the internal coordinates vector. For a 4 d.o.f. manipulator it is

$$X_g = q = \left[q_1\ q_2\ q_3\ q_4\right]^T \tag{2.4.23}$$

Thus if $\ddot{X}_g$ is given it means that $\ddot{q}$ is given and no calculation is needed. Such an approach understands that the motion is prescribed directly in terms of internal coordinates. This approach may simply be used for some configurations such as the cylindrical manipulator (Fig. 2.24).

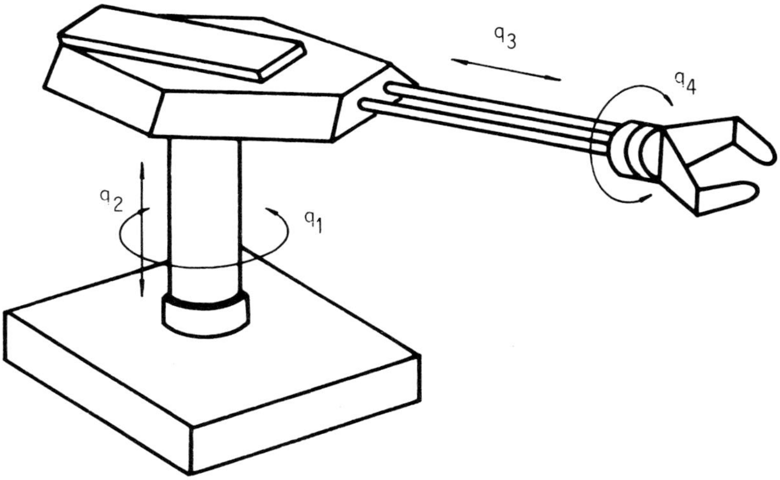

Fig. 2.24. Manipulator with four d.o.f.

The second block (4-2). A manipulator with 4 d.o.f. solves the positioning task by using three d.o.f., while the remaining one performs operations frequently sufficient for many practical manipulation tasks (Fig. 2.24). Hence we choose a generalized position vector

$$X_g = \left[x\ y\ z\ q_4\right]^T \tag{2.4.24}$$

where (x, y, z) represent the Cartesian coordinates of some point of the gripper. Thus the manipulation task is defined via positioning plus one internal coordinate which is given directly.

From (2.4.13) and (2.4.14), it follows that:

$$\begin{bmatrix} \ddot{x} \\ \ddot{y} \\ \ddot{z} \end{bmatrix} = w = \Omega' \begin{bmatrix} \ddot{q}_1 \\ \ddot{q}_2 \\ \ddot{q}_3 \\ \ddot{q}_4 \end{bmatrix} + \Theta' \qquad (2.4.25)$$

For prescribed $\ddot{X}_g = [\ddot{x}\ \ddot{y}\ \ddot{z}\ \ddot{q}_4]$, (2.4.25) represents a system of 3 equations with 3 unknowns $\ddot{q}_1$, $\ddot{q}_2$, $\ddot{q}_3$. After solving these equations the whole vector $\ddot{q} = [\ddot{q}_1\ \ddot{q}_2\ \ddot{q}_3\ \ddot{q}_4]^T$ becomes known.

This approach is closer to the essence of practical manipulation tasks. For better understanding consult the example in 2.8.1.

Other possibilities. Depending on the form of the desired trajectory, in some cases it is suitable to prescribe positioning in cylindrical or spherical coordinates (Fig. 2.25).

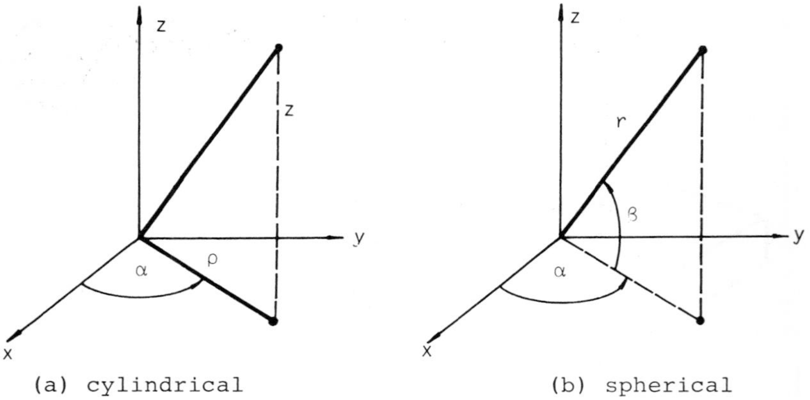

Fig. 2.25. Cylindrical and spherical coordinates

Let us first consider the cylindrical coordinate system (ρ, α, z). Projected onto the axes of such a system, acceleration $\vec{w}$ has the form

$$w_\rho = \ddot{\rho} - \rho\dot{\alpha}^2, \quad w_\alpha = \rho\ddot{\alpha} + 2\dot{\rho}\dot{\alpha}, \quad w_z = \ddot{z} \qquad (2.4.26)$$

and further, the following Cartesian projections are obtained:

$$w_x = w_\rho \cos\alpha - w_\alpha \sin\alpha, \quad w_y = w_\rho \sin\alpha + w_\alpha \cos\alpha, \quad w_z = \ddot{z} \qquad (2.4.27)$$

i.e. the acceleration vector is

$$w = \begin{bmatrix} w_x & w_y & w_z \end{bmatrix}^T \tag{2.4.28}$$

Now, vector (2.4.28) together with (2.4.27), (2.4.26) is used in eq. (2.4.25) instead of $w = [\ddot{x}\ \ddot{y}\ \ddot{z}]^T$ in order to compute $\ddot{q}_1$, $\ddot{q}_2$, $\ddot{q}_3$.

For the computation of w by using (2.4.26) and (2.4.27), it is necessary to know $\ddot{\rho}$, $\ddot{\alpha}$, $\ddot{z}$ and also ρ, $\dot{\rho}$, α, $\dot{\alpha}$. Only $\ddot{\rho}$, $\ddot{\alpha}$, $\ddot{z}$, $\ddot{q}_4$ appear as input values. So, during the recursion from one time instant to another, the values of ρ, $\dot{\rho}$, α, $\dot{\alpha}$ are calculated by integration together with integration over the generalized coordinates.

If a spherical coordinate system (r, α, β) is considered, then the acceleration projected onto the corresponding axes has the form

$$\begin{aligned} w_r &= \ddot{r} - r\dot{\beta}^2 - r\cos^2\beta\dot{\alpha}^2 \\ w_\alpha &= 2\dot{r}\dot{\alpha}\cos\beta + r\ddot{\alpha}\cos\beta - 2r\dot{\beta}\dot{\alpha}\sin\beta \\ w_\beta &= 2\dot{r}\dot{\beta} + r\ddot{\beta} + r\sin\beta\cos\beta\dot{\alpha}^2 \end{aligned} \tag{2.4.29}$$

or, in the Cartesian system

$$\begin{aligned} w_x &= w_r\cos\beta\cos\alpha - w_\alpha\sin\alpha - w_\beta\sin\beta\cos\alpha \\ w_y &= w_r\cos\beta\sin\alpha + w_\alpha\cos\alpha - w_\beta\sin\beta\sin\alpha \\ w_z &= w_r\sin\beta + w_\beta\cos\beta \end{aligned} \tag{2.4.30}$$

so the acceleration vector is

$$w = \begin{bmatrix} w_x & w_y & w_z \end{bmatrix}^T \tag{2.4.31}$$

The vector (2.4.31) together with (2.4.30), (2.4.29) is now used in (2.4.25) instead of $w = [\ddot{x}\ \ddot{y}\ \ddot{z}]^T$. As input data $\ddot{r}$, $\ddot{\alpha}$, $\ddot{\beta}$, $\ddot{q}_4$ appear and r, $\dot{r}$, α, $\dot{\alpha}$, β, $\dot{\beta}$ are obtained by integration.

2.4.4. Manipulator with five degrees of freedom

The first block (block 5-1). The first possibility is to prescribe the manipulator motion directly in terms of internal coordinates. Then

$$X_g = q = \begin{bmatrix} q_1 & q_2 & q_3 & q_4 & q_5 \end{bmatrix}^T \tag{2.4.32}$$

So, $\ddot{X}_g$ i.e. $\ddot{q}$ represent directly the input values

The second block (5-2). A manipulator with 5 d.o.f. solves the positioning task along with the task of partial orientation.

The positioning will be treated in external Cartesian coordinate system (x, y, z), so these coordinates will be included in the position vector X_g.

Let us now discuss the problem of partial orientation. Some given gripper axis (i.e. some arbitrary fixed direction on the gripper) has to coincide with the prescribed direction in external space. In order to define a direction on the gripper we use a unit vector $\vec{\tilde{h}}$, where the tilde shows that the vector is expressed via projections onto the axes of the corresponding body-fixed system. So, the unit vector $\vec{\tilde{h}}$ determines the direction with respect to the gripper b.-f. system. It is constant and represents the input value. The direction in external space will be prescribed by two angles θ and φ (Fig. 2.26). The angles determine the direction with respect to the external system. These two directions have to coincide. Let us point out that the b.-f. system and the external system need not coincide. Fig. 2.26. presents the determination of a direction with respect to the b.-f. and to the external coordinate system.

Hence, the position vector is

$$X_g = [x \; y \; z \; \theta \; \varphi]^T \tag{2.4.33}$$

For positioning we use (2.4.13), (2.4.14) i.e.

$$\begin{bmatrix} \ddot{x} \\ \ddot{y} \\ \ddot{z} \end{bmatrix} = w = \Omega' \, \ddot{q} + \Theta' \tag{2.4.34}$$

Dimensions of these matrices are: Ω' (3×5), Θ' (3×1), $\ddot{q}$ (5×1).

In the sequel we shall use transition matrices. Let A_g be the transition matrix of the gripper b.-f. system and let A be the transition

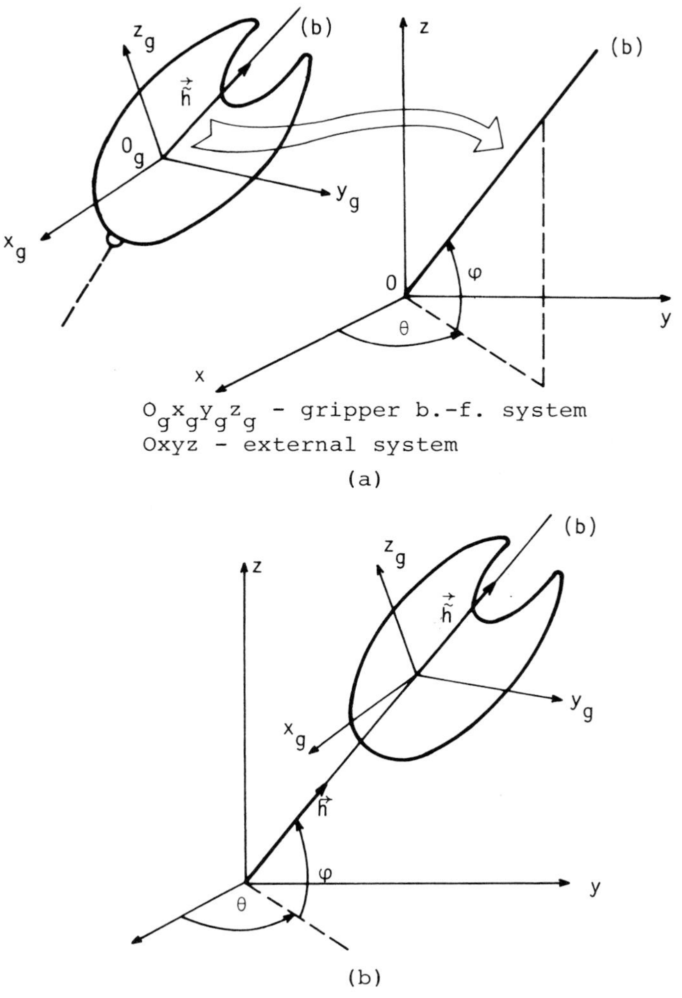

Fig. 2.26. Determination of partial orientation

matrix of the orientation coordinate system. Since the x-axis of the orientation system coincides with the direction (b) the unit vector $\vec{h}$ can be expressed in the orientation system via projections $[1\ 0\ 0]^T$. So, the vector $\vec{h}$ can be expressed in the external system as

$$h = A \cdot \begin{bmatrix} 1 \\ 0 \\ 0 \end{bmatrix} \tag{2.4.35}$$

The second derivative is

$$\ddot{h} = \ddot{A} \begin{bmatrix} 1 \\ 0 \\ 0 \end{bmatrix} \tag{2.4.36}$$

Introducing (2.4.22) into (2.4.36), one obtains

$$\ddot{h} = F_\theta \begin{bmatrix} 1 \\ 0 \\ 0 \end{bmatrix} \ddot{\theta} + F_\varphi \begin{bmatrix} 1 \\ 0 \\ 0 \end{bmatrix} \ddot{\varphi} + F_\psi \begin{bmatrix} 1 \\ 0 \\ 0 \end{bmatrix} \ddot{\psi} + G \begin{bmatrix} 1 \\ 0 \\ 0 \end{bmatrix} \tag{2.4.37}$$

The matrix product $F_\theta\ [1\ 0\ 0]^T$ represents the first column of matrix F_θ. Let us mark this column by $f_{\theta 1}$. The first column of F_φ is $f_{\varphi 1}$ and the first column of F_ψ is $f_{\psi 1}$. Let it be noticed that $f_{\psi 1} = 0$. So, from (2.4.37) it follows

$$\ddot{h} = f_{\theta 1}\ddot{\theta} + f_{\varphi 1}\ddot{\varphi} + G[1\ 0\ 0]^T \tag{2.4.38}$$

or

$$\ddot{h} = \begin{bmatrix} f_{\theta 1} & f_{\varphi 1} \end{bmatrix} \begin{matrix} \ddot{\theta} \\ \ddot{\varphi} \end{matrix} + G[1\ 0\ 0]^T \tag{2.4.39}$$

If we introduce the notation

$$v = \begin{bmatrix} f_{\theta 1} & f_{\varphi 1} \end{bmatrix}; \qquad u = G[1\ 0\ 0]^T \tag{2.4.40}$$

then

$$\ddot{h} = v \begin{bmatrix} \ddot{\theta} \\ \ddot{\varphi} \end{bmatrix} + u \tag{2.4.41}$$

Dimensions of matrices are: $\ddot{h}$ (3×1), v (3×2), u (3×1).

On the other hand, the first and the second derivative of vector $\vec{h}$ may be expressed as

$$\dot{\vec{h}} = \vec{\omega}_g \times \vec{h} \tag{2.4.42}$$

$$\ddot{\vec{h}} = \vec{\varepsilon}_g \times \vec{h} + \vec{\omega}_g \times (\vec{\omega}_g \times \vec{h}) = -\vec{h} \times \vec{\varepsilon}_g + \vec{\omega}_g \times (\vec{\omega}_g \times \vec{h}) \tag{2.4.43}$$

were index g indicates the gripper. In matrix form

$$\ddot{h} = -\underline{\underline{h}}\varepsilon_g + \alpha \tag{2.4.44}$$

where $\underline{\underline{h}}$ is analogous to (2.4.9) and $\vec{\alpha} = \vec{\omega}_g \times (\vec{\omega}_g \times \vec{h})$. The vector h expressed in external system is obtained from one of these two expressions

$$h = A_g \tilde{h} \qquad \text{or} \qquad h = A[1\ 0\ 0]^T \tag{2.4.45}$$

Expression (2.4.3b) or (2.4.13) gives

$$\varepsilon_g = A_g \Gamma \ddot{q} + A_g \Phi \tag{2.4.46}$$

Introducing (2.4.46) into (2.4.44) one obtains

$$\ddot{h} = -\underline{\underline{h}} A_g \Gamma \ddot{q} - \underline{\underline{h}} A_g \Phi + \alpha \tag{2.4.47}$$

or

$$\ddot{h} = \Gamma' \ddot{q} + \Phi' \tag{2.4.48}$$

where

$$\Gamma' = -\underline{\underline{h}} A_g \Gamma \qquad \Phi' = -\underline{\underline{h}} A_g \Phi + \alpha \tag{2.4.49}$$

Dimensions of matrices are: $\ddot{h}$ (3×1), Γ (3×5), Γ' (3×5), Φ (3×1), Φ' (3×1), $\ddot{q}$ (5×1), α (3×1), A_g (3×3).

Combining (2.4.41) and (2.4.48), it follows

$$v \begin{bmatrix} \ddot{\theta} \\ \ddot{\varphi} \end{bmatrix} + u = \Gamma' \ddot{q} + \Phi'. \tag{2.4.50}$$

(2.4.50) represents a set of three equations which should be solved for two unknowns: $\ddot{\theta}$ and $\ddot{\varphi}$. The left minimal inverse (the generalized inverse) [27] of matrix v is

$$v^{LM} = (v^T v)^{-1} v^T \tag{2.4.51}$$

Now

$$\begin{bmatrix} \ddot{\theta} \\ \ddot{\varphi} \end{bmatrix} = \underbrace{v^{LM}\Gamma'}_{\Gamma''} \ddot{q} + \underbrace{v^{LM}(\Phi' - u)}_{\Phi''} \tag{2.4.52}$$

or introducing Γ'', Φ''

$$\begin{bmatrix} \ddot{\theta} \\ \ddot{\varphi} \end{bmatrix} = \Gamma''\ddot{q} + \Phi'' \tag{2.4.53}$$

Dimensions of matrices are: v^{LM} (2×3), Γ'' (2×5), Φ'' (2×1), $\ddot{q}$ (5×1).

Now, equations (2.4.34) and (2.4.53) can be written together in the form

$$\begin{bmatrix} \ddot{x} \\ \ddot{y} \\ \ddot{z} \\ \ddot{\theta} \\ \ddot{\varphi} \end{bmatrix} = \begin{bmatrix} \Omega' \\ \Gamma'' \end{bmatrix} \ddot{q} + \begin{bmatrix} \Theta' \\ \Phi'' \end{bmatrix} \tag{2.4.54}$$

or

$$\ddot{X}_g = J\ddot{q} + A \tag{2.4.55}$$

where the Jacobian and the adjoint matrix are

$$J = \begin{bmatrix} \Omega' \\ \Gamma'' \end{bmatrix}, \qquad A = \begin{bmatrix} \Theta' \\ \Phi'' \end{bmatrix} \tag{2.4.56}$$

Dimensions of matrices are: $\ddot{X}_g$ (5×1), $\ddot{q}$ (5×1), J (5×5), A (5×1).

Further procedure is as given in the general algorithm

$$\ddot{q} = J^{-1}(\ddot{X}_g - A) \tag{2.4.57}$$

It is evident that $\ddot{x}$, $\ddot{y}$, $\ddot{z}$, $\ddot{\theta}$, $\ddot{\varphi}$ have to be the input values. θ, $\dot{\theta}$, φ, $\dot{\varphi}$ which are also needed for calculation are obtained by integration starting from the previous time instant. This integration is performed together with the integration over generalized coordinates. $\tilde{h}$ also appears as input value.

When we first introduced the angles θ, φ (in 2.4.2) we said that this approach may sometimes result in some singularity problems. From Fig. 2.19. it is evident that if $\varphi = \pi/2$ then the angle θ cannot be defined (Fig. 2.27). In that case $f_{\theta 1} = 0$ and the rank of matrix v is equal to 1. Hence the left minimal inverse v^{LM} cannot be computed because of singularity problem. This singularity is called virtual singularity.

The adjective "virtual" is used to distinguish such singularity from the previously mentioned true singularity. In the case of true singularity a manipulator really cannot move and rotate in all directions because of the loss of one d.o.f. In the case of virtual singularity no d.o.f. is lost and the manipulator can perform any motion. Such virtual singularity follows from the mathematical apparatus applied.

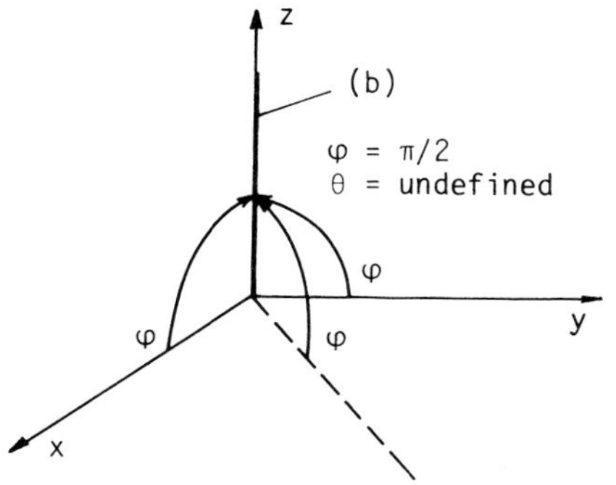

Fig. 2.27. Virtual singularity

The simplest way of avoiding the virtual singularities is to avoid positions where $\varphi = \pi/2$. If we have a trajectory which includes the position $\varphi = \pi/2$ it is enough to move slightly the trajectory so that it does not go just through the position $\varphi = \pi/2$ but passes near it.

For better understanding of this block consult the examples in 2.8.2. and 2.8.3.

At the end we mention the possibility of defining the positioning in cylindrical or spherical coordinates. It can be done in a way analogous to that explained for 4 d.o.f. manipulators.

2.4.5. Manipulator with six degrees of freedom

The first block (block 6-1). As in the case of 4 and 5 d.o.f. we may choose position vector $X_g = q$ and prescribe directly internal accelerations $\ddot{q}_1, \ldots, \ddot{q}_6$.

The second block (6-2). In this case the manipulation task is defined in the following way: we prescribe positioning and partial orientation and also one internal coordinate directly, namely $q_6(t)$. Since positioning is defined by x, y, z and partial orientation (one direction) by θ, φ, the generalized position vector is

$$X_g = \begin{bmatrix} x & y & z & \theta & \varphi & q_6 \end{bmatrix}^T \qquad (2.4.58)$$

It should be explained why this block is incorporated in the algorithm i.e. when it is suitable. To solve the desired position and partial orientation (one direction) five d.o.f. are needed. With most manipulators the sixth d.o.f. is rotational and designed so that its axis of rotation coincides with the longitudinal axis of the gripper and the working object (Fig. 2.28). So by directly prescribing the corresponding coordinate $q_6(t)$, the rotation of the working object around its axis is also prescribed (Fig. 2.28).

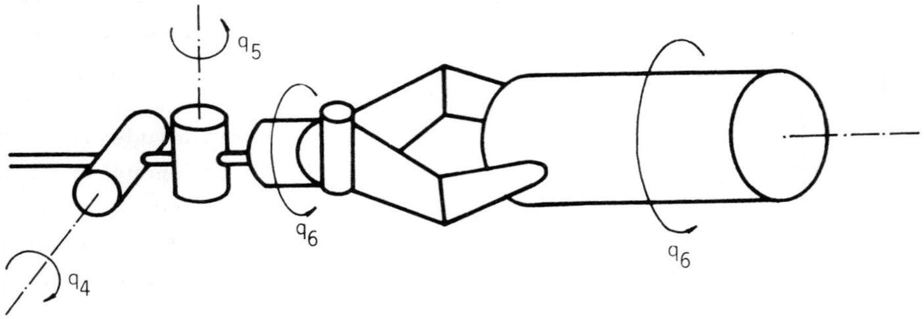

Fig. 2.28. Rotation of working object around its axis

The procedure of derivation of this block is similar to the procedure applied to 5 d.o.f. manipulators. In an analogous way we obtained equation (2.4.54) i.e.

$$\begin{bmatrix} \ddot{x} \\ \ddot{y} \\ \ddot{z} \\ \ddot{\theta} \\ \ddot{\varphi} \end{bmatrix} = \begin{bmatrix} \Omega' \\ \Gamma'' \end{bmatrix} \ddot{q} + \begin{bmatrix} \Theta' \\ \Phi'' \end{bmatrix} \tag{2.4.59}$$

But the dimensions of matrices are now different: $\ddot{q}$ (6×1), Ω' (3×6), Θ' (3×1), Γ'' (2×6), Φ'' (2×1).

If we wish to obtain the Jacobian form, then

$$\ddot{X}_g = \begin{bmatrix} \ddot{x} \\ \ddot{y} \\ \ddot{z} \\ \ddot{\theta} \\ \ddot{\varphi} \\ \ddot{q}_6 \end{bmatrix} = \underbrace{\begin{bmatrix} \Omega'_{(3\times 6)} \\ \Gamma''_{(2\times 6)} \\ \hline 0_{(1\times 5)} \quad 1 \end{bmatrix}}_{J} \ddot{q} + \underbrace{\begin{bmatrix} \Theta' \\ \Phi'' \\ \hline 0 \end{bmatrix}}_{A} \tag{2.4.60}$$

$\ddot{q}$ can now be computed by the inverse of the 6×6 Jacobian matrix

$$\ddot{q} = J^{-1}(\ddot{X}_g - A) \qquad (2.4.61)$$

Since $\ddot{q}_6$ is given we can simplify this calculation and avoid the inverse of the 6×6 matrix. Let us consider equation (2.4.59). If $\ddot{x}$, $\ddot{y}$, $\ddot{z}$, $\ddot{\theta}$, $\ddot{\varphi}$, $\ddot{q}_6$ (i.e. $\ddot{X}_g$) are input values, then (2.4.59) represents the system of 5 equations which should be solved for the 5 unknowns $\ddot{q}_1$, $\ddot{q}_2$, $\ddot{q}_3$, $\ddot{q}_4$, $\ddot{q}_5$. It is clear that the 5×5 matrix inverse now appears.

The third block (6-3). A manipulator with 6 d.o.f. solves the positioning task along with the task of total orientation.

Positioning will be treated in external Cartesian coordinate system (x, y, z), so these coordinates will be included in the position vector X_g.

Let us now discuss the problem of total orientation. When we consider the example in Fig. 2.20(b) we conclude that with the total orientation not only the direction (*) but also the direction (**) is important. So the total orientation may be considered in terms of two directions. Leter, we shall show that for the one direction given, the other is replaced by the rotation angle ψ.

Let us introduce the two directions: the main direction (b) and the auxiliary one (c). These directions are perpendicular to each other (Fig. 2.29).

Let us first define these directions with respect to the gripper. For such definition unit vectors $\vec{h}$ and $\vec{s}$ are used (Fig. 2.29). These vectors are expressed in the gripper b.-f. system. They are constant and represent the input values. In order do define the two directions (b), (c) with respect to the external system we use the three angles θ, φ, ψ (Fig. 2.29). The two directions on the gripper should coincide with the two directions in the external space. It should be pointed out that the gripper b.-f. system and the external system need not coincide.

Hence, the generalized position vector is

$$X_g = [x \; y \; z \; \theta \; \varphi \; \psi]^T \qquad (2.4.62)$$

For positioning we use (2.4.13), (2.4.14) i.e.

$$\begin{bmatrix} \ddot{x} \\ \ddot{y} \\ \ddot{z} \end{bmatrix} = w = \Omega'\ddot{q} + \Theta' \tag{2.4.63}$$

Dimensions of matrices are: $\ddot{q}$ (6×1), Ω' (3×6), Θ' (3×1).

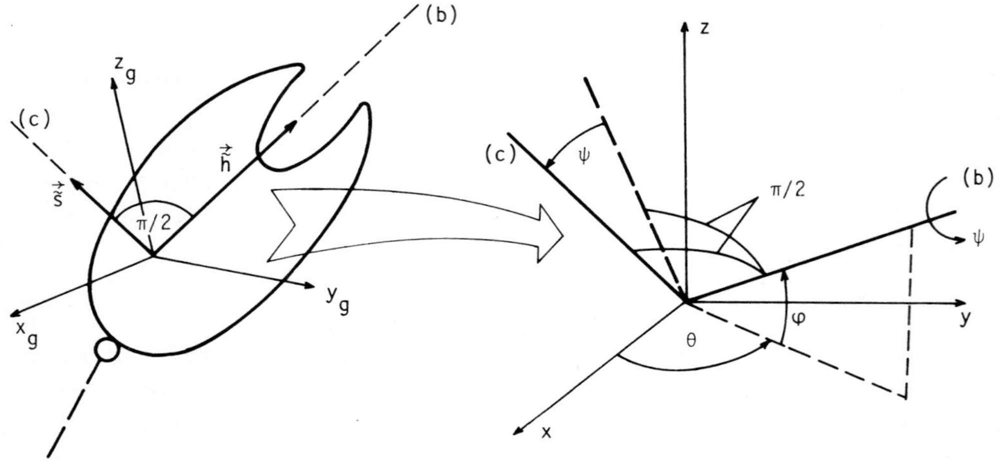

Fig. 2.29. Definition of total orientation in terms of two directions

Let A_g be the transition matrix of the gripper b.-f. system and A the transition matrix of the orientation system. Since the x-axis of the orientation system coincides with the direction (b) the unit vector $\vec{h}$ can be expressed in the orientation system via projections $[1\ 0\ 0]^T$. So, the vector $\vec{h}$ can be expressed in the external system as

$$h = A \begin{bmatrix} 1 \\ 0 \\ 0 \end{bmatrix} \tag{2.4.64}$$

From Figures 2.23 and 2.29 we conclude that the z-axis of the orientation system coincides with the direction (c). Hence the unit vector $\vec{s}$ can be expressed in the orientation system via projections $[0\ 0\ 1]^T$ and in the external system

$$s = A \begin{bmatrix} 0 \\ 0 \\ 1 \end{bmatrix} \tag{2.4.65}$$

The following transformations are similar to those derived in the second block of 2.4.4. but they will be shortly repeated here.

The second derivative of (2.4.64) gives

$$\ddot{h} = \ddot{A} \begin{bmatrix} 1 \\ 0 \\ 0 \end{bmatrix}$$

By combining this relation with (2.4.22) one obtains

$$\ddot{h} = F_\theta \begin{bmatrix} 1 \\ 0 \\ 0 \end{bmatrix} \ddot{\theta} + F_\varphi \begin{bmatrix} 1 \\ 0 \\ 0 \end{bmatrix} \ddot{\varphi} + F_\psi \begin{bmatrix} 1 \\ 0 \\ 0 \end{bmatrix} \ddot{\psi} + G \begin{bmatrix} 1 \\ 0 \\ 0 \end{bmatrix} \tag{2.4.66}$$

or

$$\ddot{h} = f_{\theta 1}\ddot{\theta} + f_{\varphi 1}\ddot{\varphi} + G[1\ 0\ 0]^T$$

$$= \begin{bmatrix} f_{\theta 1} & f_{\varphi 1} \end{bmatrix} \begin{bmatrix} \ddot{\theta} \\ \ddot{\varphi} \end{bmatrix} + G[1\ 0\ 0]^T \tag{2.4.67}$$

where $f_{\theta 1}$ is the first column of matrix F_θ and $f_{\varphi 1}$ is the first column of F_φ. The first column of F_ψ is $f_{\psi 1} = 0$. If we introduce the notation

$$v_h = \begin{bmatrix} f_{\theta 1} & f_{\varphi 1} \end{bmatrix}; \qquad u_h = G[1\ 0\ 0]^T \tag{2.4.68}$$

then

$$\ddot{h} = v_h \begin{bmatrix} \ddot{\theta} \\ \ddot{\varphi} \end{bmatrix} + u_h \tag{2.4.69}$$

Dimensions of matrices are: $\ddot{h}$ (3×1), v_h (3×2), u_h (3×1).

In a way analogous to that in the second block of 2.4.4. one obtains

$$\begin{aligned} \dot{\vec{h}} &= \vec{\omega}_g \times \vec{h} \\ \ddot{\vec{h}} &= -\vec{h} \times \vec{\varepsilon}_g + \overbrace{\vec{\omega}_g \times (\vec{\omega}_g \times \vec{h})}^{\vec{\alpha}_h} \end{aligned} \tag{2.4.70}$$

and in the matrix form

$$\ddot{h} = -\underline{\underline{h}}\varepsilon_g + \alpha_h \tag{2.4.71}$$

Since from (2.4.3b) or (2.4.13) it follows that

$$\varepsilon_g = A_g\Gamma\ddot{q} + A_g\Phi \tag{2.4.72}$$

equation (2.4.71) becomes

$$\ddot{h} = \Gamma_h'\ddot{q} + \Phi_h' \tag{2.4.73}$$

where

$$\Gamma_h' = -\underline{\underline{h}}A_g\Gamma \qquad \Phi_h' = -\underline{\underline{h}}A_g\Phi + \alpha_h \tag{2.4.74}$$

Dimensions of matrices are: $\ddot{h}$ (3×1), $\ddot{q}$ (6×1), Γ (3×6), Γ_h' (3×6), Φ (3×1), Φ_h' (3×1), α_h (3×1), A_g (3×3).

Combining (2.4.69) with (2.4.73)

$$v_h \begin{bmatrix} \ddot{\theta} \\ \ddot{\varphi} \end{bmatrix} + u_h = \Gamma_h'\ddot{q} + \Phi_h' \tag{2.4.75}$$

Since v_h is of dimensions (3×2), system (2.4.75) is solved for $[\ddot{\theta}\ \ddot{\varphi}]^T$ by using the left minimal inverse

$$v_h^{LM} = \left(v_h^T\ v_h\right)^{-1} v_h^T \tag{2.4.76}$$

Now

$$\begin{bmatrix} \ddot{\theta} \\ \ddot{\varphi} \end{bmatrix} = \Gamma_h''\ \ddot{q} + \Phi_h'' \tag{2.4.77}$$

where

$$\Gamma_h'' = v_h^{LM}\Gamma_h', \qquad \Phi_h'' = v_h^{LM}\left(\Phi_h' - u_h\right) \tag{2.4.78}$$

Dimensions of matrices are: v_h^{LM} (2×3), Γ_h'' (2×6), Φ_h'' (2×1), $\ddot{q}$ (6×1).

Now, for the vector s it follows from (2.4.65) that

$$\ddot{s} = \ddot{A} \begin{bmatrix} 0 \\ 0 \\ 1 \end{bmatrix} \tag{2.4.79}$$

By using (2.4.22) in (2.4.79) one obtains

$$\ddot{s} = F_\theta \begin{bmatrix} 0 \\ 0 \\ 1 \end{bmatrix} \ddot{\theta} + F_\varphi \begin{bmatrix} 0 \\ 0 \\ 1 \end{bmatrix} \ddot{\varphi} + F_\psi \begin{bmatrix} 0 \\ 0 \\ 1 \end{bmatrix} \ddot{\psi} + G \begin{bmatrix} 0 \\ 0 \\ 1 \end{bmatrix} \qquad (2.4.80)$$

The matrix product $F_\theta[0\ 0\ 1]^T$ represents the third column of matrix F_θ. Let this column be marked by $f_{\theta 3}$. The third column of F_φ is $f_{\varphi 3}$ and the third column of F_ψ is $f_{\psi 3}$. Thus (2.4.80) gives

$$\ddot{s} = f_{\theta 3}\ddot{\theta} + f_{\varphi 3}\ddot{\varphi} + f_{\psi 3}\ddot{\psi} + G[0\ 0\ 1]^T \qquad (2.4.81)$$

If we introduce the notation

$$v_s' = \begin{bmatrix} f_{\theta 3} & f_{\varphi 3} \end{bmatrix}; \qquad v_s'' = \begin{bmatrix} f_{\psi 3} \end{bmatrix}; \qquad u_s = G[0\ 0\ 1]^T \qquad (2.4.82)$$

then

$$\ddot{s} = v_s' \begin{bmatrix} \ddot{\theta} \\ \ddot{\varphi} \end{bmatrix} + v_s'' \ddot{\psi} + u_s \qquad (2.4.83)$$

Dimensions of matrices are: $\ddot{s}$ (3×1), v_s' (3×2), v_s'' (3×1), u_s (3×1).

Analogously to (2.4.70) one obtains

$$\dot{\vec{s}} = \vec{\omega}_g \times \vec{s}$$

$$\ddot{\vec{s}} = -\vec{s} \times \vec{\varepsilon}_g + \vec{\omega}_g \times (\vec{\omega}_g \times \vec{s}) \qquad (2.4.84)$$

and in the matrix form

$$\ddot{s} = -\underline{\underline{s}}\varepsilon_g + \alpha_s \qquad (2.4.85)$$

where $\vec{\alpha}_s = \vec{\omega}_g \times (\vec{\omega}_g \times \vec{s})$, and the vector $\vec{s}$ in the external system is obtained from one of these two expressions

$$s = A_g\, \tilde{s} \qquad \text{or} \qquad s = A[0\ 0\ 1]^T \qquad (2.4.85a)$$

If we introduce (2.4.72) into (2.4.85) it follows

$$\ddot{s} = \Gamma_s' \ddot{q} + \Phi_s' \qquad (2.4.86)$$

where

$$\Gamma_s' = -\underline{\underline{sA}}_g\Gamma; \qquad \Phi_s' = -\underline{\underline{sA}}_g\Phi + \alpha_s \tag{2.4.87}$$

Dimensions of matrices are: $\ddot{s}$ (3×1), $\ddot{q}$ (6×1), Γ (3×6), Γ_s' (3×6), Φ (3×1), Φ_s' (3×1), α_s (3×1).

Combining (2.4.83) with (2.4.86)

$$v_s' \begin{bmatrix} \ddot{\theta} \\ \ddot{\varphi} \end{bmatrix} + v_s'' \ddot{\psi} + u_s = \Gamma_s'\ddot{q} + \Phi_s' \tag{2.4.88}$$

Let us now introduce (2.4.77) into (2.4.88) to obtain

$$v_s'\Gamma_h''\ddot{q} + v_s'\Phi_h'' + v_s''\ddot{\psi} + u_s = \Gamma_s'\ddot{q} + \Phi_s' \tag{2.4.88a}$$

or

$$v_s''\ddot{\psi} = \left(\Gamma_s' - v_s'\Gamma_h''\right)\ddot{q} + \Phi_s' - v_s'\Phi_h'' - u_s \tag{2.4.89}$$

(2.4.89) represents a system of three equations which should be solved for $\ddot{\psi}$. The left minimal inverse of v_s'' is

$$v_s''^{LM} = \left(v_s''^{T} v_s''\right)^{-1} v_s''^{T} \tag{2.4.90}$$

Now

$$\ddot{\psi} = \Gamma_s''\ddot{q} + \Phi_s'' \tag{2.4.91}$$

where

$$\Gamma_s'' = v_s''^{LM}\left(\Gamma_s' - v_s'\Gamma_h''\right), \qquad \Phi_s'' = v_s''^{LM}\left(\Phi_s' - v_s'\Phi_h'' - u_s\right) \tag{2.4.92}$$

Dimensions of matrices are: $v_s''^{LM}$ (1×3), Γ_s'' (1×6), Φ_s'' (1×1), $\ddot{q}$ (6×1).

Now, equations (2.4.63), (2.4.77) and (2.4.91) can be written together in the form

$$\begin{bmatrix} \ddot{x} \\ \ddot{y} \\ \ddot{z} \\ \ddot{\theta} \\ \ddot{\varphi} \\ \ddot{\psi} \end{bmatrix} = \begin{bmatrix} \Omega' \\ \Gamma_h'' \\ \Gamma_s'' \end{bmatrix}\ddot{q} + \begin{bmatrix} \Theta' \\ \Phi_h'' \\ \Phi_s'' \end{bmatrix} \tag{2.4.93}$$

or

$$\ddot{X}_g = J\ddot{q} + A \tag{2.4.94}$$

where the Jacobian and the adjoint matrix are

$$J = \begin{bmatrix} \Omega' \\ \Gamma''_h \\ \Gamma''_s \end{bmatrix} ; \qquad A = \begin{bmatrix} \Theta' \\ \Phi''_h \\ \Phi''_s \end{bmatrix} \tag{2.4.95}$$

Dimensions of matrices are: $\ddot{X}_g$ (6×1), $\ddot{q}$ (6×1), J (6×6), A (6×1).

The accelerations can now be calculated

$$\ddot{q} = J^{-1}(\ddot{X}_g - A) \tag{2.4.96}$$

It is evident that $\ddot{x}$, $\ddot{y}$, $\ddot{z}$, $\ddot{\theta}$, $\ddot{\varphi}$, $\ddot{\psi}$ have to be the input values. θ, $\dot{\theta}$, φ, $\dot{\varphi}$, ψ, $\dot{\psi}$ are obtained by integration from the previous time instant. $\tilde{h}$, $\tilde{s}$ also appear as input.

The discussion on virtual singularities presented in the second block of 2.4.4. also holds here.

During the derivation of the adapting blocks (the second block in 2.4.4. and the second and the third block in 2.4.5.) we have used the unit vectors $\vec{\tilde{h}}$, $\vec{\tilde{s}}$ to define the directions on the gripper. These vectors are expressed in the gripper b.-f. system. They are constant and represent the input values. But we can conclude from the derivation presented that these vectors are not necessary. The vectors $\vec{\tilde{h}}$, $\vec{\tilde{s}}$ are used only for the calculation of $\vec{h}$, $\vec{s}$. From (2.4.45) and (2.4.85a) we see that the vectors $\vec{h}$, $\vec{s}$ can be obtained by using the orientation system and thus without using $\vec{\tilde{h}}$, $\vec{\tilde{s}}$. Let us explain this. If the position of a manipulator is known (e.g. known q) and if the directions (b), (c) are known in the external system (e.g. known θ, φ, ψ) then the relative orientation of these two directions with respect to the manipulator gripper is completely determined. Hence, determination of this relative orientation by defining $\vec{\tilde{h}}$, $\vec{\tilde{s}}$ is unnecesary (superfluous). So, why do we still use vectors $\vec{\tilde{h}}$, $\vec{\tilde{s}}$? It was said that the values of q, $\dot{q}$, θ, $\dot{\theta}$, φ, $\dot{\varphi}$, ψ, $\dot{\psi}$ in each time instant are obtained by integration from the previous time instant. Hence, in the initial time instant t_o all these values have to be prescribed. One possibility is to prescribe x, y, z, θ, φ,

ψ in this initial time instant and compute $q(t_o)$ (for simplicity we assume that the manipulator starts from a resting position so all velocities i.e. all first derivatives are equal to zero). But, in this approach we face the problem of calculating the internal coordinates q for known external coordinates X, i.e. $q = \eta^{-1}(X)$. In Para. 2.2. it was said that such procedure was very extensive and undesirable. Hence we use another approach. We prescribe q in the initial time instant (i.e. $q(t_o)$) and also $\vec{\tilde{h}}$, $\vec{\tilde{s}}$. Now the values of θ, φ, ψ are obtained by simple calculation. This approach using $\vec{\tilde{h}}$, $\vec{\tilde{s}}$ also offers a possibility of better visual relation with the task. Let us explain it by an example. If a manipulator has to move an object in an assembly task (Fig. 2.30) then the longitudinal axis (b) of the object and the perpendicular one (c) are essential. In order to define which directions on the gripper are important we use $\vec{\tilde{h}} = \{0, 1, 0\}$, $\vec{\tilde{s}} = \{-1, 0, 0\}$. In this way we define directly the relative orientations of directions (b), (c) with respect to the gripper. From the standpoint of visual relation with the task we consider this approach to be more convenient than the definition of relative orientations in terms of absolute position of the manipulator and the absolute orientations of (b), (c) with respect to external system. Anyhow, these two approaches are both possible, and the user will choose one of them. For better understanding of this block consult the example in 2.8.4.

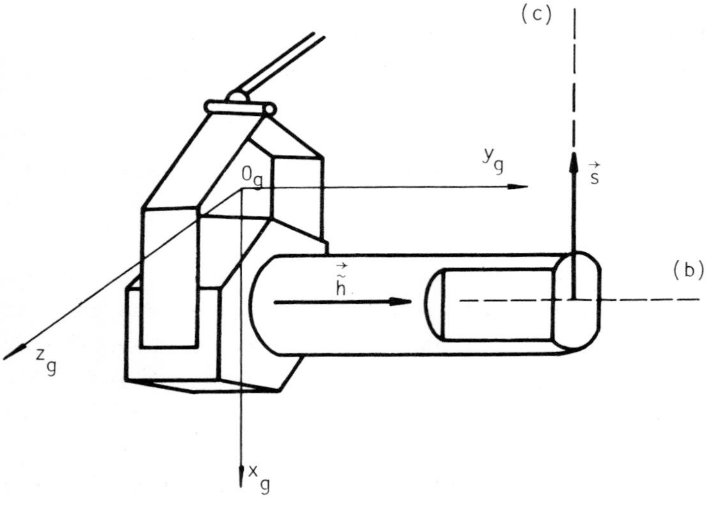

Fig. 2.30. A gripper carrying a screw

2.4.6. Velocity profiles and practical realization of adapting blocks

In the paragraphs dealing with the derivation of adapting blocks it was said that each block uses its special generalized position vector X_g which represents set of parameters defining a manipulator position. The internal and the external coordinates may appear as the elements of X_g. If a manipulator has n d.o.f. then the generalized position vector is of dimension n. Let $p_1, p_2, \ldots, p_n$ be parameters defining a manipulator position. Then the generalized position vector is

$$X_g = \begin{bmatrix} p_1 & p_2 & \cdots & p_n \end{bmatrix}^T \tag{2.4.97}$$

It was explained that an adapting block operates in such a way that for some time instant it calculates $\ddot{q}$ on the basis of known $\ddot{X}_g$. Hence $\ddot{X}_g$ is needed in each time instant. The algorithm also uses the values of the parameters and their first derivatives. It does not hold for all parameters but only for those defining the gripper orientation (e.g. θ, φ, ψ). For generality we say that X_g, $\dot{X}_g$ are also needed in each time instant. X_g and $\dot{X}_g$ in some time instant are computed by integration starting from the previous time instant. Thus, $\ddot{X}_g$ for each time instant represents the input. In order to start the time iterative procedure we also need the initial values of X_g, $\dot{X}_g$ i.e. $X_g(t_o)$, $\dot{X}_g(t_o)$.

The procedure described is completely general and can be used for any trajectory. But, for some given manipulation task it is not so easy to find the values of $\ddot{X}_g$ in a series of time instants. Hence, we develop a subroutine which prepares these data for some typical manipulation tasks. We consider manipulator motions where X_g changes from one given value to another with the triangular or trapezoidal velocity profile.

The triangular profile for one of parameters is shown in Fig. 2.31a.

We define this profile by a constant acceleration and deceleration:

$$a_i = \frac{4\Delta p_i}{T^2} = \frac{4\left(p_i^{end} - p_i^{start}\right)}{T^2} \tag{2.4.98}$$

Now

$$\ddot{p}_i(t) = \begin{cases} a_i \; ; & t < T/2 \, (t \in AB) \\ -a_i \; ; & t > T/2 \, (t \in BC) \end{cases} \tag{2.4.99}$$

The algorithm operates with discretized time. Let the interval T be divided into 60 subintervals Δt by using a series of time instants t_o, $t_1, \ldots, t_{60}$. Let $\ddot{p}_{i\ell}$ be the acceleration in time instant t_ℓ. Now, in discretized form:

$$\ddot{p}_{i\ell} = \begin{cases} a_i; & \ell = 0,\ 1, \ldots, 29 \\ -a_i; & \ell = 30,\ 31, \ldots, 59 \end{cases} \tag{2.4.100}$$

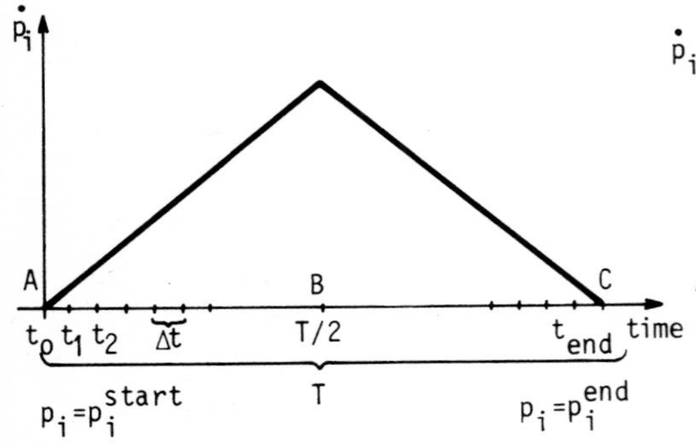

Fig. 2.31a. Triangular profile of parameter velocity

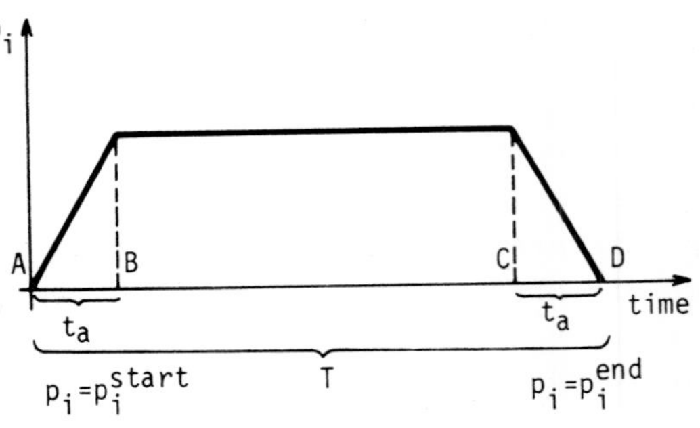

Fig. 2.31b. Trapezoidal profile of parameter velocity

Repeating the procedure (2.4.100) for each parameter one obtains $\ddot{X}_g$ for all time instants. This is done and stored in the computer memory before starting the time-iterative algorithm. Thus only the starting and the terminal position of manipulator together with the execution time have to be prescribed (i.e. X_g^{start}, X_g^{end}, T). The number of subintervals (it has been assumed to be 60 here) can be changed as wished.

Let us explain what this triangular profile means in practice. If $X_g = q$ then each internal coordinate changes between the two given values with constant acceleration and deceleration. Let us now consider another case: $X_g = [x\ y\ z\ \theta\ \varphi\ \psi]^T$. Then the triangular velocity profiles for parameters x, y, z produce straight-line motion of the gripper with the triangular profile of gripper velocity $\vec{v}_g$. Rotations θ, φ, ψ also follow such a profile.

The trapezoidal velocity profile is shown in Fig. 2.31b. Such motion is sometimes called the motion with constant velocity because the velocity is constant over the longest part of the trajectory. For such motion one has to prescribe the initial and the terminal position (i.e. p_i^{start}, p_i^{end}, i=1,...,n) together with the execution time T and the acceleration (deceleration) time t_a. The subroutine procedure computes the constant value of acceleration:

$$a_i = \frac{p_i^{end} - p_i^{start}}{t_a(T-t_a)} \qquad (2.4.101)$$

and the values of $\ddot{p}_i$, i=1,...,n:

$$\ddot{p}_i(t) = \begin{cases} a_i; & t < t_a \quad (t \in AB) \\ 0\ ; & t_a < t < T - t_a \quad (t \in BC) \\ -a_i; & t > T - t_a \quad (t \in CD) \end{cases} \qquad (2.4.102)$$

If $X_g = [x\ y\ z\ \theta\ \varphi\ \psi]$, then the trapezoidal velocity profiles for parameters x, y, z produce straight line motion of the gripper with the trapezoidal profile of gripper velocity $\vec{v}_g$.

We have discussed the two most common velocity profiles. Let us notice that the initial and terminal velocities equal zero. Some additional subroutines can be prepared for any other desired profile.

In order to obtain a more general algorithm we consider a trajectory which consists of several segments each of them being of triangular or trapezoidal type. Let there be m segments of trajectory i.e. let a manipulator move sequentially into m given points $A_1,\ldots,A_m$ starting from the initial point A_o (Fig. 2.32).

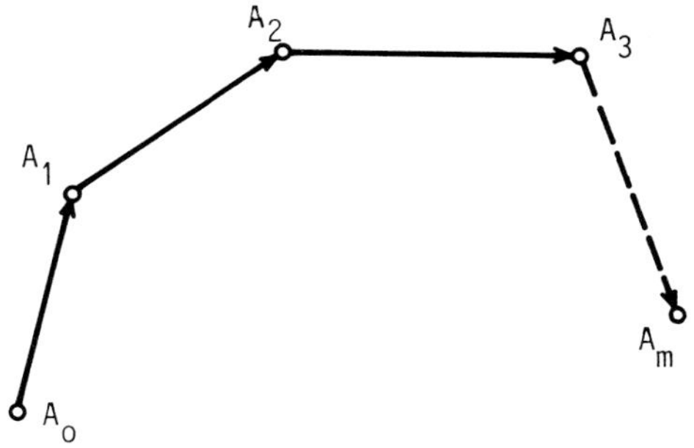

Fig. 2.32. Manipulator trajectory

So, it is first necessary to prescribe the number of points m. Then we prescribe the positions of division points $A_1, A_2,\ldots,A_m$ in terms of $X_{g1}, X_{g2},\ldots,X_{gm}$. Now if a transition $A_{k-1} \to A_k$ is of triangular type then T^k is prescribed and if it is of trapezoidal type then T^k, t_a^k are prescribed (k is not exponent but upper index). Special indicators are used to define the profile of each transition. For the trapezoidal profile the value of the indicator is 1 and for the triangular profile the value is 2. Let it be noticed that in the case of triangular and trapezoidal profiles a manipulator stops in each point $A_1,\ldots,A_m$.

There is another transformation incorporated in the algorithm. It was said that θ, φ, ψ have to be prescribed in each point A_i, in order to define the directions (b), (c) with respect to the external space. But it is sometimes easier to determine the external projections of the corresponding vectors $\vec{h}$ and $\vec{s}$ then determine the angles θ, ψ, φ. In that case we prescribe $\vec{h}$, $\vec{s}$ (with respect to external system) in each division point $A_1,\ldots,A_m$ and there is a subroutine which computes θ, ψ, φ. Prescription of $\vec{h}$ is equivalent to the prescription of a plane perpendicular to $\vec{h}$.

At the end of 2.4.5. it was explained that there are two possibilities for the initial point A_o. For that point we can prescribe X_{go} and compute the initial position $q(t_o)$ necessary for the algorithm start. In order to avoid the extensive calculation $q = \eta^{-1}(X_g)$ we suggest that the initial state be directly prescribed and then simply calculated $X_g = \eta(q)$. It is due to the fact that manipulators usually start from some well known starting position so $q(t_o)$, $\dot{q}(t_o)$ can easily be determined. If sometimes the determination of $q(t_o)$, $\dot{q}(t_o)$ is complicated then we suggest to start the algorithm from some new initial point in which the state is known and move the manipulator towards the old initial point which is now the first point in the sequence.

The trajectory described is not completely general but is general enough for the aim of the algorithm. It covers most manipulation tasks in practice. Anyhow the algorithm can be generalized by developing and incorporating some additional trajectory-input subroutines.

2.5. Calculation of Other Dynamic Characteristics

In the previous paragraphs it was shown how driving forces and torques $P(t)$ which produce the prescribed motion of a manipulator are calculated. As the output we also obtained the trajectory in the state space i.e. $q(t)$, $\dot{q}(t)$. This calculation is the basis of the algorithm for dynamic analysis of manipulator motion. But, we are usually also interested in some other dynamic characteristics. The calculation of each dynamic characteristic is based on $P(t)$, $q(t)$, $\dot{q}(t)$, $\ddot{q}(t)$ and represents only an additional block in the algorithm described. We now discuss some interesting dynamic characteristics.

2.5.1. Diagrams of torque vs. r.p.m.

One interesting characteristic which can be obtained as the output of the algorithm is the diagram of torque vs. rotation speed of motor. Since the rotation speed of a motor is usually expressed in terms of revolutions per minute (r.p.m.), we talk about torque-r.p.m. diagram. This diagram can be computed for each joint i.e. each motor (actuator). It will be shown that such diagrams are suitable for both manipulators driven by D.C. electromotors and manipulators driven by hydraulic actuators.

Let us first consider D.C. actuators. The dynamic analysis algorithm computes the torque P_i and the internal (generalized) velocity $\dot{q}_i$ for each joint S_i and each time instant. The r.p.m. for a joint is $n_i = \frac{60}{2\pi}\dot{q}_i$ ($\dot{q}_i$ is expressed in rad/s), so each time instant gives one point of the P_i-n_i diagram. This diagram is valid for the shaft of the joint considered. But we are usually interested in the diagram for the motor itself. Then the reducer in the joint must be taken into account. Let us consider a reducer with the speed reduction ratio equal to N. Then the motor r.p.m. is

$$n_i^m = N_i n_i = N_i \frac{60}{2\pi} \dot{q}_i \tag{2.5.1}$$

If the reducer has the mechanical efficiency $\eta(N, n)$, then the torque reduction ratio is $N \cdot \eta(N, n)$ and so the motor torque is

$$P_i^m = \frac{P_i}{N_i \cdot \eta_i} \tag{2.5.2}$$

Hence the motor diagram $P_i^m - n_i^m$ is obtained by calculating (2.5.1) and (2.5.2) in each time instant $t_o, t_1, t_2, \ldots$ One example of a torque - r.p.m. diagram is shown in Fig. 2.33. Only a qualitative presentation of the diagram is given.

Such diagrams are very useful during the synthesis and choice of D.C. servosystems. The producer gives the $P^m_{max} - n^m$ motor characteristic in the catalog, where P^m_{max} is the maximal motor torque at motor r.p.m.$=n^m$. By comparing the necessary characteristic, obtained by means of the algorithm described, with the one from catalogue, one can decide whether the chosen motor suits its application. The use of these characteristics will be discussed further in 2.6.1. and 4.5.

In the case of hydraulic actuator the procedure is similar but usually without reducers. In the case of a rotational hydraulic actuator we obtain the diagram of torque vs. rotational velocity in the joint i.e. P-$\dot{q}$. If a hydraulic actuator with a cylinder (translational motion of the piston) is used then we obtain the characteristic of force vs. linear velocity of the piston (there is no r.p.m.). In the case of a hydraulic actuator the maximal characteristic P_{max} - $\dot{q}$ is different from the one holding for D.C. motors.

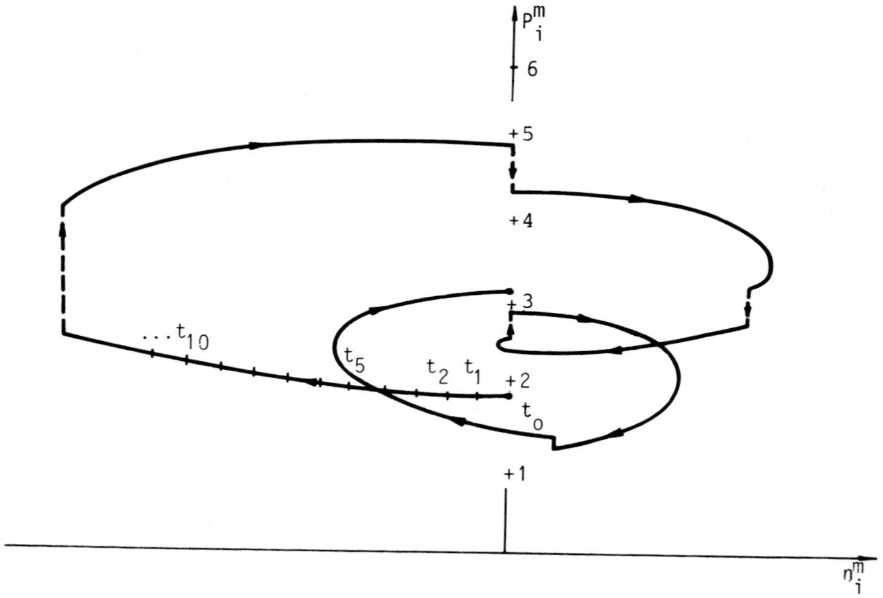

Fig. 2.33. A torque - r.p.m. diagram

A few things should be pointed out. If a rotational actuator drives a rotational joint then the connection is usually direct (or by means of a reducer). Another simple case appears when hydraulic actuator with cylinder drives a linear joint. But, if an actuator with the hydraulic cylinder drives a rotational joint then there is usually a nonlinear dependence between the piston motion and the rotation in joint. This should be kept in mind.

2.5.2. Calculation of the power needed and the energy consumed

The next interesting characteristics (e.g. for the choice of actuators) are the power needs in each joint. For a joint S_i the power needed in some time instant is obtained as $Q_i = P_i \cdot \dot{q}_i$. But the power produced by the motor has to be larger because of the power loss in the reducer. So the necessary motor power in this time instant is $Q_i^m = P_i \cdot \dot{q}_i / \eta_i$, where η_i is the mechanical efficiency of the reducer. Hence, the time history of the power needed is obtained. It should be pointed out that this is the output mechanical power of the motor.

The energy consumption may also be easily computed. Let us consider a joint S_i and the corresponding actuator. Let $E_i^{(k)}$ be the energy consumed in the first k time steps, and let ΔE_i^k be the energy consumed in the k-th time step (time interval Δt_k). The total energy consumed by the i-th joint actuator is calculated in such a way that, during the time-iterative procedure of dynamic analysis, summation of the energy at each step Δt_k is found:

$$E_i^{(k)} = E_i^{(k-1)} + \Delta E_i^k \tag{2.5.3}$$

To calculate ΔE_i^k, we adopt the medium drive value on the interval

$$P_{imed}^k = \frac{1}{2}\left(P_i^{k-1} + P_i^k\right), \tag{2.5.4}$$

where the upper index indicates the k-th time instant. Now

$$\Delta E_i^k = P_{imed}^k \cdot \Delta q_i^k \tag{2.5.5}$$

where $\Delta q_i^k = q_i^k - q_i^{k-1}$.

It should be stressed that this discussion on energy has dealt with the mechanical power and mechanical energy only. If we want to calculate the energy which has to be taken from an energy source (e.g. from an electric battery) then we should take care of the energy lost in actuators. Thus, the dynamic models of actuators have to be introduced. These models are discussed in 2.9. In this paragraph we give only some ideas of such energy consumption calculation. If a manipulator is driven by D.C. electromotors then the energy loss follows from resistance and friction effects. Let us consider one joint and the corresponding motor. The power required from a source can be computed as $Q = u\, i_r$ where u is control voltage and i_r is rotor current. Now, in a time step Δt_k the energy increment is

$$\Delta E_i^k = u_i^k i_{r_i}^k \Delta t_k = Q_i \Delta t_k$$

where the lower index represents the number of joint and the upper index indicates the k-th time instant.

In the case of a hydraulic actuator one should take care about the loss due to friction and leakage.

The energy consumed by the whole manipulator is the sum of energies consumed by actuators.

2.5.3. Calculation of reactions in joints and stresses in segments

Reaction forces and moments and, especially, stresses in manipulator segments represent very useful data for manipulator design process. We first derive the procedure for the calculation of reactions in manipulator joints. At the beginning, it is necessary to discuss in more details the case of linear joints. Such a joint was shown in Fig. 2.5. and the vector $\vec{r}'_{ii}$ was introduced. The vector $\vec{r}_{i-1,i}$ was considered constant. Such joint representation was satisfactory for the formation of mathematical model. But, if we wish to calculate reactions and stresses and, leter, elastic deformations, it is necessary to distinguish two different cases of linear joints. These two cases are presented in Fig. 2.34. Let us introduce the indicator p_i, which determines to which of the two cases linear joint S_i belongs:

$$p_i = \begin{cases} 1, & \text{if joint } S_i \text{ is of type (a) (Fig. 2.34)} \\ 0, & \text{if joint } S_i \text{ is of type (b) (Fig. 2.34)} \end{cases} \tag{2.5.6}$$

We now introduce

$$\begin{aligned} \vec{r}'_{ii} &= \vec{r}_{ii} + q_i s_i p_i \vec{e}_i \\ \vec{r}'_{i-1,i} &= \vec{r}_{i-1,i} - q_i s_i (1-p_i) \vec{e}_i \end{aligned} \tag{2.5.7}$$

Thus, the translation in joint S_i is taken into account either through $\vec{r}'_{ii}$ (in case (a)) or through $\vec{r}'_{i-1,i}$ (case (b)).

Let us now consider the i-th segment (Fig. 2.35). Let $\vec{F}_{Si}$ be the total force in the joint S_i acting on the i-th segment, and let $\vec{M}_{Si}$ be the total moment in the same joint acting on the i-th segment. Notice that now $\vec{F}_{Si+1}$ and $\vec{M}_{Si+1}$ act on the (i+1)-th segment, and $-\vec{F}_{Si+1}$, $-\vec{M}_{Si+1}$ act on the i-th segment.

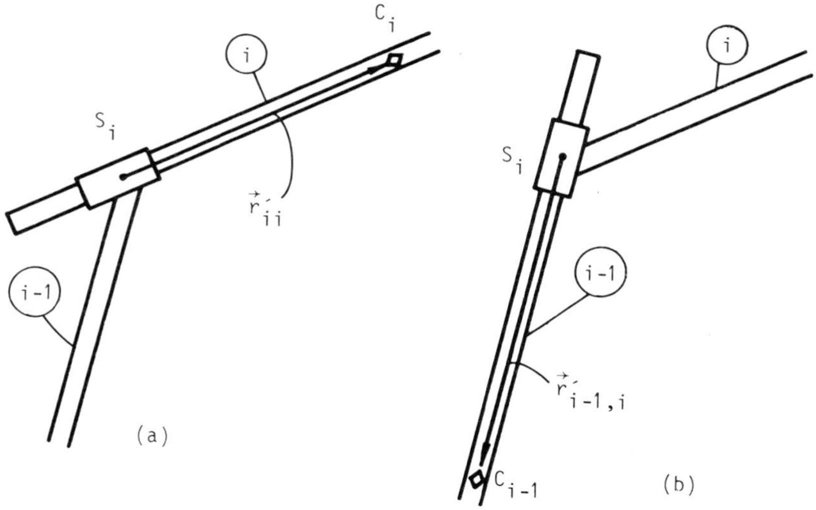

Fig. 2.34. Two types of linear joint

From the theorem of center of gravity motion it follows

$$m_i\vec{w}_i = \vec{F}_{Si} - \vec{F}_{Si+1} + m\vec{g} \quad (2.5.8)$$

i.e.

$$\vec{F}_{Si} = \vec{F}_{Si+1} - m_i\vec{g} + m_i\vec{w}_i \quad (2.5.9)$$

Now let us apply the theorem of angular momentum with respect to the point S_i.

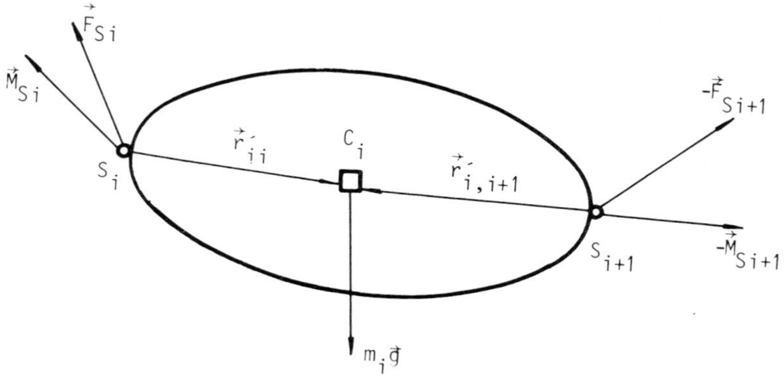

Fig. 2.35. One segment and the forces and moments acting on it

$$A_i\left[\tilde{J}_i\vec{\varepsilon}_i - \left(\tilde{J}_i\vec{\omega}_i\right)\times\vec{\omega}_i\right] + \vec{r}'_{ii}\times m_i\vec{w}_i =$$

$$= \vec{M}_{Si} - \vec{M}_{Si+1} - \left(\vec{r}'_{ii}-\vec{r}'_{i,i+1}\right)\times\vec{F}_{Si+1} + \vec{r}'_{ii}\times m_i\vec{g} \qquad (2.5.10)$$

i.e.

$$\vec{M}_{Si} = \vec{M}_{Si+1} + \left(\vec{r}'_{ii}-\vec{r}'_{i,i+1}\right)\times\vec{F}_{Si+1} + A_i\left[\tilde{J}_i\vec{\varepsilon}_i - \left(\tilde{J}_i\vec{\omega}_i\right)\times\vec{\omega}_i\right] +$$

$$+ \vec{r}'_{ii}\times m_i\vec{w}_i - \vec{r}'_{ii}\times m_i\vec{g} \qquad (2.5.11)$$

where A_i is the transition matrix of the i-th segment.

Since the motion and, thus, $\vec{w}_i$, $\vec{\varepsilon}_i$ are known, equations (2.5.9) and (2.5.11) offer the possibility of recursive calculation of total forces and moments in manipulator joints. The boundary conditions for this backward recursion are

$$\vec{F}_{Sn+1} = \vec{F}_{end} \qquad (2.5.12a)$$

$$\vec{M}_{Sn+1} = \vec{M}_{end} \qquad (2.5.12b)$$

and they are shown in Fig. 2.36. $\vec{F}_{end}$ and $\vec{M}_{end}$ are the force and the moment the manipulator produces if in contact with some object on the ground (Fig. 2.36a). If the last segment of the manipulator is free (not in contact with the ground), $\vec{F}_{end} = 0$, $\vec{M}_{end} = 0$ (Fig. 2.36b).

Recursive expressions (2.5.9) and (2.5.11) may become even more convenient if written in b.-f. systems.

Let us now consider a rotational joint S_i. In that joint, there is the driving torque $\vec{P}_i^M$ (with the direction along $\vec{e}_i$), reaction moment $\vec{M}_{Ri}$ (perpendicular to $\vec{e}_i$), and the reaction force $\vec{F}_{Ri}$ (Fig. 2.37a).

If the joint S_i is linear, then there is a driving force $\vec{P}_i^F$ (along $\vec{e}_i$), the reaction force $\vec{F}_{Ri}$ (perpendicular to $\vec{e}_i$), and the reaction moment $\vec{M}_{Ri}$ acting on the i-th segment (Fig. 2.37b).

Thus, the total force in joint S_i is:

$$\vec{F}_{Si} = \vec{F}_{Ri} + s_i\vec{P}_i \qquad (2.5.13)$$

and the total moment

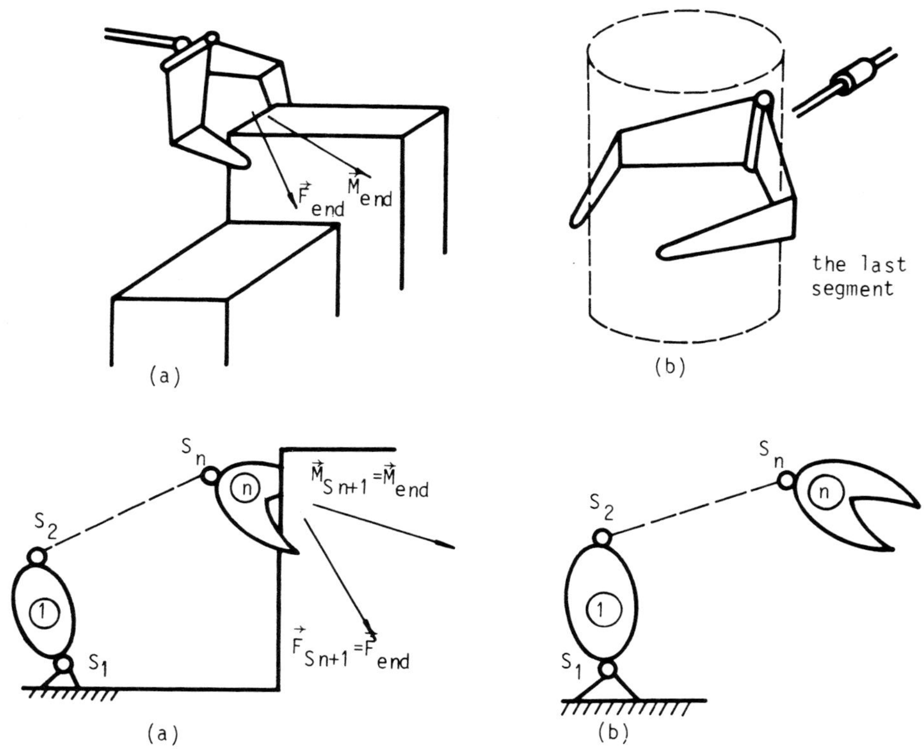

Fig. 2.36. Boundary conditions on the last segment

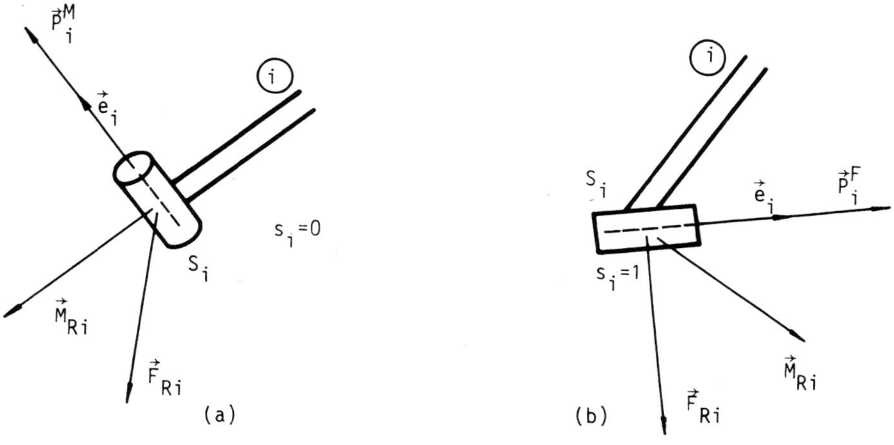

Fig. 2.37. Reactions and drives in joints

$$\vec{M}_{Si} = \vec{M}_{Ri} + (1-s_i)\vec{P}_i \tag{2.5.14}$$

The reaction force can now be computed as

$$\vec{F}_{Ri} = \vec{R}_{Si} - s_i\vec{P}_i \qquad (2.5.15)$$

and the reaction moment

$$\vec{M}_{Ri} = \vec{M}_{Si} - (1-s_i)\vec{P}_i \qquad (2.5.16)$$

where $\vec{P}_i$, $\vec{F}_{Si}$, $\vec{M}_{Si}$ are already computed.

When the total joint forces and moments are calculated, it is possible to compute stresses in manipulator segments. We are interested in the maximal bending and maximal torsion stresses.

Let us consider a segment on which the forces and moments are acting according to Fig. 2.38. and let us use the notation from that figure.

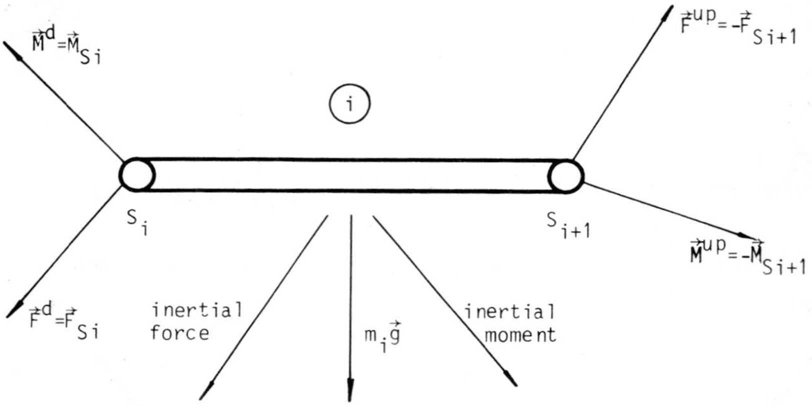

Fig. 2.38. Forces and moments acting on the i-th segment

Stresses in a segment depend on its form. They are usually calculated for segments which have some typical form. The algorithm described here contains special subroutines for two standard forms of segments. These are a circular tube (Fig. 2.39 (a)) and a rectangular tube (Fig. 2.39 (b)). For any other form a special subroutine can be programmed and added to the algorithm. For these standard segment forms the algorithm itself adopts body-fixed coordinate system with the z-axis along the segment (Fig. 2.39). Indices i which denote the i-th segments are omitted for shorter writing.

It is suitable to express the forces and moments in the segment b.-f. system. For the segment "i" it is

segment mass m_i is divided into two parts $\mu_i^{\ell} = m_i/2$ and $\mu_i^{up} = m_i/2$, μ_i^{ℓ} being concentrated on the lower side of cane (point S_i) and μ_i^{up} on the upper side (point S_{i+1}). The masses of motors and reducers, if they are placed in joints, are added to these joint masses μ_i. It is clear that the existence of some masses which are really concentrated in joints reduces the error which appears because of segment mass division.

In 2.5.3. we distinguished two sorts of linear joint and introduced the corresponding indicators p_i (relation (2.5.6)).

Now, the length $\vec{\ell}_i$ of a segment (between the two joints) is

$$\vec{\ell}_i = \overrightarrow{S_i S_{i+1}} = \vec{r}_{ii} + s_i p_i q_i \vec{e}_i + \vec{r}_{i,i+1} + s_{i+1}(1-p_{i+1}) q_{i+1} \vec{e}_{i+1} \qquad (2.5.24)$$

For standard form segments, with the z_i-axis of b.-f. system placed along the cane, it holds

$$\vec{\ell}_i = \left\{0,\ 0,\ |\ell_i|\right\} \qquad (2.5.25)$$

For the gripper and its point A the length is defined as

$$\vec{\ell}_n = \overrightarrow{S_n A} \qquad (2.5.26)$$

<u>Calculation of deformations</u>. Linear deviation $\vec{u}_i$ consists of three or two components (Fig. 2.42) depending on whether the segment "i" is elastic or rigid:

$$\vec{u}_i = \begin{cases} \vec{u}_{i-1} + \vec{u}_i^{e\ell} + \vec{\varphi}_{i-1} \times \vec{\ell}_i, & k_{ei} = 1 \\ \vec{u}_{i-1} + \vec{\varphi}_{i-1} \times \vec{\ell}_i, & k_{ei} = 0 \end{cases} \qquad (2.5.27)$$

$\vec{u}_i^{e\ell}$ is the component resulting from the elastic deformation of segment "i".

For the angular deviation one obtains

$$\vec{\varphi}_i = \begin{cases} \vec{\varphi}_{i-1} + \vec{\varphi}_i^{e\ell}, & k_{ei} = 1 \\ \vec{\varphi}_{i-1}, & k_{ei} = 0 \end{cases} \qquad (2.5.28)$$

where the component $\vec{\varphi}_i^{e\ell}$ results from the elastic deformation of segment "i".

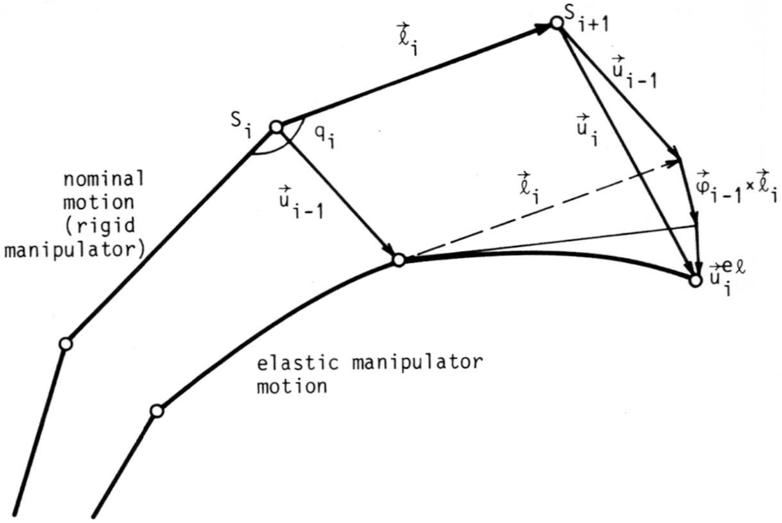

Fig. 2.42. Components of linear deviation

Relations (2.5.27) and (2.5.28) make possible the recursive calculation of deviations $\vec{u}_i$ and $\vec{\varphi}_i$ starting with $\vec{u}_o = 0$, $\vec{\varphi}_o = 0$. But, the deformations $\vec{u}_i^{e\ell}$, $\vec{\varphi}_i^{e\ell}$ of the segment "i" still remain to be found.

The segment "i" will be considered as a cantilever beam having the lower end S_i fixed and the upper end S_{i+1} free. The action of the next segment "i+1" is replaced by total joint force $-\vec{F}_{Si+1}$ and moment $-\vec{M}_{Si+1}$. The sign "-" follows from the previous assumption that $\vec{F}_{Si+1}$, $\vec{M}_{Si+1}$ act on the next segment "i+1" (see 2.5.3). Thus, the following forces and moments (Fig. 2.43) act on the free end of segment "i": joint force $\vec{F}_{Si+1}$, moment $\vec{M}_{Si+1}$, gravity force $\mu_i^{up}\vec{g}$ ($\vec{g}$ = {0, 0, -9.81}), and nominal inertial force $-\mu_i^{up}\vec{w}_{Si+1}$ ($\vec{w}_{Si+1}$ is the nominal acceleration of the point S_{i+1}).

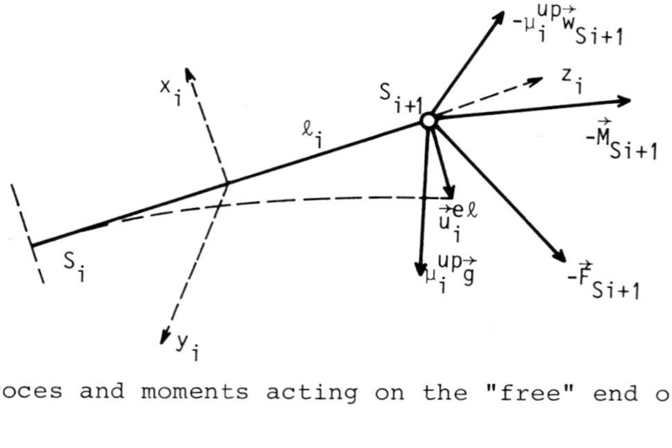

Fig. 2.43. Froces and moments acting on the "free" end of a segment

Now, the linear deviation $\vec{u}_i^{e\ell}$ is:

$$\vec{u}_i^{e\ell} = \alpha_i\left(-\vec{F}_{Si+1}+\mu_i^{up}\vec{g}-\mu_i^{up}\vec{w}_{Si+1}\right) + \beta_i\left(-\vec{M}_{Si+1}\right), \tag{2.5.29}$$

and the angular deviation

$$\vec{\varphi}_i^{e\ell} = \gamma_i\left(-\vec{F}_{Si+1}+\mu_i^{up}\vec{g}-\mu_i^{up}\vec{w}_{Si+1}\right) + \delta_i\left(-\vec{M}_{Si+1}\right) \tag{2.5.30}$$

α_i, β_i, γ_i, δ_i are matrix influence coefficients. These coefficients will now be discussed but, before that, we conclude that equations (2.5.27) and (2.5.28) combined with (2.5.29) and (2.5.30) allow recursive calculations of $\vec{u}_1, \vec{\varphi}_1, \ldots, \vec{u}_{n-1}, \vec{\varphi}_{n-1}$ and finally the deviations of gripper point A i.e. $\vec{u}_A$, $\vec{\varphi}_A$.

<u>Influence coefficients</u>. Let us consider a segment "i" and its b.-f. system $O_i x_i y_i z_i$ (Fig. 2.43). Since the segment is considered as a cantilever beam, in the b.-f. system it holds

$$\tilde{\vec{u}}_i^{e\ell} = \tilde{\alpha}_i\left(-\underset{\sim}{\vec{F}}_{Si+1}+\mu_i^{up}\left(\tilde{\vec{g}}-\underset{\sim}{\vec{w}}_{Si+1}\right)\right) + \tilde{\beta}_i\left(-\underset{\sim}{\vec{M}}_{Si+1}\right) \tag{2.5.31}$$

and

$$\tilde{\vec{\varphi}}_i^{e\ell} = \tilde{\gamma}_i\left(-\underset{\sim}{\vec{F}}_{Si+1}+\mu_i^{up}\left(\tilde{\vec{g}}-\underset{\sim}{\vec{w}}_{Si+1}\right)\right) + \tilde{\delta}_i\left(-\underset{\sim}{\vec{M}}_{Si+1}\right) \tag{2.5.32}$$

where

$$\tilde{\alpha}_i = \begin{bmatrix} \alpha_{xi} & 0 & 0 \\ 0 & \alpha_{yi} & 0 \\ 0 & 0 & \alpha_{zi} \end{bmatrix}, \qquad {}_i = \begin{bmatrix} 0 & \beta_{xi} & 0 \\ -\beta_{yi} & 0 & 0 \\ 0 & 0 & 0 \end{bmatrix},$$

$$\tilde{\gamma}_i = \begin{bmatrix} 0 & -\gamma_{xi} & 0 \\ \gamma_{yi} & 0 & 0 \\ 0 & 0 & 0 \end{bmatrix}, \qquad \tilde{\delta}_i = \begin{bmatrix} \delta_{xi} & 0 & 0 \\ 0 & \delta_{yi} & 0 \\ 0 & 0 & \delta_{zi} \end{bmatrix} \tag{2.5.33}$$

α_{xi} is the influence coefficient for the bending deflection (linear deviation) along the axis x_i, under the action of the force at the point S_{i+1}. Similarly, α_{yi} is defined in terms of the y_i axis.

α_{zi} is the influence coefficient for extension along the z_i axis, under the action of the force at S_{i+1}. It can usually be neglected by taking $\alpha_{zi} = 0$.

β_{xi} is the influence coefficient for bending deflection along the x_i axis due to the moment acting at S_{i+1}. Likewise for the y_i axis.

γ_{xi} is the influence coefficient for the bending angle around x_i axis due to the force acting at S_{i+1}. Likewise for y_i axis.

δ_{xi} is the influence coefficient for the bending angle around x_i axis due to the moment acting at S_{i+1}. Likewise for y_i axis.

δ_{zi} is the influence coefficient for the torsion around the z_i axis due to the moment acting at S_{i+1}.

In order to transform equations (2.5.31), (2.5.32) into external system i.e. to obtain equations (2.5.29), (2.5.30) we transform the influence coefficients by using

$$\begin{aligned} \alpha_i &= A_i \tilde{\alpha}_i A_i^{-1}, \qquad \beta_i = A_i \tilde{\beta}_i A_i^{-1}, \\ \gamma_i &= A_i \tilde{\gamma}_i A_i^{-1}, \qquad \delta_i = A_i \tilde{\delta}_i A_i^{-1} \end{aligned} \tag{2.5.34}$$

where A_i is the transition matrix of the segment "i".

For the case of force acting of S_{i+1} the influence coefficients for deflection and extension are

$$\alpha_{xi} = \frac{\ell_i^3}{3E_i I_{yi}}, \quad \alpha_{yi} = \frac{\ell_i^3}{3E_i I_{xi}}, \quad \alpha_{zi} = \frac{\ell_i}{E_i A_i} \tag{2.5.35}$$

where E_i is Young's modulus for the adopted material, I_{xi} and I_{yi} are axial moments of inertia of the cross-section and A_i is the cross-section area.

If the moment is acting the influence coefficients for segment bending deflection are

$$\beta_{xi} = \frac{\ell_i^2}{2E_i I_{yi}} , \qquad \beta_{yi} = \frac{\ell_i^2}{2E_i I_{xi}} \tag{2.5.36}$$

Let us consider angular deviations. If a force is acting at S_{i+1} the influence coefficients for bending angles are

$$\gamma_{xi} = \frac{\ell_i^2}{2E_i I_{xi}} , \qquad \gamma_{yi} = \frac{\ell_i^2}{2E_i I_{yi}} \tag{2.5.37}$$

and if a moment is acting then the coefficients for bending and torsion angles are

$$\delta_{xi} = \frac{\ell_i}{E_i I_{xi}} , \qquad \delta_{yi} = \frac{\ell_i}{E_i I_{yi}} , \qquad \delta_{zi} = \frac{\ell_i}{G_i I_{oi}} \tag{2.5.38}$$

where I_{oi} is the polar moment of inertia of the cross-section, and G_i is the torsion modulus. G_i is given by

$$G_i = \frac{E_i}{2(1+\nu_i)} \tag{2.5.39}$$

where ν_i is Poisson's coefficient.

Let us now discuss the influence coefficients for two standard form cross-sections. We consider a circular tube (Fig. 2.39a) and a rectangular tube (Fig. 2.39b). For the circular tube it holds

$$\begin{aligned} I_{xi} &= I_{yi} = \frac{\pi R_i^4}{4}(1-\psi_i^4) , \qquad \psi_i = r_i/R_i \\ I_{oi} &= \frac{\pi R_i^4}{2}(1-\psi_i^4) \end{aligned} \tag{2.5.40}$$

and for the rectangular tube

$$\begin{aligned} I_{xi} &= \frac{1}{12} H_{xi} H_{yi}^3 (1-\psi_{xi}\psi_{yi}^3) , \qquad \psi_{xi} = \frac{h_{xi}}{H_{xi}} , \qquad \psi_{yi} = \frac{h_{yi}}{H_{yi}} \\ I_{yi} &= \frac{1}{12} H_{yi} H_{xi}^3 (1-\psi_{yi}\psi_{xi}^3) \end{aligned} \tag{2.5.41}$$

The torsion of rectangular cross-section will not be discussed here, so the readers are referred to the literature.

There are also some interesting characteristics which are connected with control problems. Some of these characteristics which can be obtained by means of the dynamic analysis algorithm are explained in 2.9.

2.6. Tests of Dynamic Characteristics

This chapter deals with the computer-aided algorithm for the dynamic analysis of manipulator motion. The algorithm computes some important dynamic characteristics. The knowledge of these characteristics is very useful in the design process and in the application of the device. But in order to obtain a more applicable algorithm, we supplement the algorithm with the possibility of testing the relevent dynamic characteristics. Now, the design procedure can be shortly described. We chose some values of manipulator parameters and run the algorithm which calculates the relevant dynamic characteristics and tests whether they satisfy the conditions imposed. So, we immediately have the answer as to whether the chosen parameters are correct. If they are not, a correction is made.

It should be mentioned that, in principle, each dynamic characteristic which is calculated can be tested. Here, we mention only the most relevant tests.

2.6.1. Tests of a D.C. electromotor

Suppose that we have chosen some D.C. electromotor as the actuator for a joint S_i. We have also adopted some value of the reduction ratio. All the calculations refer to some defined manipulation task with a prescribed execution time.

Test 1. We can find the $P^m_{max} - n^m$ characteristic of the chosen motor in the catalog. It is the diagram of maximal torque depending on motor r.p.m. If the diagram is not given directly, it can be constructed from the data given in the catalog. In manipulator systems, we often use permanent magnet D.C. motors. For such motors the $P^m_{max} - n^m$ characteristic has a polygonal form (straight lines in Fig. 2.44). In the catalog we find sometimes the value of maximal motor torque (corresponding to the point A in Fig. 2.44) and the maximal rotation speed (point B in Fig. 2.44). These two values define the maximal characteristic (we assume a straight line). Let the value of maximal torque for $n^m \to 0$ (point A) be marked by P_M^m and let the value of rotation speed for $P^m \to 0$ (point B) be marked by n_M^m. The torque P_M^m is often called stall torque, and the rotation speed n_M^m is called no-load speed. When this speed is expressed in terms of r.p.m. then it is marked by n_M^m and if it is expressed in terms of rad/s then we mark it by ω_M^m. The straight line maximal characteristic between P_m^M and n_M^m will be proved in paragraph 4.5.1. It should be said that there is sometimes a difference between the real value of maximal torque (P_{Mr}^m) and its theoretical value (P_M^m); the real value of P_{Mr}^m is less than P_M^m. In such a case the maximal characteristic $P_{max}^m - n^m$ has an upper bound P_{Mr}^m (point C in Fig. 2.44a). This characteristic defines the feasible domain. The real $P^m - n^m$ characteristics must be wholly within this domain. The calculation of $P^m - n^m$ characteristic was explained in 2.5.1. In each iteration a new point of the diagram is obtained. The algorithm checks whether it is within the permissible domain. If it is, a new iteration starts, and if it is not, the algorithm signals that there is a violation of the constraint.

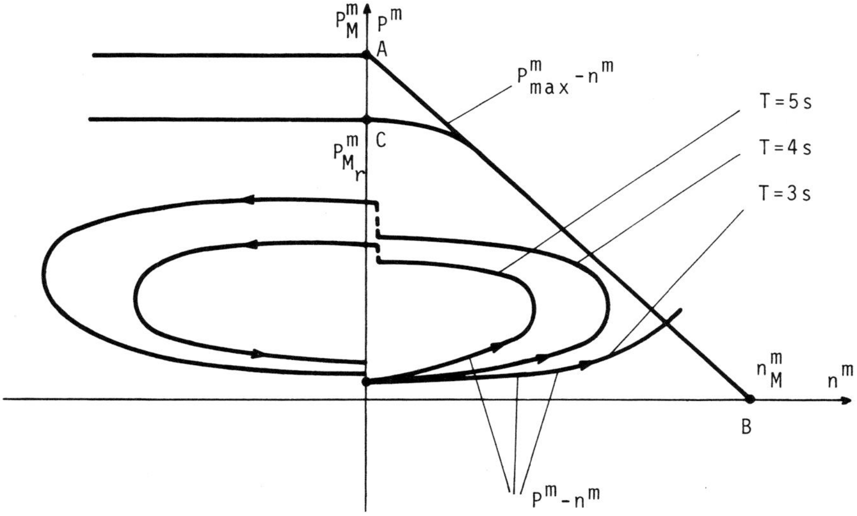

Fig. 2.44a. $P^m - n^m$ diagrams

The constraint considered follows from the mathematical model of D.C. actuator. The model will be completely dicussed in 2.9.2. and we here give only a simplified derivation of torque-speed constraint. According to Ohm's law it is

$$u = Ri + C_E\dot{q}^m \tag{2.6.1}$$

where u is input voltage, R is rotor resistance, i is rotor current, C_E is the constant of electromotor force, and $\dot{q}^m$ is rotation speed (rad/s). If we neglect rotor acceleration effects and friction term, motor output torque is

$$P^m = C_M i \tag{2.6.2}$$

where C_M is the constant of torque. Combining (2.6.1) and (2.6.2) one obtains

$$P^m = \frac{C_M}{R} u - \frac{C_M C_E}{R} \dot{q}^m \tag{2.6.3a}$$

If rotation speed is expressed in terms of r.p.m. ($n^m = \frac{60}{2\pi} \dot{q}^m$) then

$$P^m = \frac{C_M}{R} u - \frac{C_M C_E 60}{R\ 2\pi} n^m \tag{2.6.3b}$$

Let us introduce the constraint of maximal input voltage u_{max}. Then, from (2.6.3) it follows

$$\frac{P^m_{max}}{P^m_M} + \frac{n^m}{n^m_M} = 1 \tag{2.6.4}$$

where $P^m_M = \frac{u_{max} C_M}{R}$ is stall torque and $n^m_M = \frac{60}{2\pi} \frac{u_{max}}{C_E}$ is no-load speed. This constraint of maximal input voltage can be represented by a strainght line in $P^m - n^m$ plane ((1.) in Fig. 2.44b). We use this constraint in the quadrants I and III of the $P^m - n^m$ plane. For the quadrants II and IV we introduce the constraint of maximal rotor current in order to keep this current smaller than the stall current value: $|i| < i_M = \frac{u_{max}}{R}$. This constraint is represented by a horizontal line (2.) in $P^m - n^m$ plane (Fig. 2.44b). Finally, we introduce the constraint of maximal allowed speed n_{max} ((3.) in Fig. 2.44b), for instance $n_{max} = n^m_M$.

If viscous friction is not neglected then no-load speed becomes $\omega^m_M = u_{max}/(C_E + \frac{RB_C}{C_M})$. B_C is the viscous friction coefficient. This modified constraint is represented by dotted line in Fig. 2.44b.

An example is shown in Fig. 2.44a. Straight lines represent the constraint $P^m_{max} - n^m$. It can be concluded that the $P^m - n^m$ diagrams spread

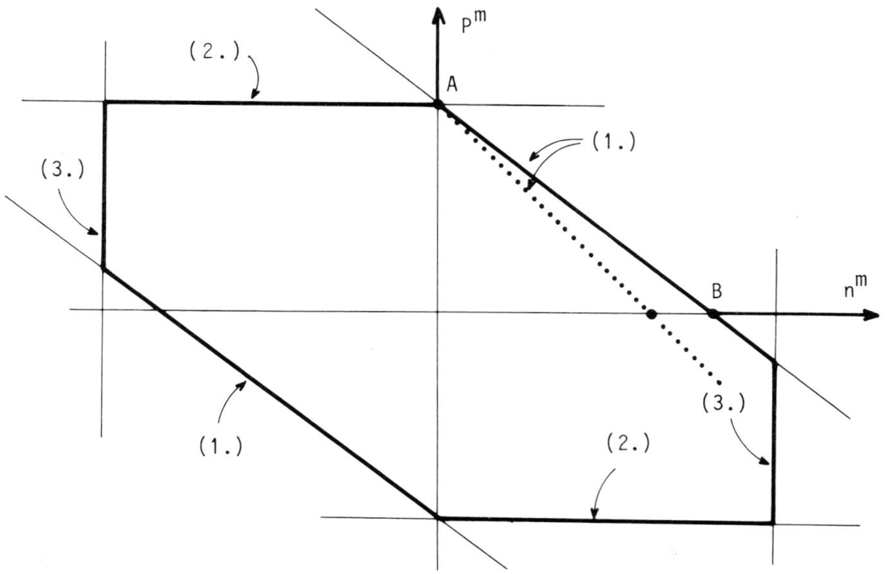

Fig. 2.44b. Constraints of D.C. motor

when the working speed increases, i.e., the execution time T decreases. For T = 5s and T = 4s the diagrams are wholly within the permissible domain. It means that the actuator chosen can produce manipulator work at this speed. For T = 3s the diagram extends beyond the permissible domain i.e. the constraint is violated. It means that the motor cannot produce manipulator work at that speed. The reduction ratio N influences strongly these $P^m - n^m$ characteristics: the larger N, the lower and wider the diagram. $P^m_{max} - n^m$ is the characteristic of the motor and does not depend on N. Thus, the violation of the constraint can sometimes be corrected by choosing some more appropriate value of the reduction ratio.

Test 2. Another test can be made from the standpoint of necessary motor power. The algorithm computes the power needed in each joint i.e. $Q_i^m = P_i\dot{q}_i/\eta_i$. By comparing this function (for the joint considered) with the maximal power which can be produced by the chosen motor we conclude whether the motor is chosen correctly. The loss of power due to reducer is taken into account.

This was a rather simplified approach to tests of power requirements. The complete theory of power testing is a complex problem and allows the development of the general methodology for motor choice. This theory will be explained in detail in paragraph 4.5.1. which deals with the choice of D.C. actuators. Here we give only a short presentation of power test procedure. Let us first give some definitions: besides power $Q_i = P_i\dot{q}_i$ which is a product of torque and rotation speed, we introduce a notion of dynamic power (or acceleration power) DQ as a product of torque and acceleration i.e. $DQ_i = P_i\ddot{q}_i$. In such forms these variables (Q, DQ) refer to joint shaft (i.e. outside the reducer), but, if we are interested in power which is to be produced by actuator then $Q_i^m = Q_i/\eta_i$ and $DQ_i^m = DQ_i/\eta_i$ are used since these forms refer to motor shaft (before reducer). Now, a very useful characteristic is a diagram connecting power and dynamic power. Each time instant gives one point having the coordinates Q^m and DQ^m. In this way, at the end of manipulation task we obtain Q^m-DQ^m characteristic (Fig. 2.45). This characteristic has to be within the feasible domain which is defined by a straight line connecting the maximal values Q_M^m and DQ_M^m (points B and A respectively, in Fig. 2.45). If the diagram violates this constraint then the test is negative. Q_M^m is the maximal motor power and can be expressed in the form $Q_M^m = P_M^m\omega_M^m/4 = (P_M^m)^2R_r/4C_MC_E$ where P_M^m is the stall torque of motor, ω_M^m is no-load speed, R_r is rotor resistance, and finally C_M and C_E are constants of torque and electromotor force. Maximal value DQ_M^m can be obtained by $DQ_M^m = Q_M^m/T_{em}$. T_{em} is called electromechanical constant and has the form $T_{em} = J_rR_r/C_MC_E$, where J_r is the rotor moment of inertia. It should be said that the real value of maximal dynamic power DQ_{Mr}^m can be less than theoretical value DQ_M^m

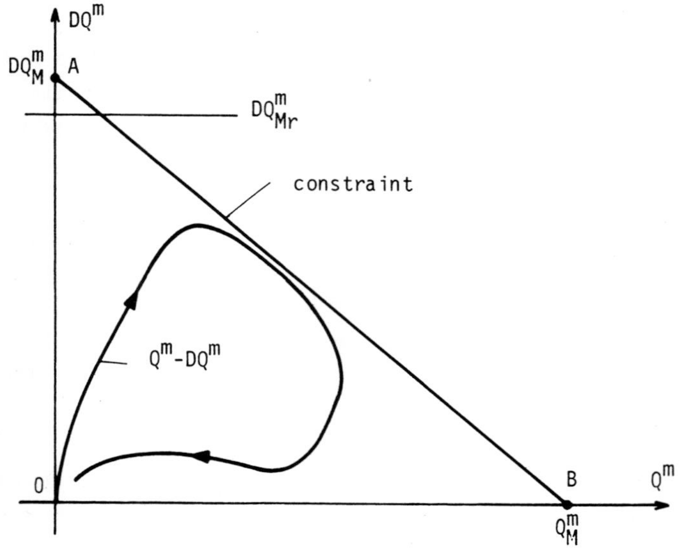

Fig. 2.45. Q^m-DQ^m diagram

calculated (Para. 2.5.) and, thus, which can be printed. We use the indicators LB_i to define which of these dynamic characteristics will be printed.

The indicators LC_i define the set of tests. Thus, we choose the tests by using these indicators.

On the basis of these two sets of indicators (LB and LC) the algorithm itself decides what should be computed in order to give the required prints and to perform the required tests.

The configuration input contains the definition of geometry and inertial properties. Some other parameters are also defined, for instance the cross section inertial moment if stress or deformation analysis is required. This input block allows the choice of standard-form segments. A set of indicators is used to determine the segment which has a standard form and which form it is. For a standard-form segment the algorithm itself computes the segment mass, inertial tensor, cross section inertial moments etc. on the basis of the input dimensions of standard form and the input data on material.

If the actuators tests or torque - r.p.m. prints are required, then the catalog characteristics of actuator and reducers are needed. Special indicators are used to define which actuators have to be tested.

If stress or elastic deformation analysis (the corresponding tests or prints) is required, some properties of segments material must be known (e.g. Young's modulus, maximal permitted stress, etc.). A set of indicators is used to define which segments are considered elastic, and another set to define which segments have to be stress-tested.

Manipulation task input has been discussed in 2.4. and the corresponding input file will be explained through examples in 2.8.

2.8. Examples

In this paragraph we present several examples illustrating the operation of the algorithm for dynamic analysis. We demonstrate the use of adapting blocks, the calculation of various dynamic characteristics and some testings. The first example (2.8.1) deals with a 4 d.o.f.

manipulator. In (2.8.2) and (2.8.3) two 5 d.o.f. manipulators are considered. Finally, in (2.8.4) we present an example of a manipulator with 6 d.o.f.

2.8.1. Example 1

We consider a cylindrical manipulator UMS-2V - variant with 4 d.o.f. The external look and manipulator data are presented in Fig. 2.48a. This figure also shows the choice of generalized coordinates (internal coordinates) q_1, q_2, q_3, q_4 and the adopted b.-f. systems. The kinematic scheme of manipulator is shown in Fig. 2.48b. The minimal configuration consists of one rotational (q_1) and two linear (q_2, q_3) degrees of freedom. With this 4 d.o.f. variant, the gripper is connected to the minimal configuration by means of a rotational joint (q_4).

Manipulation task. The manipulator carries a 3kg mass working object. The moments of inertia of the working object are $I_{x4} = I_{y4} = I_{z4} = 0.01$ kgm^2 (with respect to the corresponding b.-f. system). The object is to be moved along the trajectory A_o A_1 A_2 A_3 (Fig. 2.48b). Every part of the trajectory ($A_o \to A_1$, $A_1 \to A_2$, $A_2 \to A_3$) is a straight line. Object rotation for the angle $\pi/2$ has to be performed on the trajectory part $A_o \to A_1$, and the backward rotation $(-\frac{\pi}{2})$ on the part $A_1 \to A_2$. The complete scheme of manipulation task is shown in Fig. 2.48b.

We have to notice a few things. If a cylindrical manipulator has to reach the points A_o, A_1, A_2, A_3, it usually follows the trajectory represented by a dashed line in Fig. 2.49. This is done because of simplier control synthesis. In this example we have chosen straight line motion betwen two points (full line in Fig. 2.49) in order to demonstrate the algorithm possibilities. Triangular velocity profile is adopted.

Adapting block 4-2 is suitable for this manipulation task because it uses the position vector $X_g = [x\ y\ z\ q_4]^T$. Now, let us discuss the input values. The manipulator has to move to points A_o, A_1, A_2, A_3 one after another, so m = 3. In the starting point A_o we give the initial state $q(t_o) = [0\ -0.1\ -0.2\ 0]^T$, $\dot{q}(t_o) = 0$. In the point A_1 we give the value of position vector $X_g(A_1) = X_{g1} = [0.57\ 0.1\ 0.6\ \pi/2]^T$ and also the time interval in which motion from the previous point is performed $T_{(A_o \to A_1)} = T^1 = 1.5s$. Analogous values have to be given for points A_2,

MANIPULATOR DATA				
SEGMENT-i	1	2	3	4
m_i[kg]	10	7	4.15	0.5
I_{xi}[kgm^2]	-	-		0.01
I_{yi}[kgm^2]	-	-	0.318	0.001
I_{zi}[kgm^2]	0.029	0.055	-	0.001

Empty spaces denote insignificant data

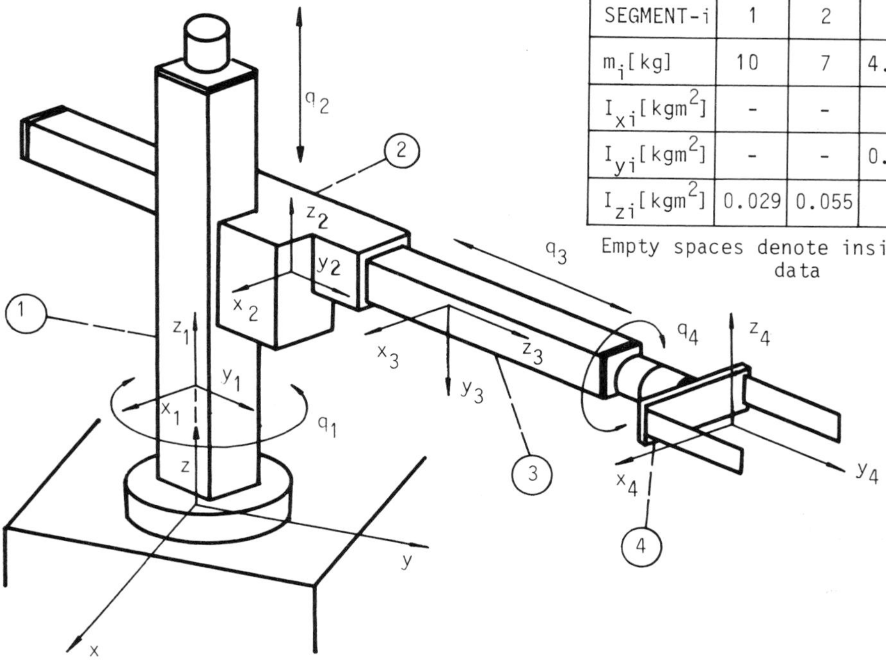

Fig. 2.48a. Cylindrical manipulator UMS-2V

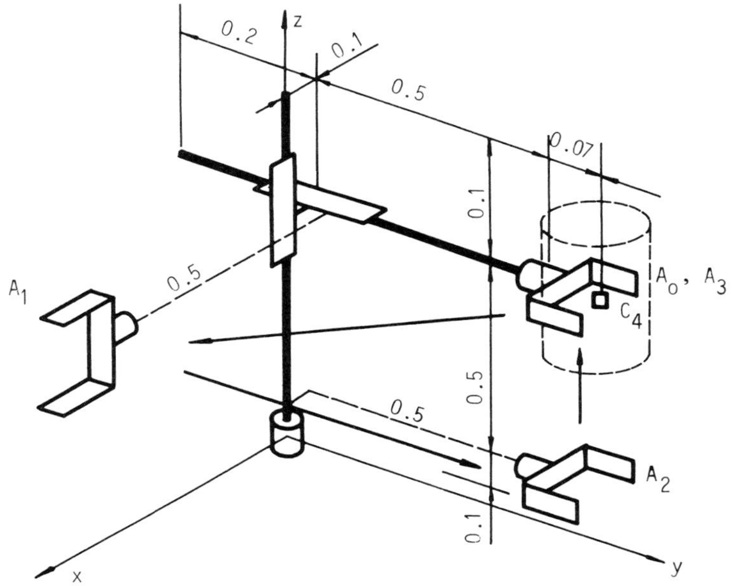

Fig. 2.48b. Scheme of manipulation task

A_3. Thus the input list for the definition of manipulation task is:

m - number of points to be reached	3
Indicator of adaption block	2
Indicators for the choice of velocity profiles	222
Initial position in terms of $q(t_o)$	0 -0.1 -0.2 0
$X_{g1} = \left[x\ y\ z\ q_4\right]_{A_1}$	0.57 0.1 0.6 $\pi/2$
$T^1 = T\left(A_o \to A_1\right)$	1.5
$X_{g2} = \left[x\ y\ z\ q_4\right]_{A_2}$	-0.1 0.57 0.1 0
$T^2 = T\left(A_1 \to A_2\right)$	1.5
$X_{g3} = \left[x\ y\ z\ q_4\right]_{A_3}$	-0.1 0.57 0.6 0
$T^3 = T\left(A_2 \to A_3\right)$	1.5

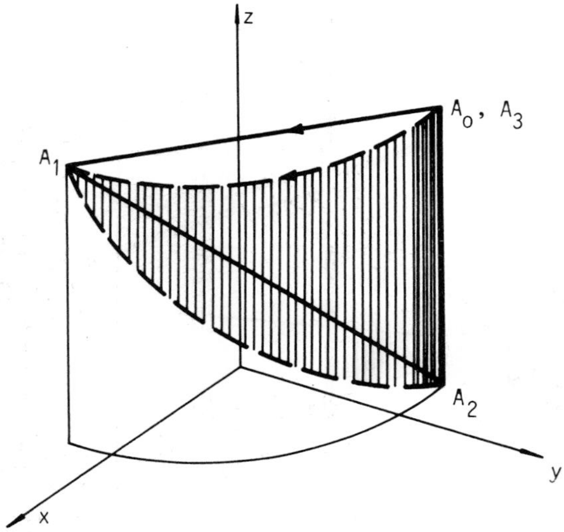

Fig. 2.49. Two variants of manipulator motion

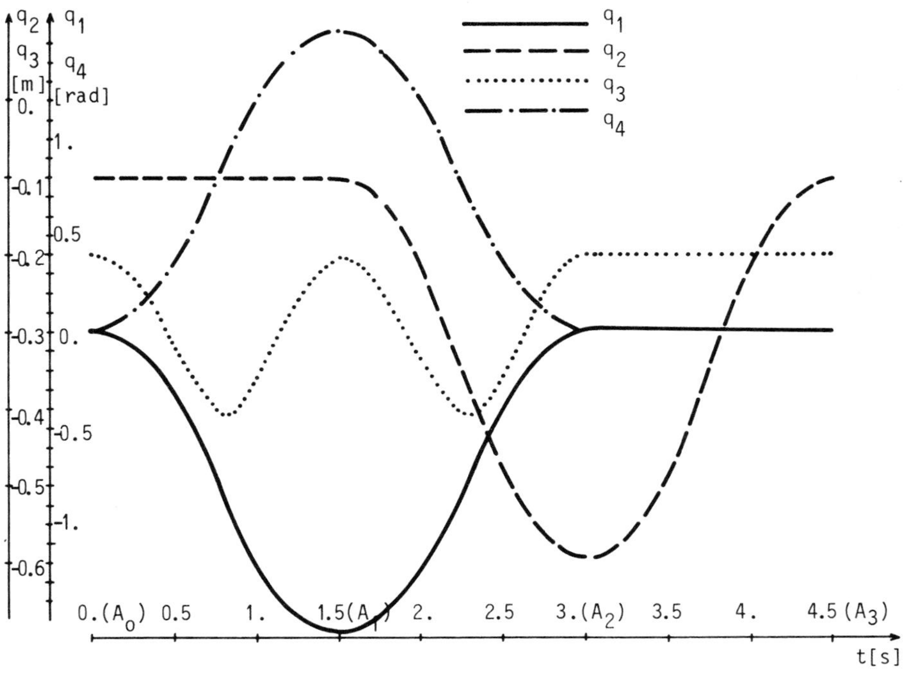

Fig. 2.50. Internal coordinates

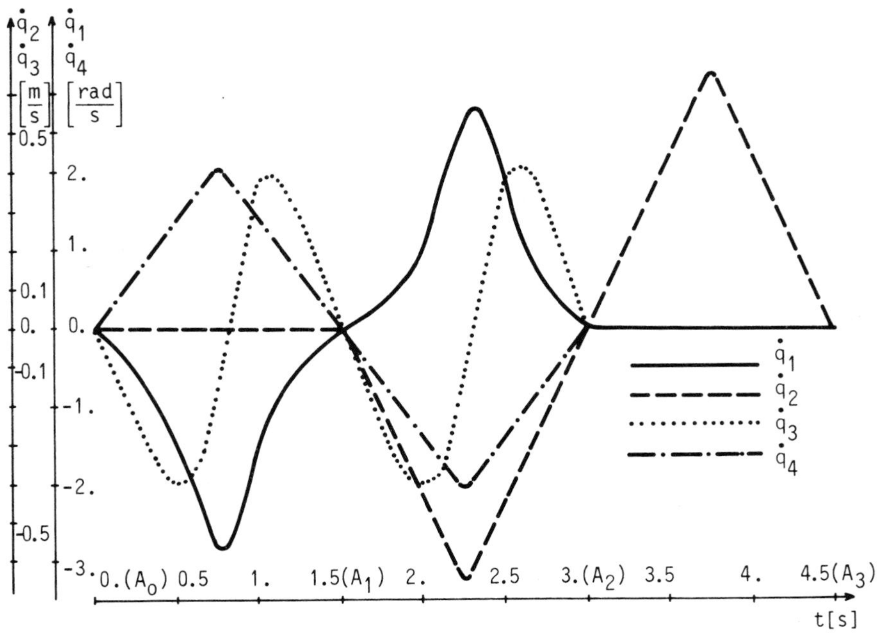

Fig. 2.51. Internal velocities

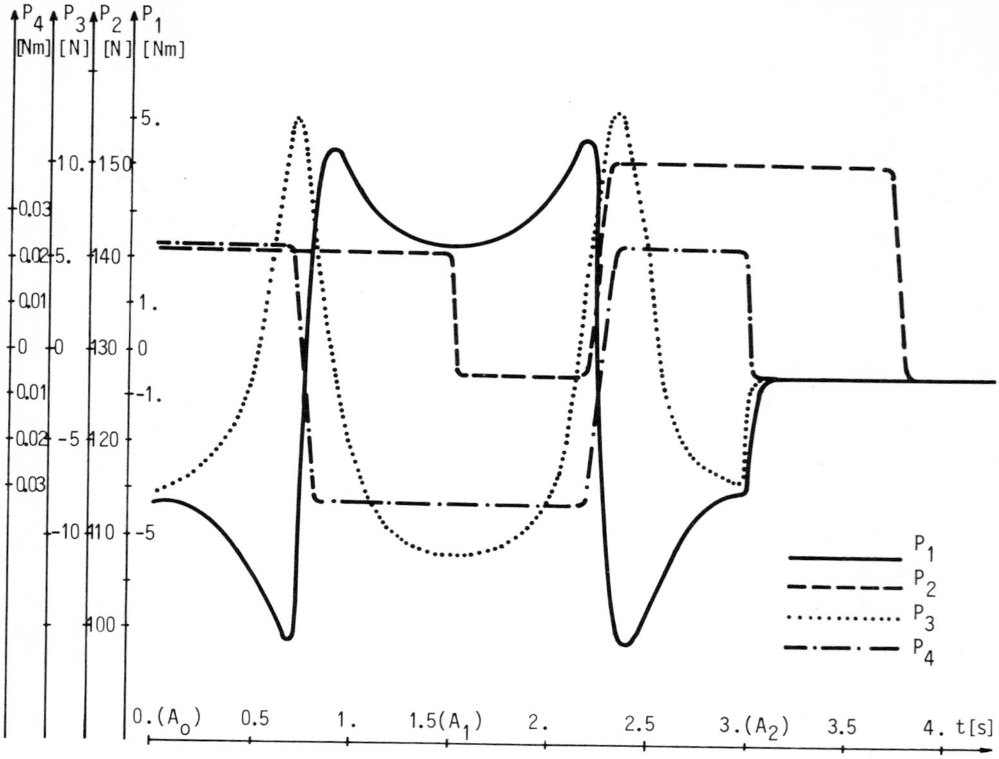

Fig. 2.52. Time histories of driving forces and torques

Results. We now present some results obtained by means of the dynamic analysis algorithm. The trajectory in the state space is shown first. Fig. 2.50. presents the time history of internal coordinates q and Fig. 2.51. presents the same for internal velocities $\dot{q}$. The next figure (Fig. 2.52) shows the corresponding time history of the driving forces and torques in manipulator joints.

2.8.2. Example 2

Let us consider the arthropoid manipulator having 5 degrees of freedom. It has been designed for manipulation with heavy loads. The minimal configuration consists of three rotational d.o.f. (q_1, q_2, q_3) and the gripper is connected to the minimal configuration by means of two rotational joints (q_4, q_5). The external look, manipulator data, the choice of generalized (internal) coordinates, and the adopted b.-f. systems are shown in Fig. 2.53.

<u>Manipulation task.</u> The manipulator has to move a 250 kg mass object along the trajectory A_o A_1 A_2 A_3 (Fig. 2.54). Every part of the trajectory ($A_o \to A_1$, $A_1 \to A_2$, $A_2 \to A_3$) is a straight line. The velocity profile on each part is triangular. The complete scheme of manipulation task, i.e. the initial position, the trajectory of object motion and the changes in object orientation, is shown in Fig. 2.54. It can be concluded that in this task the partial orientation only is necessary. Thus, this manipulation task consists of positioning along with partial orientation, so five d.o.f. are enough.

It is evident from the manipulation task scheme (Fig. 2.54) that the direction (b) is the most important. In order to define the direction (b) with respect to the gripper we use the unit vector $\vec{\tilde{h}} = \{0, 1, 0\}$ (expressed in the gripper b.-f. system $O_6x_6y_6z_6$). The two angles θ, φ define the direction (b) with respect to the external system Oxyz. We use the adapting block 5-2, so the position vector is $X_g=[x\ y\ z\ \theta\ \varphi]^T$.

The nubmer of points to be reached is m = 3. The initial position is defined by $q(t_o) = [0 \quad -\pi/6 \quad -4\pi/6 \quad -\pi/6 \quad 0]^T$.

Now, the input list defining the manipulation task is:

m	3
Indicator of adapting block	2
Indicators for profiles	222
Initial position $q(t_o)$	0 $-\pi/6$ $-4\pi/6$ $-\pi/6$ 0
$X_{g1} = [x\ y\ z\ \theta\ \varphi]_{A_1}$	1.5 1.5 0 0 0
$T^1 = T(A_o \to A_1)$	3.
$X_{g2} = [x\ y\ z\ \theta\ \varphi]_{A_2}$	0 2 0.8 $\pi/2$ 0
$T^2 = T(A_1 \to A_2)$	3.
$X_{g3} = [x\ y\ z\ \theta\ \varphi]_{A_3}$	1.5 0 0 0 0
$T^3 = T(A_2 \to A_3)$	4.5

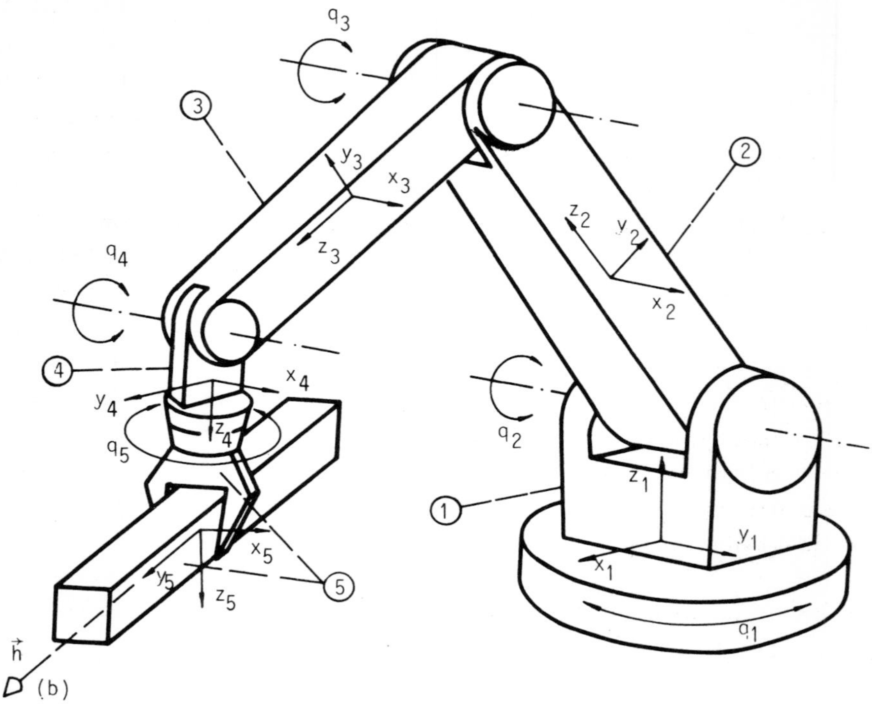

Manipulator data					
Segment-i	1	2	3	4	5*
m_i [kg]	-	125	98	10	270
I_{xi} [kgm^2]	-	31	26	0.05	38
I_{yi} [kgm^2]	-	31	26	0.05	3
I_{zi} [kgm^2]	15	2.8	2.8	0.05	38
Length[m]	0.4	1.5	1.5	0.2	0.2

* working object included

Fig. 2.53. An arthropoid manipulator

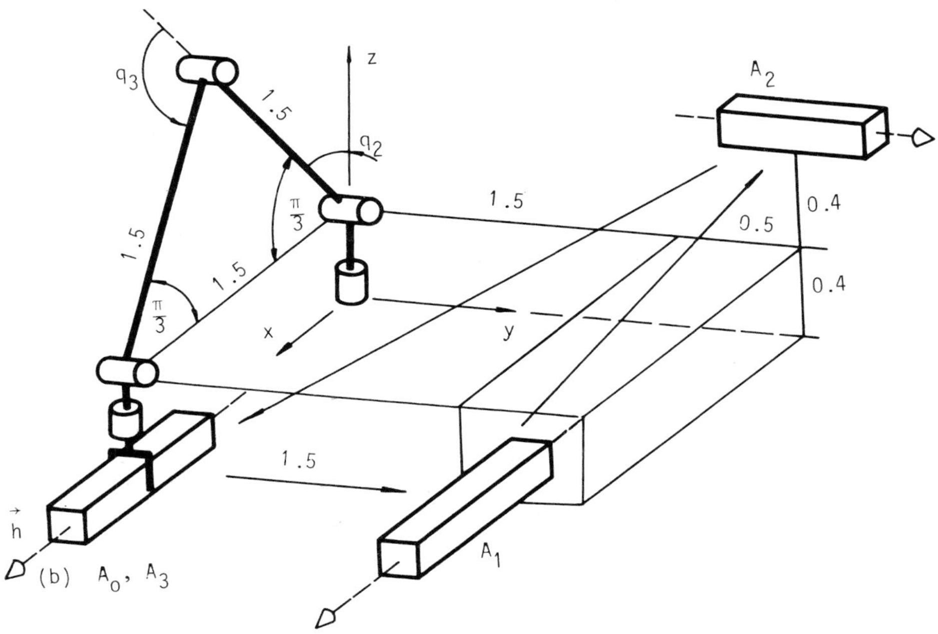

Fig. 2.54. Scheme of manipulation task

<u>Results.</u> Fig. 2.55. presents time-histories of the internal coordinates q and Fig. 2.56. presents the corresponding time histories of driving torques in manipulator joints.

In performing the task the manipulator consumed 15660 J of energy.

From the manipulator configuration and the manipulation task (Figs. 2.53, 2.54) it is clear that the joint S_4 does not play an active role. Hence we may consider a manipulator with no drive in that joint ($P_4=0$). In that case the manipulator gripper would behave like some kind of a pendulum. In order to avoid large oscillations we may apply some passive amortization. Thus the driving actuator for the joint S_4 is not necessary.

Let us now discuss the driving torques in the joints S_2 and S_3. These torques are very large due to large manipulator weight and the heavy payload. In such cases the compensation is usually applied. The hydro or pneumatic compensators may be used or sometimes even active compensation.

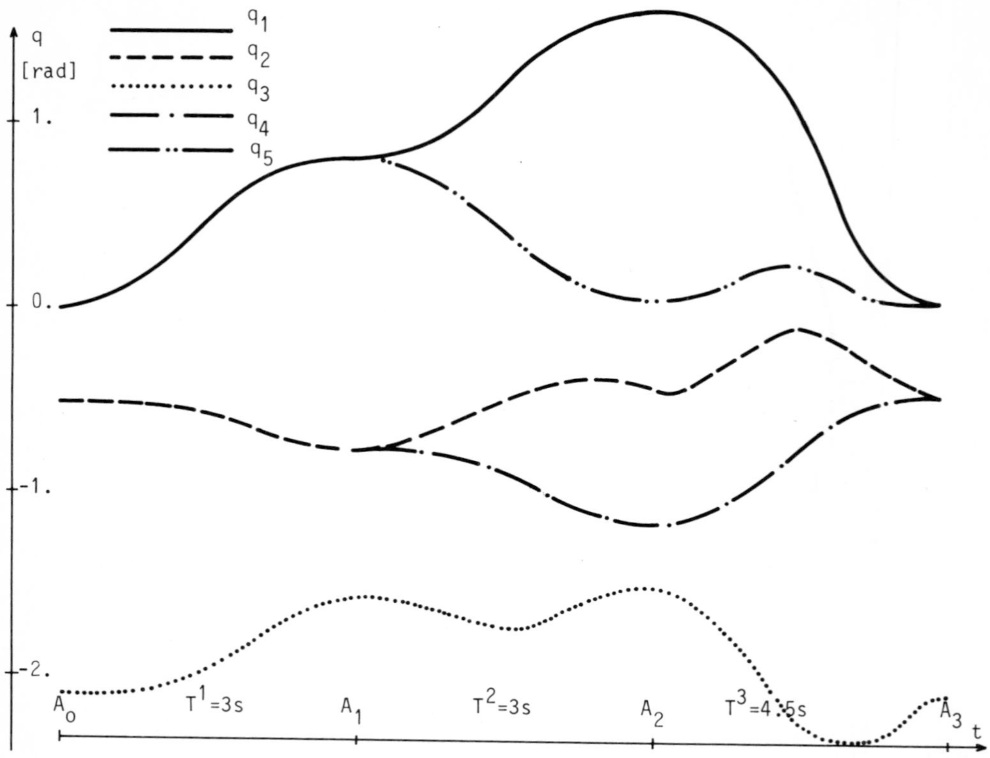

Fig. 2.55. Internal coordinates

2.8.3. Example 3

This example deals with the anthropomorphic manipulator UMS-1V - variant with 5 degrees of freedom. The manipulator is shown in Fig. 2.57.

The joints S_2 and S_3 are powered by 23 FRAME MAGNET MOTORS, 2315-P20-0, produced by INDIANA GENERAL. The reduction ratio is N = 100, and the reducer mechanical efficiency η = 0.8. The maximal characteristic (P^m_{max} - n^m) i.e. the maximal torque depending on motor rotation speed is obtained by an experiment and it is shown to be almost a straight line (Fig. 2.58). The characteristic differs from a straight line only in the region of slow speeds. If this region is not especially interesting from the standpoint of constraint violation, we may use a straight line aproximation and in such a way save some computer memory. Here, we work with the original characteristic without approximation.

The manipulator has to move a container with liquid along the trajecto-

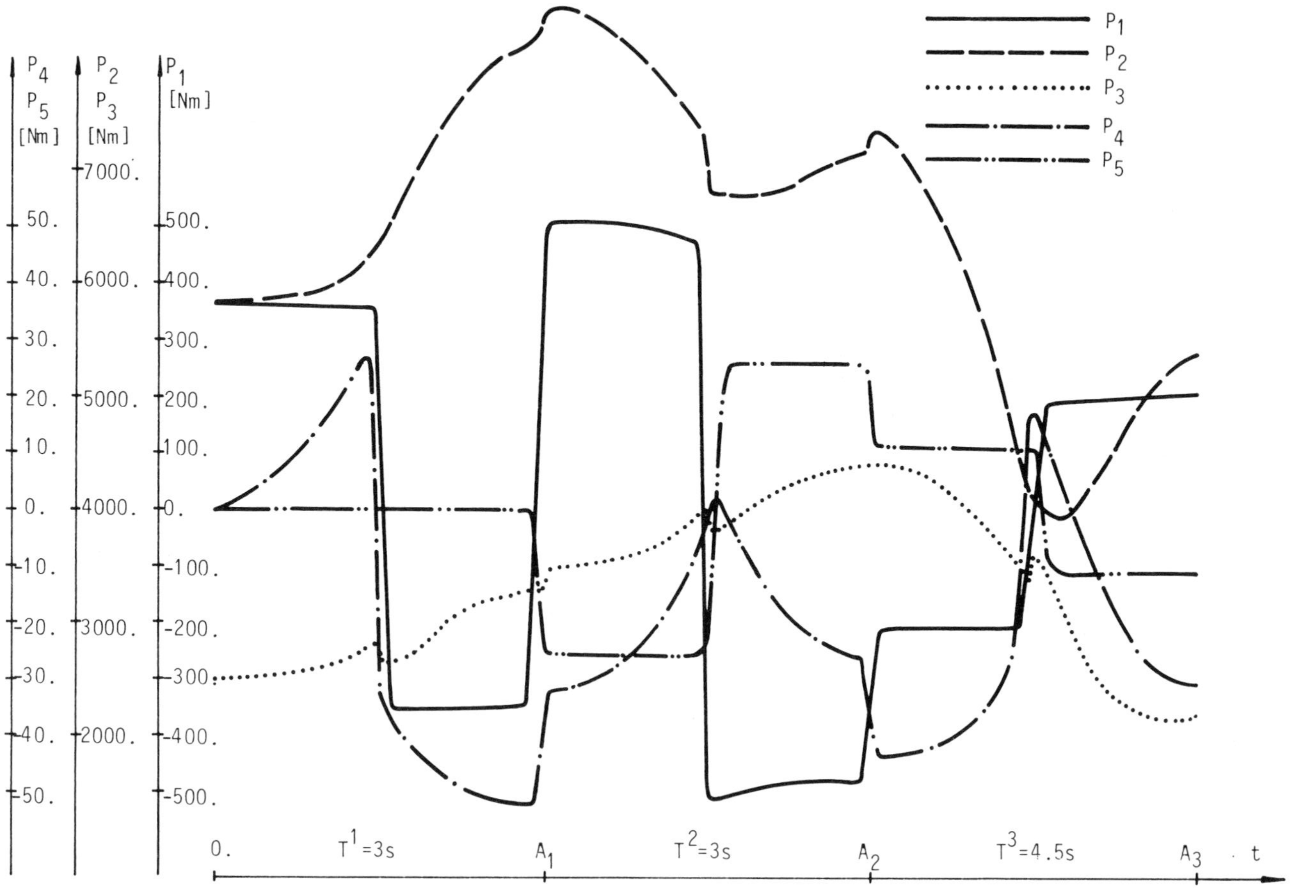

Fig. 2.56. Driving torques

ry $A_oA_1A_2$ (Fig. 2.59) keeping all the time the container axis vertical. It is the task of positioning along with the task of partial orientation. We first adopt the triangular velocity profile on each straight-line part of the trajectory. Let $T(A_o \to A_1) = T(A_1 \to A_2)$ (i.e. $T^1 = T^2$). Now the total execution time is $T = T^1 + T^2$.

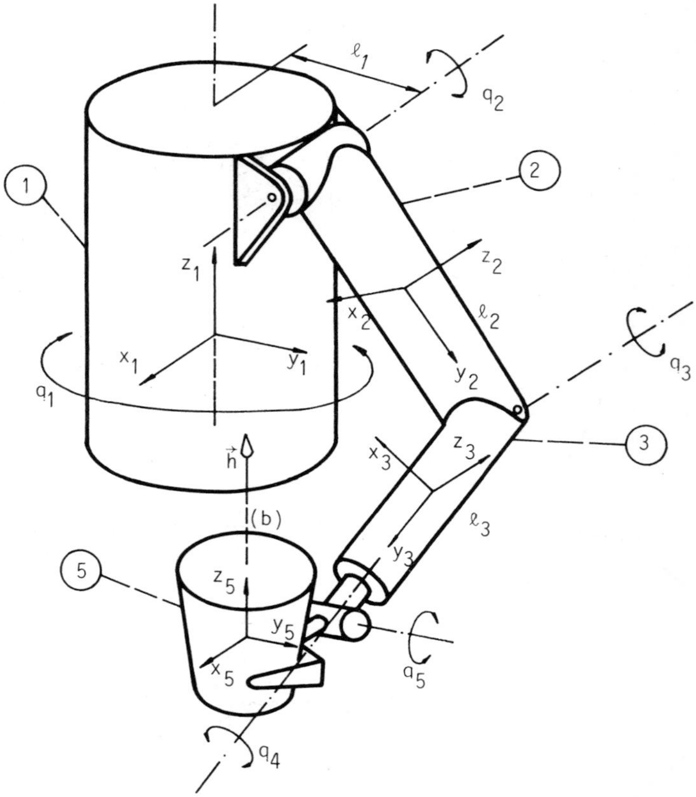

SEGMENTS	1	2	3	4	5*
LENGTH ℓ_i [m]	0.1	0.3	0.3	0.05	0.05
m_i [kg]	-	2.7	2.	0.5	5.
I_{xi} [10^{-2}kgm^2]	-	0.5	0.35	0.1	1.
I_{yi} [10^{-2}kgm^2]	-	0.1	0.1	0.1	1.
I_{zi} [10^{-2}kgm^2]	3.	0.5	0.35	0.1	1.

* working object included
empty spaces denote insignificant data

Fig. 2.57. Manipulator UMS-1V

Let us now discuss some of the results. For the total execution time T = 3.2s the tests of driving motors are positive which means that the chosen motors can produce the manipulator work at that speed. Fig. 2.60. presents the time histories of internal coordinates q and Fig. 2.61. presents the same for internal velocities $\dot{q}$. Finally, Fig. 2.62. presents the corresponding driving torques P.

Fig. 2.63. shows the torque-r.p.m. diagrams ($P^m - n^m$) for the motor in joint S_2 together with the maximal motor characteristic ($P^m_{max} - n^m$). It can be seen that for the execution time T = 3.2s, the diagram is wholly within the permissible domain (defined by maximal torque characteristic). For the faster manipulator work, T = 2.8 s, the diagram extends beyond the permissible domain which means that the chosen motor cannot produce this faster work. For T = 2.4 s the constraint is violated much earlier.

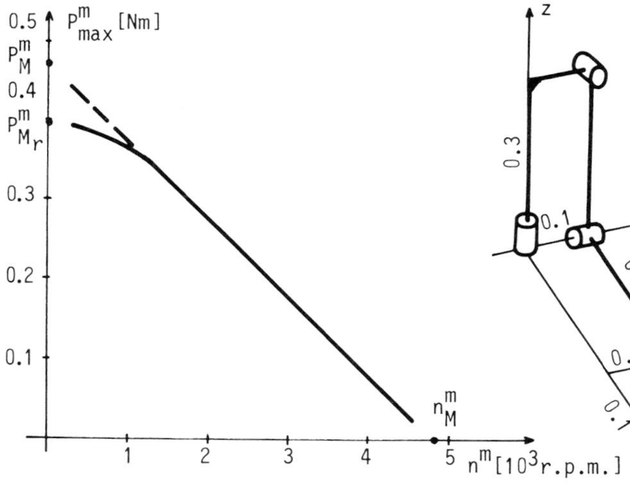

Fig. 2.58. Maximal characteristic of motor torque

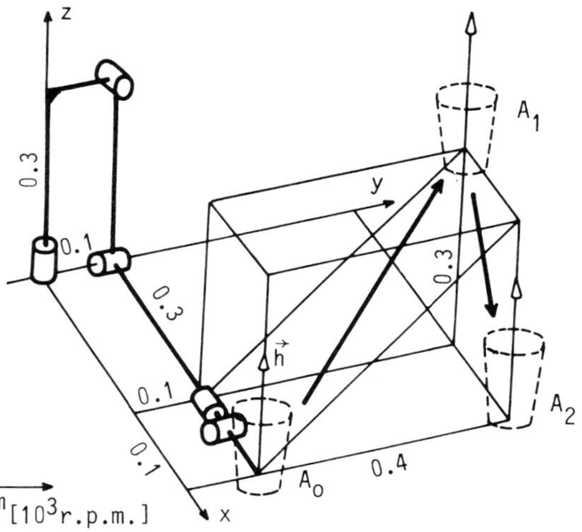

Fig. 2.59. Scheme of manipulation task

We now consider a different velocity profile. Let the motion A_oA_1 be performed with trapezoidal profile having T^1 = 1 s and the acceleration time t_a = 0.2s, and let the motion A_1A_2 be performed with triangular profile having T^2 = 1.6 s. Time histories of internal coordinates are shown in Fig. 2.64, and internal velocities and driving torques in Figs. 2.65, 2.66. Finally Fig. 2.67. shows the $P^m - n^m$ diagrams for the motor in joint S_2. We can see that the trapezoidal velocity profile enables faster moving up ($A_o \to A_1$) compared with the triangular profile.

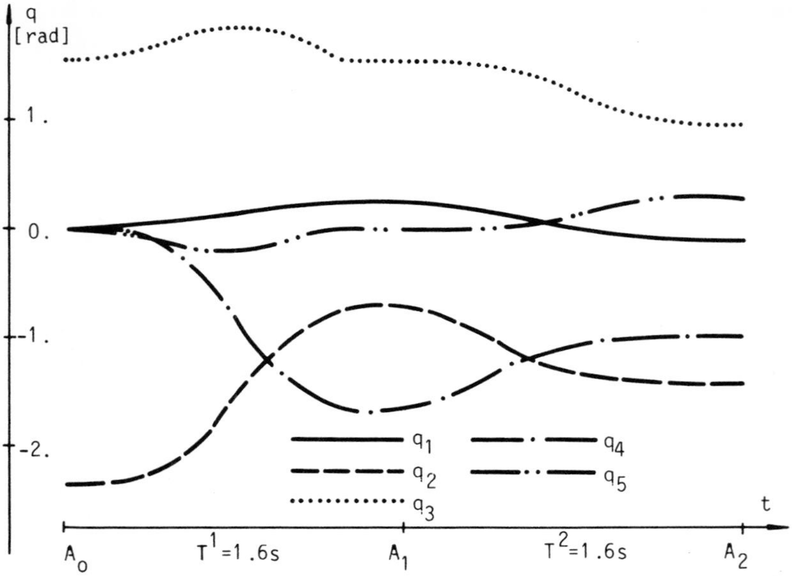

Fig. 2.60. Time histories of internal coordinates q

With the triangular profile $T(A_o \rightarrow A_1) = 1.2$ s is not posible and with the trapezoidal profile even $T(A_o \rightarrow A_1) = 1$ s is posible. One can conclude that with this manipulator and this manipulation task the trapezoidal profile is more convenient, because the actuator capabilities are used in a more efficient way (Fig. 2.67).

2.8.4. Example 4

In this example we present the manipulator GORO-80 having six rotational degrees of freedom. Fig. 2.68a, shows the external look, Fig. 2.68b. shows the kinematical scheme, and in Fig. 2.68c. there is a table with the manipulator data.

The manipulation task is shown in Fig. 2.69. The initial position of manipulator (A_o in Fig. 2.69) is given in 2.68b. Working object has to be inserted into a hole as shown in Fig. 2.69. First, the object is moved from A_o into a position A_1. Keeping in mind the form of the object and the hole, it is clear that the total orientation is necessary. It is shown in Fig. 2.69 via two directions (b) and (c) i.e. via two vectors $\vec{h}$ and $\vec{s}$. Now insertion is performed. The working object is moved from A_1 to A_2 along the direction (b) without any change in orientation. Each straight line motion is performed with the triangular velocity

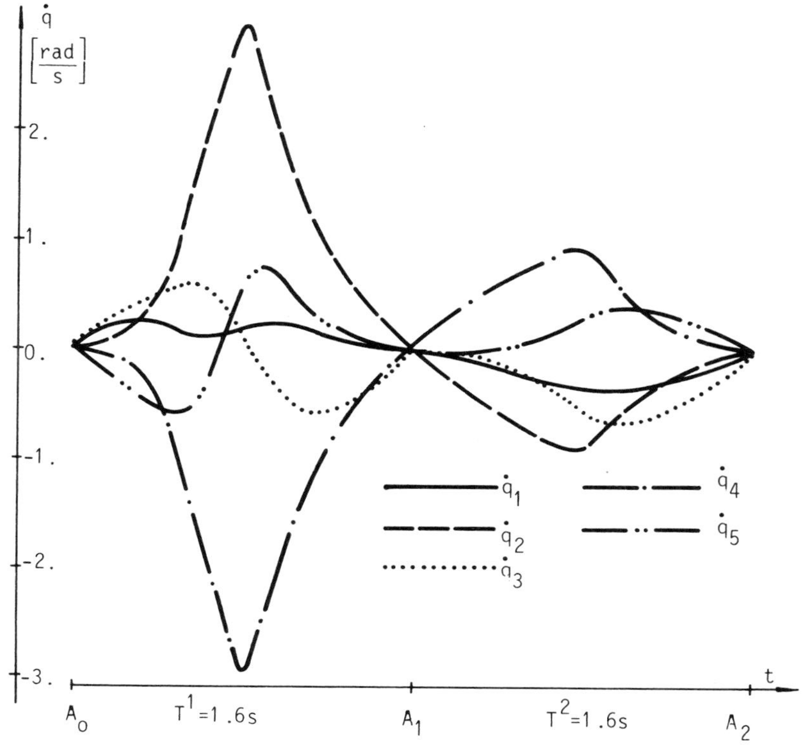

Fig. 2.61. Time histories of internal velocities $\dot{q}$

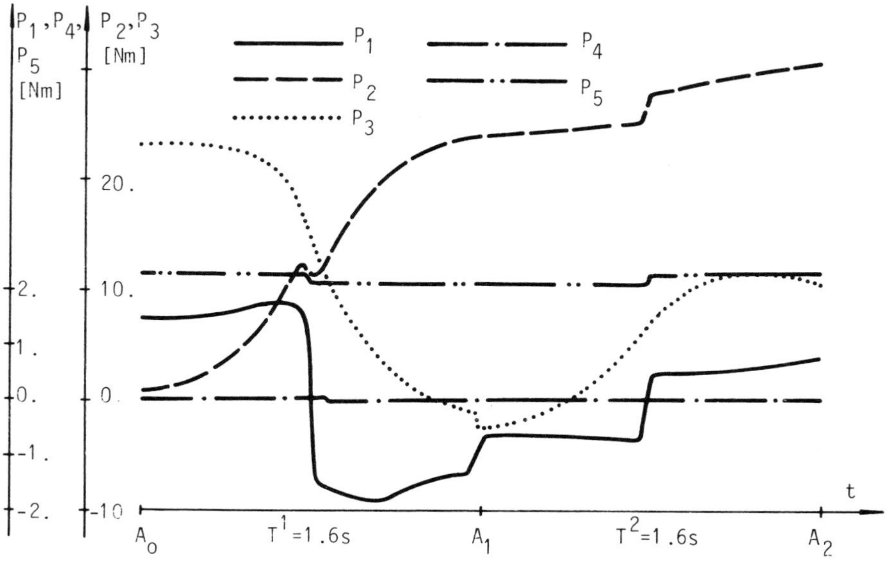

Fig. 2.62. Time histories of driving torques P

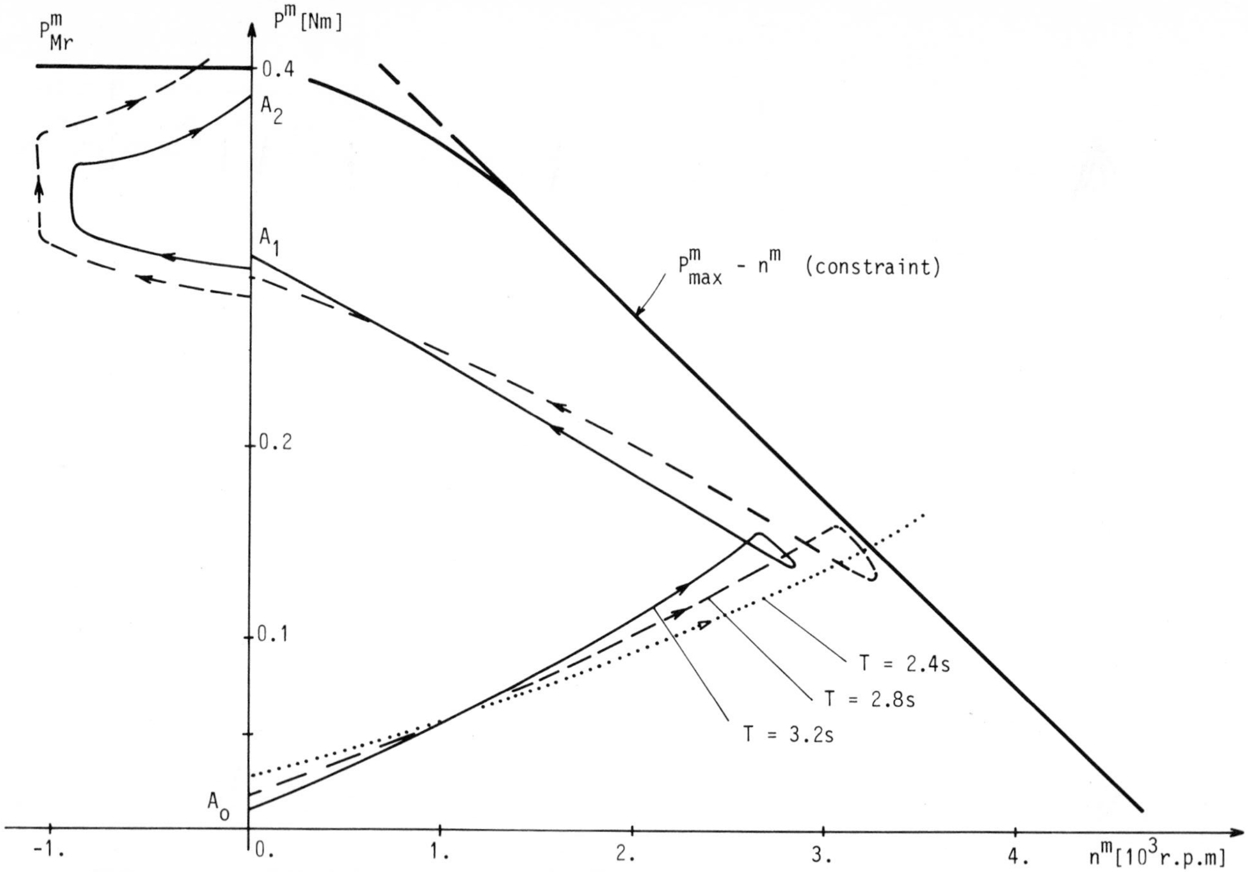

Fig. 2.63. Diagrams $P^m - n^m$ (torque. vs. r.p.m.) for joint S2

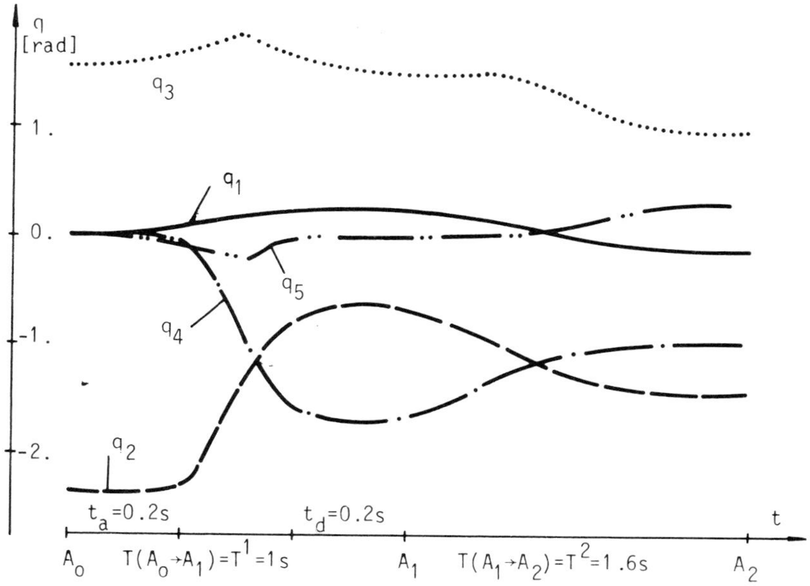

Fig. 2.64. Internal coordinates q

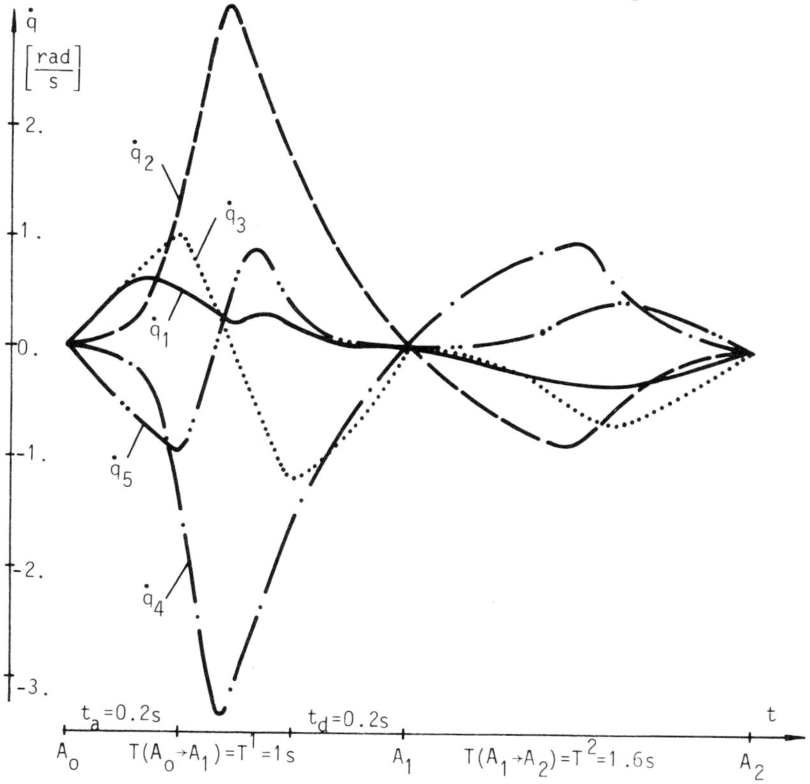

Fig. 2.65. Internal velocities $\dot{q}$

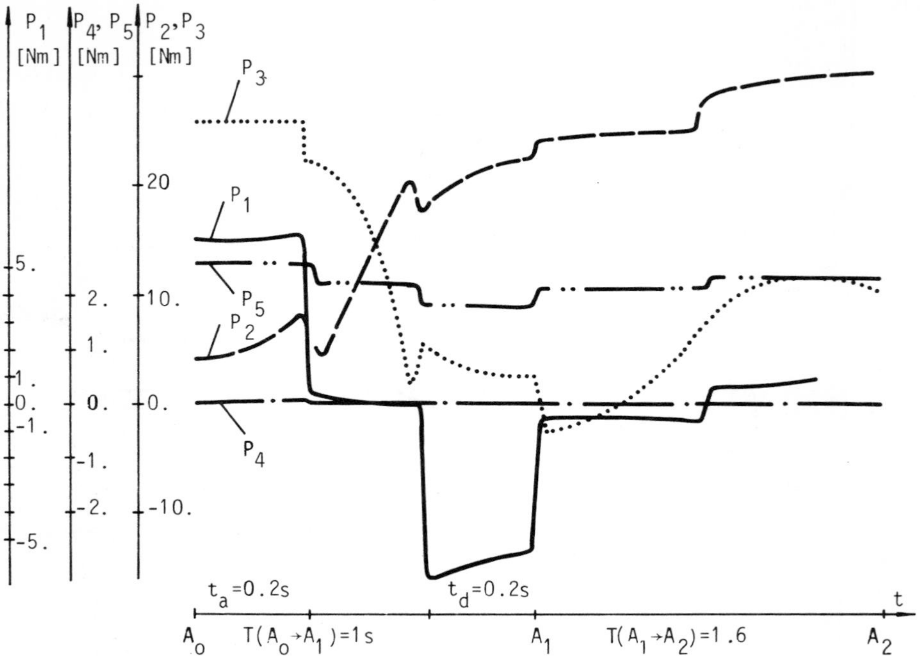

Fig. 2.66. Driving torques P

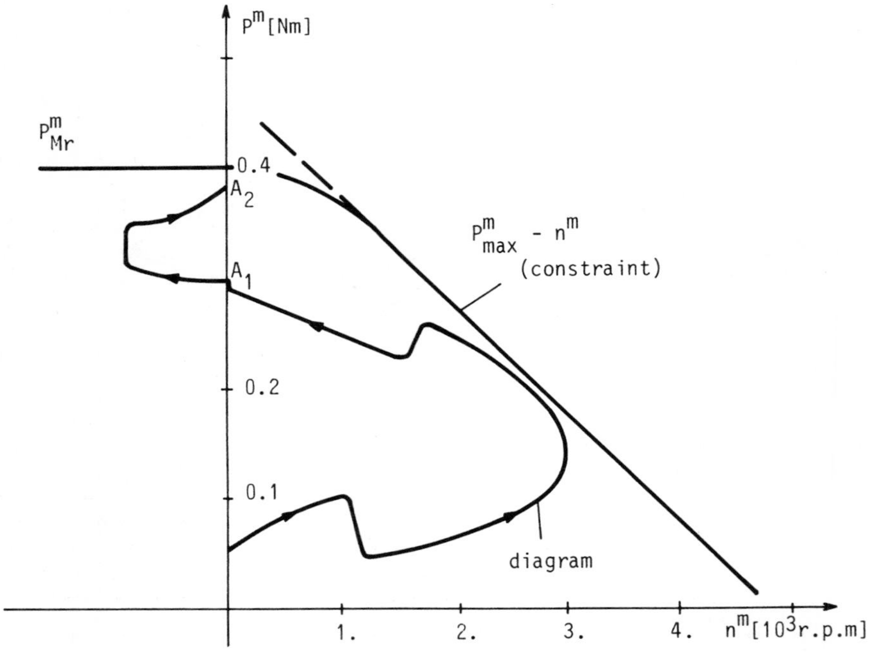

Fig. 2.67. Diagrams torque - r.p.m. for joint S_2

profile. The execution time is $T = T^1 + T^2$ with $T^1 = 2$ s and $T^2 = 1$ s.

Some of the results are shown in Figs. 2.70. and 2.71. Time histories of internal coordinates q and torques P are given in the figures. In performing the task the manipulator consumed 1263 J of energy. No resistance to the insertion is considered.

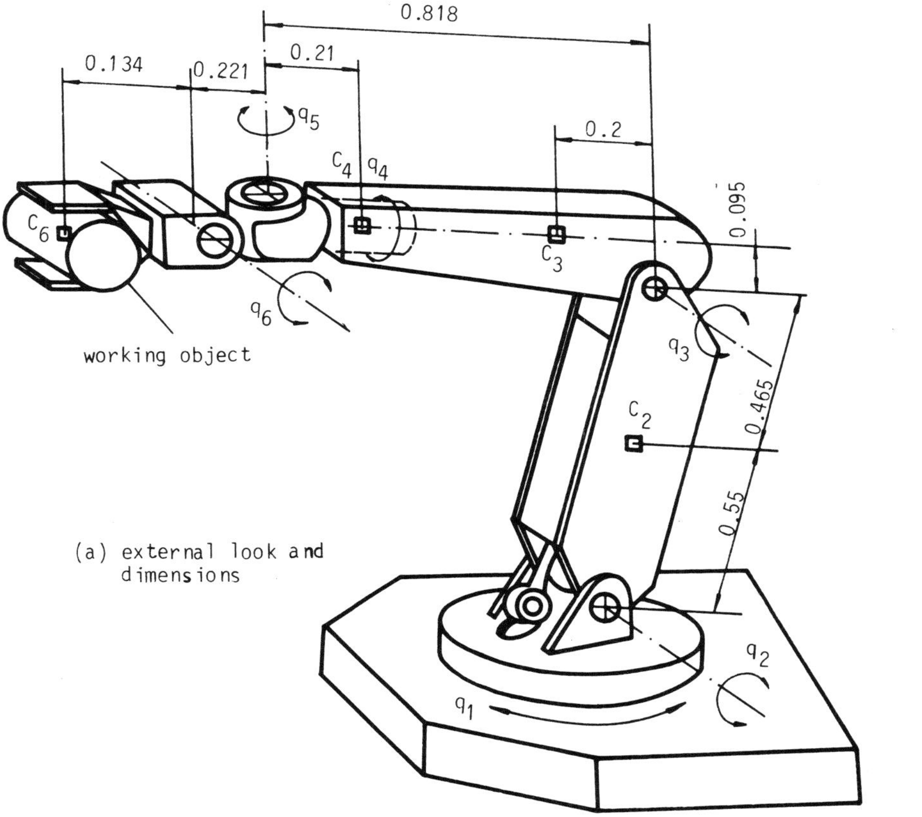

(a) external look and dimensions

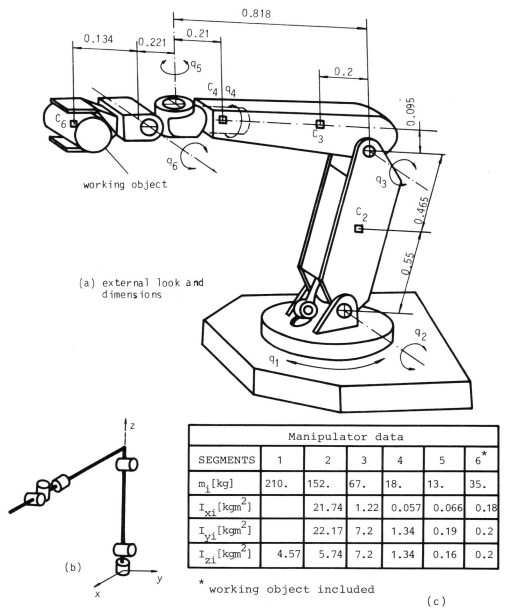

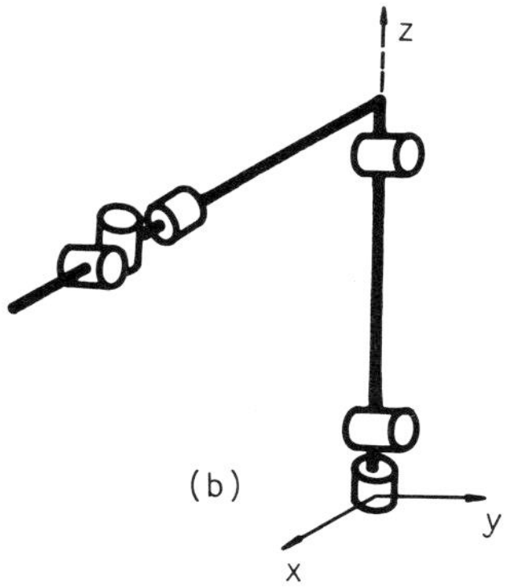

(b)

Manipulator data						
SEGMENTS	1	2	3	4	5	6*
m_i[kg]	210.	152.	67.	18.	13.	35.
I_{xi}[kgm^2]		21.74	1.22	0.057	0.066	0.18
I_{yi}[kgm^2]		22.17	7.2	1.34	0.19	0.2
I_{zi}[kgm^2]	4.57	5.74	7.2	1.34	0.16	0.2

* working object included

(c)

Fig. 2.68. Arthropoid manipulator GORO-80

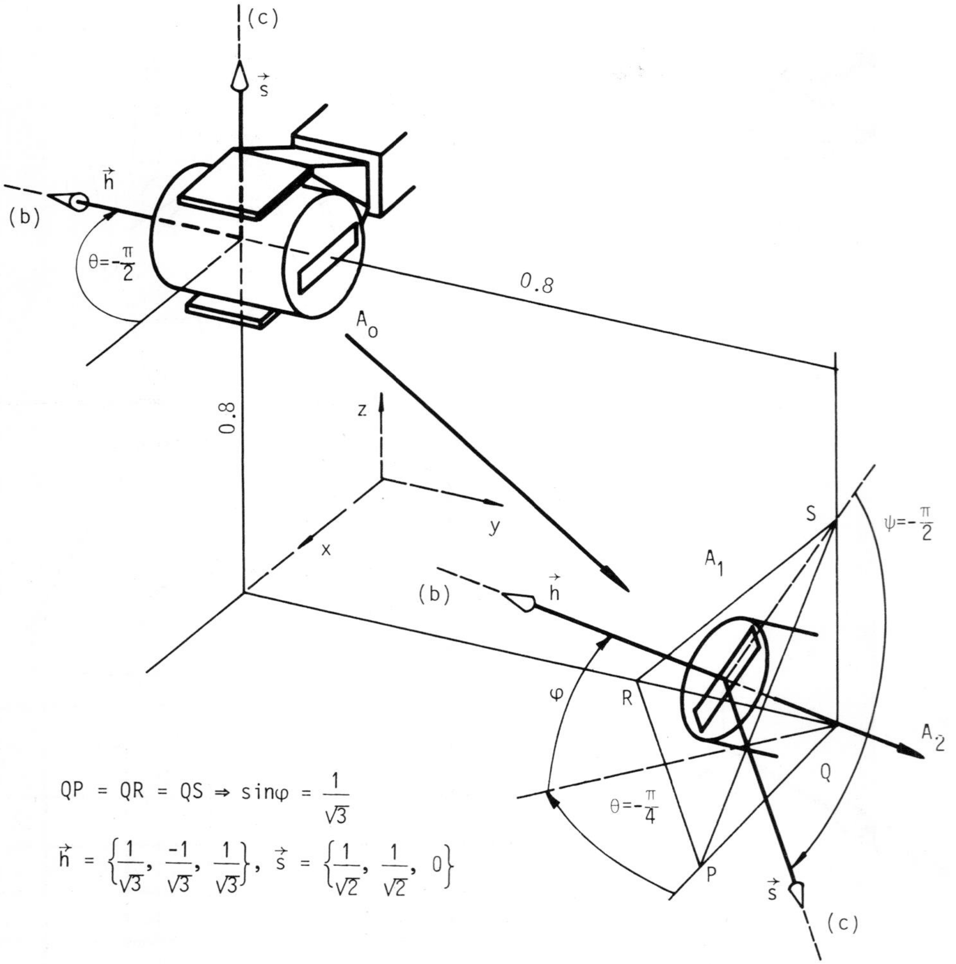

Fig. 2.69. Scheme of manipulation task

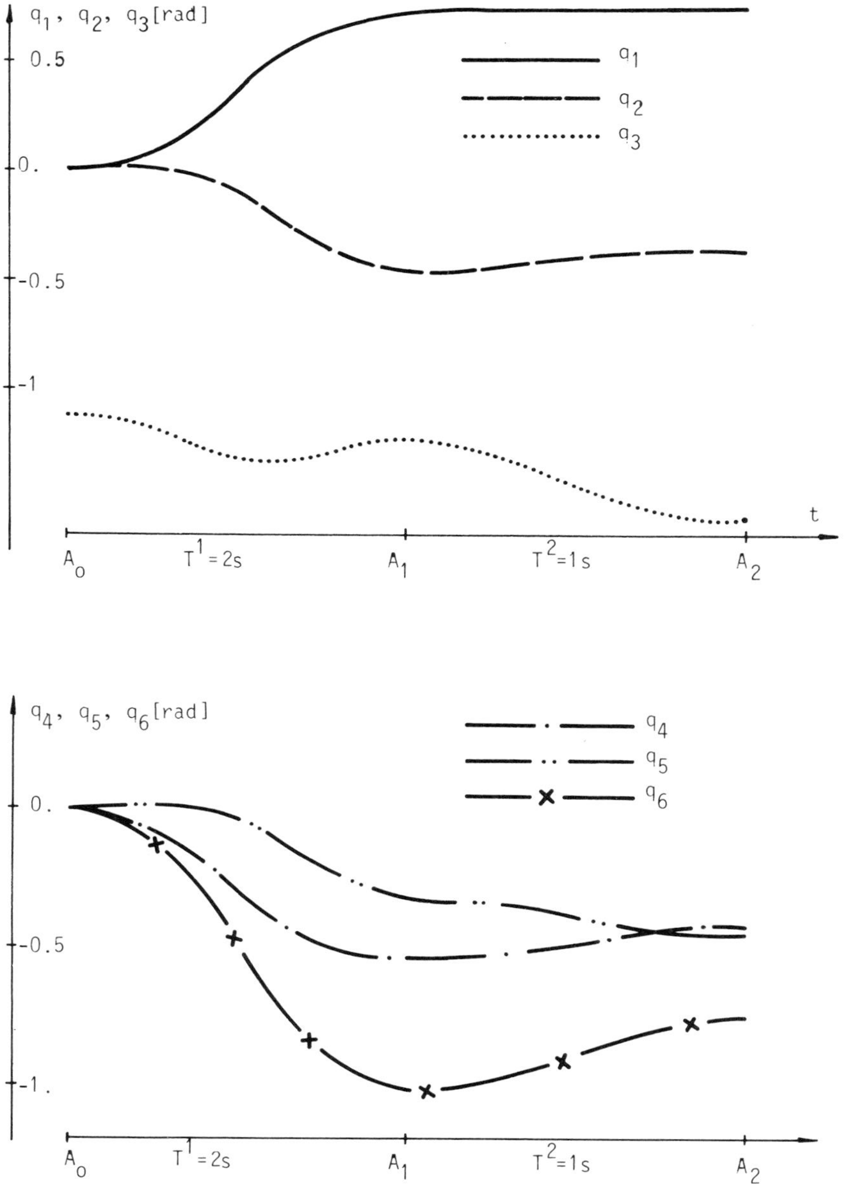

Fig. 2.70. Internal coordinates

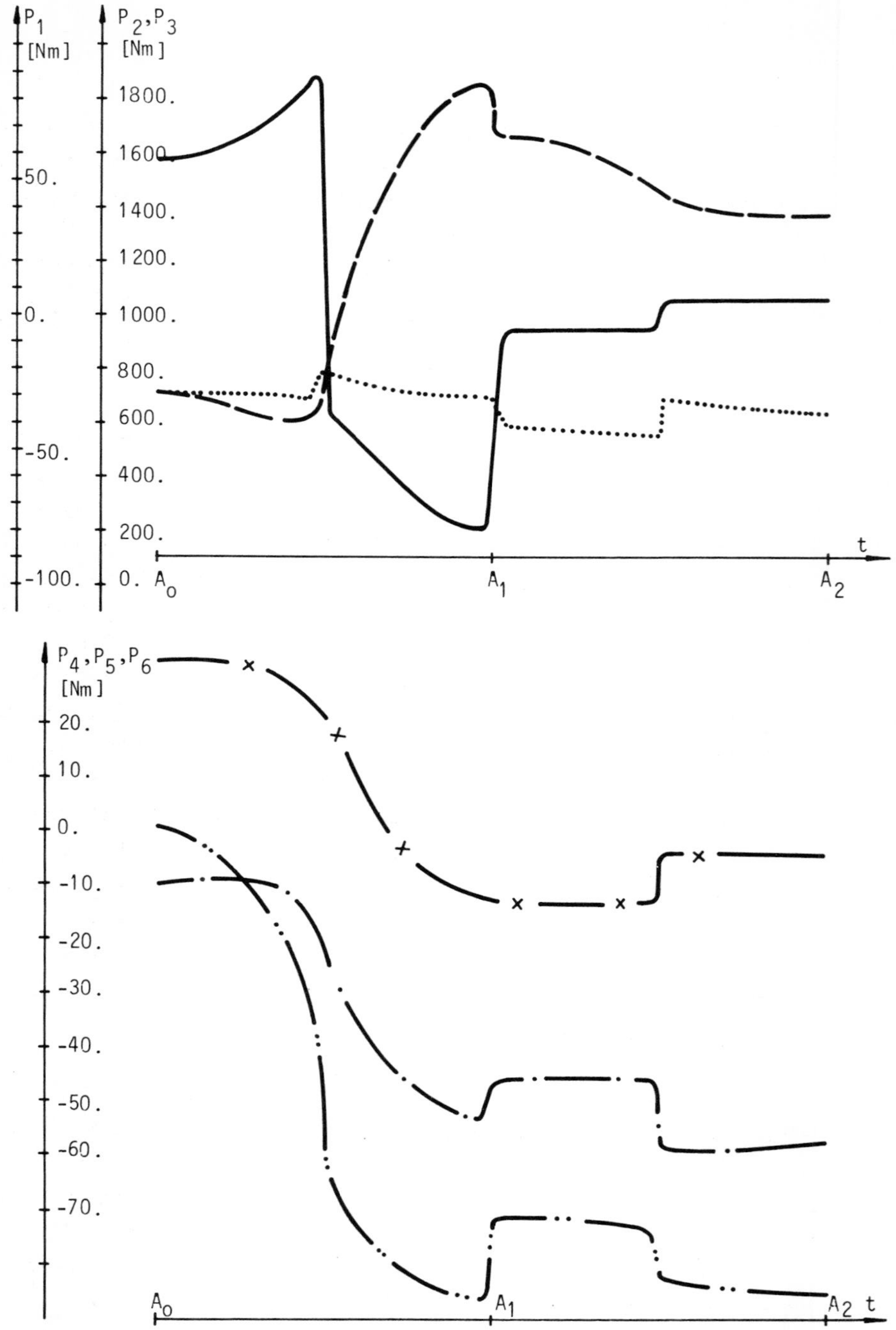

Fig. 2.71. Driving torques

2.9. Synthesis of Nominal Dynamics of Manipulation Movements

In this paragraph we consider some control problems. The dynamic analysis algorithm described in previous paragraphs is not intended for control synthesis. Hence, we develop here a block which solves only some control problems such as synthesis of nominal control. Such a block can easily be incorporated in the algorithm described. The reason for this expansion of the dynamic analysis algorithm is our aim to obtain the algorithm which will consider the whole manipulator dynamics including the dynamics of actuator units.

Now it is necessary to define some notions which appear in control theory. In the general case, the control task can be defined as a task of transferring the system state from any initial state into a defined point in the state space during a finite time interval. The initial state can usually only belong to a bounded zone in the system state space X^I. It is not necessary to transfer the system state into a point but a bounded zone in the state space around the desired point X^F. The system is observed during a determined time interval τ and it is required that transfer of the state from the zone of initial conditions X^I into the zone X^F is performed during a defined time interval τ_s, $\tau_s \leqslant \tau$. It is also required that during transfer from zone X^I to zone X^F the system state belongs to a certain bounded zone X^t. At the stage of nominal dynamics the task of control synthesis is the following: programmed control $u(t)$ should be synthesized for $\forall t \in T$, which should transfer the system S state from a defined initial state $x(0) \in X^I$ into a desired state $x(\tau_s) \in X^F$ during the time interval $\tau_s \leqslant \tau$, where the nominal trajectories of the system state coordinates should satisfy the conditions $x(t, x(0)) \in X^t$ for $\forall t \in T$.

Having in mind the fact that for most robots and especially manipulators in industrial practice the working conditions are known in advance and that the functional tasks, performed by the system, are repeated in cycles, so that they can be foreseen, it has been proposed that the control be synthesized in two stages [28, 29]. At the first stage of control synthesis, the nominal programmed control is synthesized, which produces the prescribed system motion from a certain chosen initial state under the supposition that no perturbations are acting on the system.

At the second stage of control synthesis, control of the tracking of the nominal trajectories is synthesized when the initial state deviates from the nominal initial state (but belongs to a bounded zone of initial states) and perturbations of the initial conditions type act on the system. At this stage, decentralized control is applied.

As this monograph only treats the active mechanisms dynamics, this paragraph will only consider the problem of the synthesis of nominal dynamics, which is the first stage in the synthesis of the control system.

The complete dynamic model of the manipulation robot will therefore be constructed, including both the mechanism and the actuator system. The dynamic analysis algorithm described earlier is then supplemented with a control block in order to permit the calculation of nominal control inputs for the given motion. The example of synthesis at the basis of functional subsystems for positioning and orientation is given.

2.9.1. The complete dynamic model

In this paragraph, the complete mathematical model of the manipulator dynamics will be derived. This model includes the model of the manipulator mechanical part as well as the model of the driving actuators.

The manipulator is considered as an active mechanism of the open chain type, with n d.o.f. (Fig. 2.3). It is assumed that the manipulator joints have one d.o.f. each, which can be rotational or linear. There is a driving actuator acting at each joint.

Manipulator mechanical part. This term refers to the manipulator mechanism as a mechanical system. If we introduce the generalized coordinates $q = [q_1 \cdots q_n]^T$ as in previous paragraphs, the mechanical part dynamics can be described by means of a system of n second-order differential equations in matrix form:

$$S_M: \quad W(q)\ddot{q} = P + U(q, \dot{q}), \quad q(t_o) = q^o, \quad \dot{q}(t_o) = \dot{q}^o, \tag{2.9.1}$$

where $P = [P_1 \cdots P_n]^T$ is the n-dimensional vector of the driving forces and torques in the joints. P_i is a force if S_i is a linear joint, or a torque if S_i is a rotational joint.

By introducing a 2n-dimensional state vector

$$\xi = \begin{bmatrix} \xi_1 \\ \xi_2 \end{bmatrix}; \quad \xi_1 = q, \quad \xi_2 = \dot{q}, \tag{2.9.2}$$

the system (2.9.1) can be reduced to canonical form

$$\dot{\xi} = \begin{bmatrix} \xi_2 \\ W(\xi_1)^{-1} U(\xi_1, \xi_2) \end{bmatrix} + \begin{bmatrix} 0 \\ W^{-1}(\xi_1) \end{bmatrix} P, \quad \xi(t_o) = \xi^o. \tag{2.9.3}$$

Introducing

$$K_M(\xi) = \begin{bmatrix} \xi_2 \\ W(\xi_1)^{-1} U(\xi_1, \xi_2) \end{bmatrix}, \qquad D_M(\xi) = \begin{bmatrix} 0 \\ W^{-1}(\xi_1) \end{bmatrix} \tag{2.9.4}$$

the model (2.9.3) of the mechanical part becomes

$$S_M: \; \dot{\xi} = K_M(\xi) + D_M(\xi) P, \quad \xi(t_o) = \xi^o \tag{2.9.5}$$

Mathematical model of the driving actuators. Driving actuators will be considered such that the model of the i-th actuator can be written in the form

$$S_A^i: \; \dot{x}_i = C_i x_i + f_i P_i + d_i N(u_i), \quad x_i(t_o) = x_i^o \tag{2.9.6}$$

where x_i is an n_i-dimensional state vector of the i-th subsystem S_A^i, C_i, f_i and d_i are constant matrices of the model, P_i is the driving torque (or force) of the i-th actuator (scalar value), u_i is the control input of the i-th actuator (scalar value) and $N(u_i)$ is a nonlinearity of the saturation type.

Further, let k_i elements of vector x_i coincide with the elements of vector ξ, i.e., let the k_i state coordinate of the i-th actuator S_A^i be already contained in the state vector of the mechanical part S_M. For instance, the generalized coordinate q_i and the generalized velocity $\dot{q}_i$ are usually included in the state vector x_i of the i-th joint actuator. So, usually $k_i = 2$ and $\sum_{i=1}^{n} k_i = 2n$ = the dimension of the vector ξ. In general, $\dot{q}_i$ and q_i need not be included in x_i, but a nonlinear dependence exists.

The subsystems S_A^i, $i=1,\dots,n$ can be united into a system of dimension $N = \sum_{i=1}^{n} n_i$, i.e.,

$$S_A: \quad \dot{x} = Cx + FP + Du, \; x(t_o) = x^o, \tag{2.9.7}$$

where $x = [x_1^T \dots x_n^T]^T$, $P = [P_1 \dots P_n]^T$, $u = [u_1 \dots u_n]^T$, $C = \mathrm{diag}[C_1 \dots C_n]$, $F = \mathrm{diag}[f_1 \dots f_n]$, $D = \mathrm{diag}[d_1 \dots d_n]$. P is the vector of the drives and u is the control vector. Thus, the model of driving actuators is written in the form (2.9.7). In the model (2.9.7), care should be taken with the nonlinearity of the saturation type.

Complete model. The models S_M: (2.9.1) and S_A: (2.9.7) can now be united into a complete dynamic model. Let it be assumed that q_i and $\dot{q}_i$ are included in x_i, i.e., $k_i = 2$, $i=1,\dots,n$, and let us introduce the transformation matrices T_i, $i=1,\dots,n$ (dimension $1\times n_i$) such that $\ddot{q}_i = T_i\dot{x}_i$. Now from (2.9.1) it follows that

$$P = W\ddot{q} - U = WT\dot{x} - U, \tag{2.9.8}$$

where $T = \mathrm{diag}[T_1 \dots T_n]$ is an $n\times N$ matrix. Substituting $\dot{x}$ in (2.9.7) into (2.9.8), one obtains

$$P = (E_n - WTF)^{-1}[WT(Cx + Du) - U], \tag{2.9.9}$$

where E_n is an $n\times n$ unit matrix.

Now, substituting P in (2.9.9) into (2.9.7), the complete system is obtained in the form

$$\dot{x} = \hat{C}(x) + \hat{D}(x)u, \tag{2.9.10}$$

where the $N\times N$ matrix $\hat{C}$ and the $N\times n$ matrix $\hat{D}$ are

$$\hat{C} = Cx + F(E_n-WTF)^{-1}(WTCx-U), \quad \hat{D} = D + (E_n-WTF)^{-1}WTD. \tag{2.9.11}$$

This form of manipulator mathematical model is used in those books of this series which discuss the control problems of industrial manipulation.

2.9.2. Mathematical models of the actuator systems

Paramanent magnet D.C. motors are widely used as the actuators for industrial manipulators. The scheme of such a motor is shown in Fig. 2.72. Let i_r be the rotor current. Then the 3-dimensional state vector for such a system is

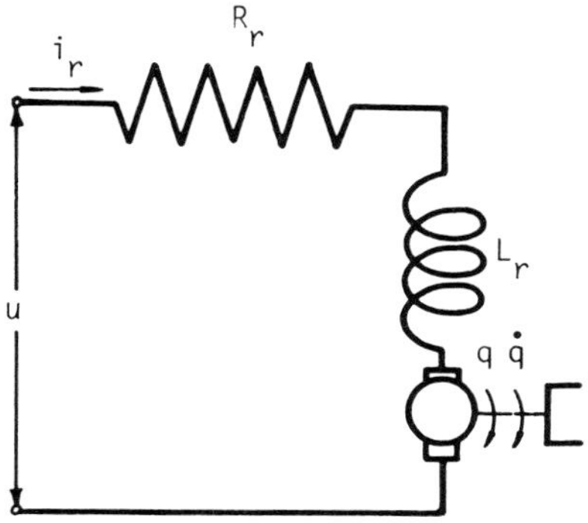

Fig. 2.72. Scheme of a D.C. motor

$$x_i = \left[q_i \ \dot{q}_i \ i_{r_i}\right]^T \tag{2.9.12}$$

if the i-th joint actuator is considered. In this case $k_i = 2$. The mathematical model of actuator can be written in the following form

$$S_A^i: \quad x_i = C_i \dot{x}_i + f_i P_i + d_i N(u_i), \tag{2.9.13}$$

where P_i is the motor torque and u_i is the control voltage. The amplitude of the control is constrained:

$$N(u_i) = \begin{cases} -u_{imax} & \text{for} \quad u_i < -u_{imax} \\ u_i & \text{for} \quad -u_{imax} < u_i < u_{imax} \\ u_{imax} & \text{for} \quad u_i > u_{imax} \end{cases} \tag{2.9.14}$$

For the third order model the system matrices are

$$C_i = \begin{bmatrix} 0 & 1 & 0 \\ 0 & -\frac{B_C}{J_r} & \frac{C_M}{J_r} \\ 0 & -\frac{C_E}{L_r} & -\frac{R_r}{L_r} \end{bmatrix}; \quad f_i = \begin{bmatrix} 0 \\ -\frac{1}{J_r} \\ 0 \end{bmatrix}; \quad d_i = \begin{bmatrix} 0 \\ 0 \\ \frac{1}{L_r} \end{bmatrix}, \tag{2.9.15}$$

where C_M and C_E are the constants of moment and electromotor force, L_r and R_r are the rotor inductivity and resistance, J_r is the rotor moment of inertia, and B_C is the viscous friction coefficient. Index "i" is omitted.

If the rotor inductivity is neglected, then the actuator model reduces to the second-order form ($n_i = 2$). The state vector is $x_i = [q_i \; \dot{q}_i]^T$ and the system matrices of the second-order model are

$$C_i = \begin{bmatrix} 0 & 1 \\ 0 & -\frac{C_E C_M}{J_r R_r} \end{bmatrix}; \quad f_i = \begin{bmatrix} 0 \\ -\frac{1}{J_r} \end{bmatrix}; \quad d_i = \begin{bmatrix} 0 \\ \frac{C_M}{J_r R_r} \end{bmatrix} \tag{2.9.16}$$

Here, B_C is also neglected.

The rotor current is then

$$i_{r_i} = \left(u_i - C_{E_i} \dot{q}_i\right)/R_{r_i} . \tag{2.9.17}$$

The viscous friction is also neglected. For this model $n_i = k_i = 2$ i.e. all state variables of the actuator are included in the state vector of the manipulator mechanical part.

<u>The electrohydraulic actuator</u> consists of a servovalve and a cylinder. The scheme of such an actuator is given in Fig. 2.73.

The 5-dimensional state vector is

$$x_i = \left[\ell_i \; \dot{\ell}_i \; \Delta p_i \; V'_{si} \; \dot{V}'_{si}\right] \tag{2.9.18}$$

if the i-th joint actuator is considered. ℓ is the displacement of the cylinder piston. $\Delta p = p^1 - p^2$, where p^1 and p^2 are pressures on the piston sides. V'_s is the flow due to servovalve piston motion (theoretic flow). Index "i" is omitted. If the actuator drives a linear joint, then assuming $\ell_i = q_i$, $\dot{\ell}_i = \dot{q}_i$, it follows that $k_i = 2$. If such an actuator drives a rotational joint, then there is usually nonlinear dependence between the actuator state variables ℓ_i, $\dot{\ell}_i$ and the joint variables q_i, $\dot{q}_i$.

Now, the fifth-order mathematical model can be written in the form (2.9.13). P_i represents the actuator force and u_i now represents the control current. The system matrices are

$$C_i = \begin{bmatrix} 0 & 1 & 0 & 0 & 0 \\ 0 & -\frac{B_c}{m} & \frac{A}{m} & 0 & 0 \\ 0 & -\frac{4\beta A}{V} & -\frac{4\beta(k_c+C_\ell)}{V} & \frac{4\beta}{V} & 0 \\ 0 & 0 & 0 & 0 & 1 \\ 0 & 0 & 0 & -c_1 & -c_2 \end{bmatrix} \; ; \quad f_i = \begin{bmatrix} 0 \\ -\frac{1}{m} \\ 0 \\ 0 \\ 0 \end{bmatrix} \; ; \quad d_i = \begin{bmatrix} 0 \\ 0 \\ 0 \\ 0 \\ k_q \end{bmatrix} \tag{2.9.19}$$

where B_c is the viscous friction coefficient, m is the mass of the cylinder piston, A is the piston area, β is the compressibility coefficient of the fluid depending on the percentage of air in the oil, V is the total volume including the volume of the valve, the cylinder and the pipes, k_c is the slope of the servovalve flow-pressure characteristic in the working point, $C_\ell = C_{i\ell} + C_{e\ell}/2$, where $C_{i\ell}$ and $C_{e\ell}$ are the coefficients of internal and external leakage, c_1 and c_2 are the coefficients depending on the servovalve frequency characteristic and k_q is a servovalve coefficient.

If the servovalve bandwidth is large enough, we can assume that its dynamics do not influence the behaviour of the whole system. The actuator model can then be reduced to the third-order form (n_i=3) with the state vector $x_i = [\ell_i \; \dot{\ell}_i \; \Delta p_i]^T$ and the system matrices

$$C_i = \begin{bmatrix} 0 & 1 & 0 \\ 0 & -\frac{B_c}{m} & \frac{A}{m} \\ 0 & \frac{4\beta}{V} A & -\frac{4\beta}{V}(k_c+C_\ell) \end{bmatrix} \; ; \quad f_i = \begin{bmatrix} 0 \\ -\frac{1}{m} \\ 0 \end{bmatrix} \; ; \quad d_i = \begin{bmatrix} 0 \\ 0 \\ \frac{4\beta}{V}\frac{k_q}{C_\ell} \end{bmatrix} \tag{2.9.20}$$

2.9.3. Algorithm for the synthesis of nominal dynamics

The notion of dynamic analysis will now be broadened. For this reason the complete model will be considered and the algorithm will include the calculation of the control inputs which have to produce the prescribed motion of the manipulator.

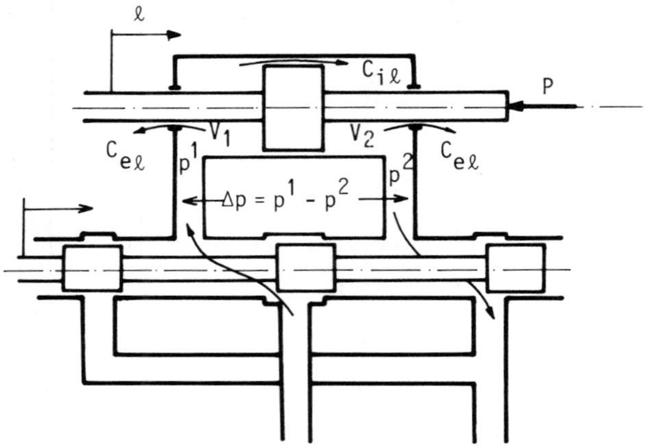

Fig. 2.73. Scheme of an electrohydraulic actuator

We shall now describe an algorithm for the analysis of manipulators with D.C. motors. In a similar way, the algorithm for manipulators with hydraulic actuators can be derived. The third-order model of D.C. motor will be used. It is assumed that the system starts with nominal initial conditions.

Let us consider a time instant t^* when the state $(q, \dot{q})$ of the mechanical part is known. The accelerations $\ddot{q}$ can be computed from (2.2.7). The driving forces and torques P are then computed from the S_M model (2.9.1), i.e.,

$$P = W\ddot{q} - U. \tag{2.9.21}$$

The control inputs should be derived from the actuator models S_A^i, $i = 1,\ldots,n$ i.e. (2.9.13). The matrix equation (2.9.13) with (2.9.15) consits of three scalar equations. The rotor current i_{r_i} is computed form the second scalar equation. The derivative $\frac{d}{dt} i_{r_i}$ can be found in the form $\frac{d}{dt} i_{r_i} = (i_{r_i}(t^*) - i_{r_i}(t^* - \Delta t))/\Delta t$. The control input u_i is now computed from the third scalar equation.

If a second-order actuator model S_A^i is used then the procedure is simplified. The second-order matrix model (2.9.13) with (2.9.16) consists of two scalar equations. For known $(q, \dot{q})$ and computed $\ddot{q}$ and P, the control input u_i can be obtained from the second scalar equation.

The block-scheme of the dynamic analysis algorithm is given in Fig. 2.74. The scheme is similar to the scheme in Fig. 2.2. but with the addition of the control block.

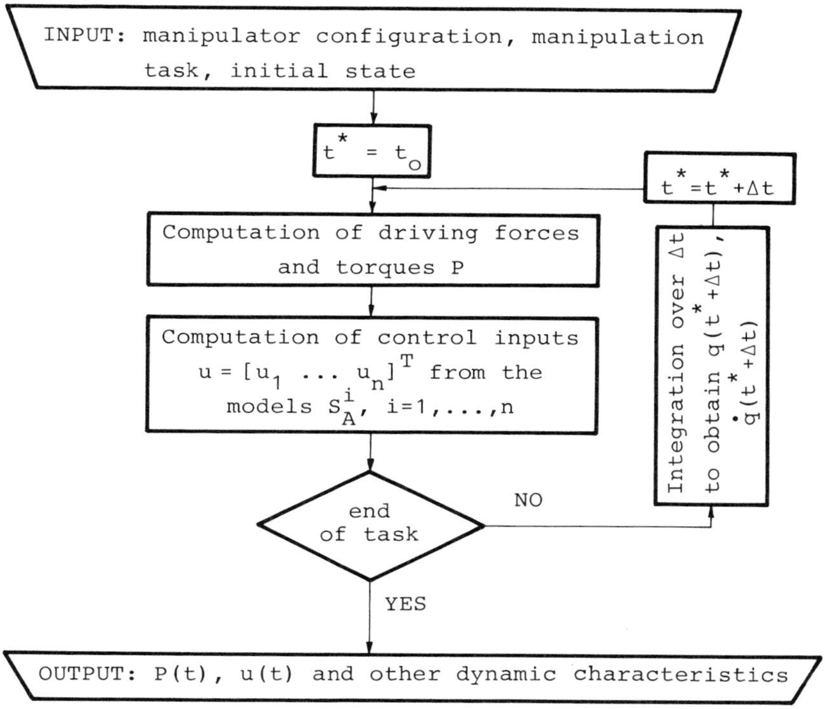

Fig. 2.74. Block-scheme of the algorithm for synthesis of nominal dynamics

2.10. Extension of Dynamic Model by Including Friction Effects

In Para. 2.3. we have considered the dynamic model of manipulator mechanism (mechanical part of manipulator). In 2.9. we have shown how the complete model is formed which includes both the mechanical part and the driving actuators. Let us now discuss these models from the standpoint of friction effects. We notice that the actuator model (Para. 2.9) contains the term which introduces viscous friction (for instance coefficient B_C in (2.9.15)). Any additional viscous fiction in the mechanical part can easily be taken into account by introducing the viscous friction term in each joint:

$$M_{vf_i} = B_i\dot{q}_i \tag{2.10.1}$$

where B_i is the friction coefficient. Thus, the dynamic model (2.3.2) from which the driving torques are computed becomes

$$P = W\ddot{q} - U + B\dot{q} \tag{2.10.2}$$

where $B = \mathrm{diag}[B_i]$.

The problem of static friction is more complex, and there are two possible approaches. The first follows from energy analysis i.e. the friction is taken into account through the energy loss it produces. This approach has already been used in the discussion on reducers. In 2.5.1. we have introduced reduction ratio N_i such that

$$\text{for speed: } n_i = n_i^m/N_i, \tag{2.10.3a}$$

$$\text{for torque: } P_i = P_i^m N_i \eta_i \tag{2.10.3b}$$

where η_i is the mechanical efficiency of reducer. In that way the ratio for speed N_i is greater than the torque ratio $(N_i\eta_i)$. In general, the mechanical efficiency is a function of speed and this dependence can be found in catalogues. An example is given in Fig. 2.75. In an analoguous way we introduce mechanical efficiencies for bearings etc. This approach is simple and gives satisfactory results.

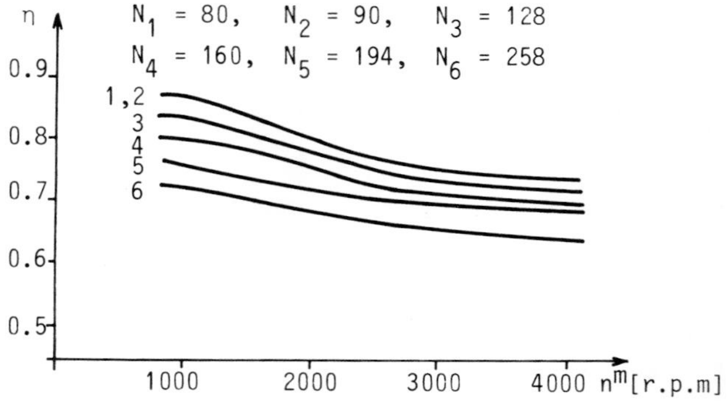

Fig. 2.75. Mechanical efficiency as a function on rotation speed

There is still one disadvantage of the methodology described: the value of friction torque can not be calculated exactly enough and especially the starting friction torque. The second approach is oriented to calculation of this friction torque. It will be described through the example of rolling bearings. First we restrict our consideration to calculation of torques for known motion since the inverse calculation in the presence of static friction would be rather complex. In Para. 2.5.3. it was shown how the total force $\vec{F}_{Si}$ and total moment $\vec{M}_{Si}$ in joint S_i are calculated. We now consider one mechanism joint, a rotational one, consisting of two rolling bearings (Fig. 2.76).

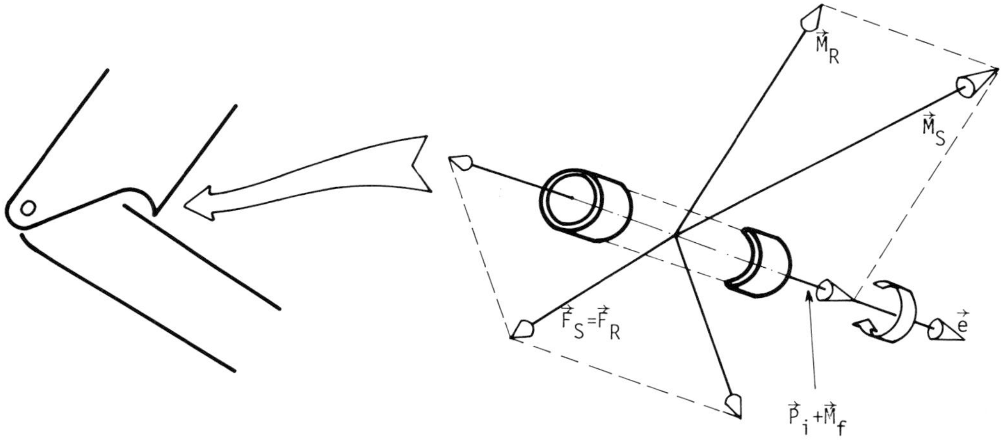

Fig. 2.76. A rotational joint

Note that the joint index "i" is omitted in Fig. 2.76. and will also be omitted in the following discussion.

After calculation of $\vec{M}_S$ and $\vec{F}_S$ the reaction moment is found in the form

$$\vec{M}_R = \vec{e}\times(\vec{M}_S\times\vec{e}) \tag{2.10.4}$$

The projection of $\vec{M}_S$ onto the axis $\vec{e}$ represents the sum of driving torque $\vec{P}_i$ and friction torque $\vec{M}_f$. Let M_f be the module of vector $\vec{M}_f$ ($\vec{M}_f = M_f\vec{e}$). Now:

$$P_i = (\vec{M}_S\cdot\vec{e}) - M_f\,\mathrm{sgn}(\dot{q}) \tag{2.10.5}$$

Let the reaction moment $\vec{M}_R$ be substituted by two forces (Fig. 2.77).

From Fig. 2.77. it follows that

$$\vec{N}_M = \frac{\vec{M}_R \times \vec{e}}{\ell/2} \ . \tag{2.10.6}$$

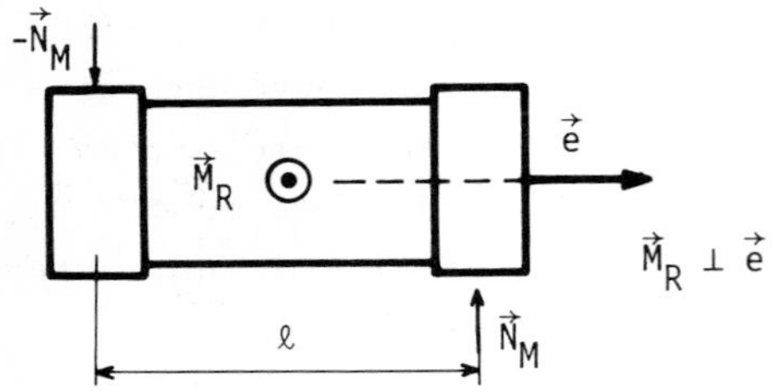

Fig. 2.77. Decomposition of reaction moment

Now the total loads of bearings (1) and (2) (see Fig. 2.77) are

$$\vec{F}_{L1} = \frac{1}{2}\vec{F}_R - \vec{N}_M \tag{2.10.7a}$$

$$\vec{F}_{L2} = \frac{1}{2}\vec{F}_R + \vec{N}_M \tag{2.10.7b}$$

We compute the equivalent loads for bearings [30]: F_{eL1}, F_{eL2}. Then the friction torques M_{f1} and M_{f2} can be computed by [31]

$$M_{f1} = \mu F_{eL1}\frac{d}{2}, \qquad M_{f2} = \mu F_{eL2}\frac{d}{2}, \qquad M_f = M_{f1}+M_{f2} \tag{2.10.8}$$

where d is bearing bore diameter and μ is friction coefficient. The values of μ which can be found in [31] are approximate but allow estimates to be made of friction torque in different types of bearings. Thus, after calculation of M_S, F_S and calculation of M_f by using (2.10.4), (2.10.6) - (2.10.8) we can finally compute the driving torque from (2.10.5).

Linear joints are treated in an analoguous way.

Conclusion

This chapter dealt with the computer oriented procedure for the dynamic analysis of manipulator motion. The computer-aided method for the formation of dynamic model was considered first. This method is based on Appel's equations and ects with generalized coordinates. The method is completely general for it operates for an arbitrary manipulator configuration. In Para. 2.4 the discussion on functional motion was presented. The solutions proposed give the posibility for manipulation task prescription in terms of suitable external coordinates and simple handling with the algorithm. Such an algorithm for dynamic analysis of manipulator motion calculates all important dynamic characteristics such as: trajectories in state space, driving forces and torques, reactions in joints, stresses in manipulator segments, some necessary driving units characteristics, energy consumption, etc. The algorithm is supplemented with a control block in order to calculate the nominal (program) control inputs. The algorithm also contains the tests of dynamic characteristics. In such a way, the algorithm described becomes a very useful tool for designers in the process of manipulator design. It enables the calculation of all relevant dynamic characteristics in advance i.e. before the device is built. It allows a designer to analyse a great number of different configurations and choose the most appropriate one keeping in mind the application of the robotic system. Chapter 3 will discuss the problems of manipulators with closed chain configurations. Finally Chapter 4 is devoted to computer-aided design of industrial manipulators. This design procedure is based on the dynamic analysis algorithm described in Chapters 2 and 3.

References

[1] Stepanenko Yu., "Method of Analysis of Spatial Level Mechanisms", (in Russian), Mekhanika mashin, Vol. 23, Mascow, 1970.

[2] Juričić D., Vukobratović M., "Mathematical Modelling of a Bipedal Walking System", ASME Publ. 72-WA/BHF-13.

[3] Vukobratović M., Stepanenko Yu., "Mathematical Models of General Anthropomorphic Systems", Math. Biosciences, Vol. 17, 1973.

[4] Stepanenko Yu., Vukobratović M., "Dynamics of Articulated Open--Chain Active Mechanisms", Math. Biosciences, Vol. 28, 1/2, 1976.

[5] Vukobratović M., "Computer Method for Mathematical Modelling of Active Kinematic Chains via Generalized Coordinates", Journal of IFToMM Mechanisms and Machine Theory, Vol. 13, No 1, 1978.

[6] Vukobratović M., Legged Locomotion Robots and Anthropomorphic Mechanisms, Monograph, 1975, Institute "M.Pupin", Beograd, Yugoslavia.

[7] Orin D., Vukobratović M., R.B. Mc Ghee., G.Hartoch, "Kinematic and Kinetic Analysis of Open-Chain Linkages Utilizing Newton-Euler Methods", Math. Biosciences, Vol. 42, 1978.

[8] Vukobratović M., "Synthesis of Artificial Motion", Journal of Mechanism and Machine Theory, Vol. 13, No 1, 1978.

[9] Vukobratović M., Potkonjak V., "Contribution to the Forming of Computer Methods for Automatic Modelling of Active Spatial Mechanisms Motion", PART1: "Method of Basic Theorems of Mechanics", Journal of Mechanisms and Machine Theory, Vol. 14, No 3, 1979.

[10] Vukobratović M., Potkonjak V., "Contribution to the Computer Methods for Generation of Active Mechanism models via Basic Theorems of Mechanics" (in Russian), Teknitcheskaya kibernetika ANSSSR, No 2, 1979.

[11] Paul R., Modelling, Trajectory Calculation and Servoing of a Computer Controlled Arm, (in Russian) "Nauka", Moscow, 1976.

[12] Luh, J.Y.S., Walker M.W., Paul R.P.C., "On-Line Computational Scheme for Mechanical Manipulators", Trans. of ASME Journal of Dynamic Systems, Measurement and Control, Vol. 102, June 1980.

[13] Vukobratović M., "Computer Method for Mathematical Modelling of Active Kinematic Chains via Euler's Angles", Journal of IFToMM Mechanisms and Machine Theory, Vol. 13, No 1, 1978.

[14] Vukobratović M., Potkonjak V., "Contribution to Automatic Forming of Active Chain Models via Lagrangian Form", Journal of Applied Mechanics, No 1, 1979.

[15] Hollerbach J.M., "A Recursive Formulation of Lagrangian Manipulator Dynamics", Proc. of JACC, Aug. 13-15, 1980, San Francisco.

[16] Potkonjak V., Vukobratović M., "Two Methods for Computer Forming of Dynamic Equations of Active Mechanisms", Journal of Mechanism and Machine Theory, Vol. 14, No 3, 1979.

[17] Vukobratović M., Nikolić I., "Further Development of Methods for Automatic Forming of Dynamic Models of Manipulators", Proc. of International Conference on Systems Engineering, Coventry, England 1980.

[18] Popov E.P., Vereschagin A.F., Ivkin A.M., Leskov A.G., Medvedov V.S., "Design of Robot Control Using Dynamic Models of Manipulator Devices", Proc. of VI IFAC Symp. on Automatic Control in Space, (in Russian), Erevan, USSR, 1974.

[19] Popov E.P., Vereschagin A.F., Zenkevich S.A., Manipulation Robots: Dynamics and Algorithms, (in Russian), "Nauka", Moscow, 1978.

[20] Medvedov V.S., Leskov A.G., Yuschenko A.S., Control Systems of Manipulation Robots, (in Russian), "Nauka", Moscow, 1978.

[21] Kulakov F.M., Supervisory Control of Manipulation Robots, (in Russian), "Nauka", Moscow, 1980.

[22] Vukobratović M., Potkonjak V., Dynamics of Manipulation Robots: Theory and Application, Monograph, Springer-Verlag, 1982.

[23] Vukobratović M., Potkonjak V., Hristić D., "Dynamic Method for the Evaluation and Choice of Industrial Manipulators", Proc. of 9th International Symp. on Industrial Robots, Washington, 1979.

[24] Vukobratović M., Potkonjak V., "Contribution to Computer-Aided Design of Industrial Manipulators Using Their Dynamic Properties", Journal of IFToMM, Mechanisms and Machine Theory, Vol. 16, No 2, 1982.

[25] Pars L., Analytical Mechanics, (in Russian), "Nauka", Moscow, 1971.

[26] Vukobratović M., Potkonjak V., "Transformation Blocks in The Dynamic Simulation Typical Manipulator Configurationa", IFToMM Journal of Mechanism and Machine Theory, No 3, 1982.

[27] Cadzow J.A., Martens H.R., Discrete-Time and Computer Control Systems, Prentice-Hall, Electrical Engineering Series, 1970.

[28] Vukobratović M., Stokić D., "Contribution to the Decoupled Control of Large-Scale Mechanical Systems", Automatica, Vol. 16, No 1, 1980.

[29] Vukobratović M., Stokić D., Control of Manipulation Robots: Theory and Application, Monograph, Springer-Verlag, 1982.

[30] International Organization for Standardisation, Recommendation R281.

[31] Tribology Handbook, Newnes-Butterworths, London, 1973.

[32] Truckenbrodt A., "Dynamics and Control Methods for Moving Flexible Structures and Their Application to Industrial Robots", Proc. of 5th World Congres on Theory of Machines and Mechanisms, publ. ASME, 1979.

[33] Sunada H.W., Dynamic Analysis of Flexible Spatial Mechanisms and Robotic Manipulators", Ph. D. Thesis, University of California, Los Angelos, 1981.

Appendix: Theory of Appel's Equations

This appendix is intended to explain the Appel's approach to the description of mechanical system dynamics. Since this theory is rarely taught even within high-level courses in mechanics, we give here a short derivation of the most important expressions and conclusions. The Appel's equations have been introduced to describe the dynamics of non-holonomic systems. Here, in robot dynamics, we do not encounter non-holonomic systems but we still use Appel's equations, since they happen to be very convenient for linked systems. These conveniences will be discussed later. We shall therefore derive these equations for holonomic systems and only notice the problems of non-holonomic dynamics.

This book is intended for engineers, designers, researches, and students interested in the field of robotics. It understands the knowledge obtainable at some standard courses in mechanics but since some readers may happen to be unfamiliar with certain notions of theoretical mechanics, we start our presentation with some definitions.

We consider a system of N particles denoted by $\ell = 1,2,\dots,N$. Let a particle ℓ have a mass m_ℓ and a position vector $\vec{r}_\ell$. Now, the position of the system is defined by $\vec{r}_1,\dots,\vec{r}_N$. Let each vector be expressed in terms of three Cartesian projections. Let these 3N Cartesian projections be denoted by $\underbrace{x_1,\ x_2,\ x_3}_{\vec{r}_1},\dots,\underbrace{x_{3N-2},\ x_{3N-1},\ x_{3N}}_{\vec{r}_N}$. This notation is shown to be more suitable than the x, y, z notation. Now the system position is defined by the 3N - dimensional vector $x = (x_1,\dots,x_{3N})$.

Let us now consider the constraints imposed on the system. The constraints of the form

$$f(x,\ t) = 0 \tag{A.1a}$$

or

$$f(x) = 0 \tag{A.1b}$$

are called the geometric or holonomic constraints. If a system moves and is subject only to the constraints of type (A.1) we then talk about holonomic systems.

The constraints having the non-integrable form

$$\sum_{\nu=1}^{3N} A_\nu(x,\, t)\; dx_\nu + Adt = 0 \tag{A.2a}$$

or

$$\sum_{\nu=1}^{3N} A_\nu(x) dx_\nu = 0 \tag{A.2b}$$

are called non-holonomic constraints. If they are imposed on a system we then talk about non-holonomic systems.

The constraints in which the time t does not enter explicitly (the forms (A.1b) and (A.2b)) are called scleronomic or stationary constraints. If there exists explicit dependence on t, then the constraints are rheonomic or non-stationary.

Let us repeat that Appel's equations are intended for the description of non-holonomic systems (and also for holonomic systems since they are special cases of non-holonomic ones). But, we restrict our consideration to holonomic systems.

We consider a system of particles which moves subject to k holonomic constraints

$$f_\mu(x,\, t) = 0, \qquad \mu = 1,\ldots,k. \tag{A.3}$$

According to D'Alembert's principle the dynamics of such a holonomic system can be described by the differential equations

$$\vec{F}_\ell - m_\ell \ddot{\vec{r}}_\ell + \vec{R}_\ell = 0 \qquad \ell = 1,\ldots,N \tag{A.4}$$

where $\vec{F}_\ell$ is the resultant of active forces acting on particle m_ℓ and $\vec{R}_\ell$ represents the resultant of reaction forces due to constraints imposed. The number of equations needed to solve the dynamics of such a system is larger than the number of system degrees of freedom. The system has $n = 3N - k$ degrees of freedom and we need $3N + k$ scalar equations (3N scalar differential equations contained in (A.4) and k equations of constraints (A.3)). Our intention is to find the minimal set of equations which will describe the system dynamics. The number of such equations should be equal to the number of degrees of freedom i.e. n. We shall derive such a set of equations in Appel's form.

In order to derive the Appel's equations we use the method of virtual displacements. We first give a simplified explanation of this notion. Let x be the position of the system at some time instant t. Let x' be another possible position (possible in the sense of being consistent with the constraints at the time instant t). To move from x to x' we need only give the system a displacement $\delta x = x' - x$. This motion performs in the time instant t and does not require any time interval. We call δx a virtual displacement to distinguish it from a true displacement (denoted by dx) which occurs in a time interval dt where forces and constraints could be changing. The symbol δ has the usual properties of the differential d.

Let us give the system a virtual displacement and multiply each of equations (A.4) by the virtual displacement $\delta\vec{r}_\ell$ of the corresponding particle.

$$\left(\vec{F}_\ell - m_\ell\ddot{\vec{r}}_\ell + \vec{R}_\ell\right)\delta\vec{r}_\ell = 0, \qquad \ell = 1,\ldots,N \tag{A.5a}$$

If we make a sum of equations (A.5a) we obtain

$$\sum_{\ell=1}^{N}\left(\vec{F}_\ell - m_\ell\ddot{\vec{r}}_\ell + \vec{R}_\ell\right)\delta\vec{r}_\ell = 0 \tag{A.5b}$$

Since we consider only ideal constraints it follows $\Sigma\vec{R}_\ell\cdot\delta\vec{r}_\ell = 0$ (due to $\vec{R}_\ell\perp\delta\vec{r}_\ell$). Thus:

$$\sum_{\ell=1}^{N}\left(\vec{F}_\ell - m_\ell\ddot{\vec{r}}_\ell\right)\delta\vec{r}_\ell = 0 \tag{A.6}$$

This equation unites the D'Alembert's principle and the principle of virtual displacements. Hence, it is often called the D'Alambert-Lagrange's equation.

Let us introduce Cartesian projections with the following notation

$$\vec{r}_1 = (x_1, x_2, x_3),\ldots,\vec{r}_N = \left(x_{3N-2}, x_{3N-1}, x_{3N}\right)$$

$$\vec{F}_1 = (P_1, P_2, P_3), \ldots, \vec{F}_N = \left(P_{3N-2}, P_{3N-1}, P_{3N}\right) \tag{A.7}$$

and also the new numeration of masses:

$$\begin{aligned} &\text{for particle 1:} \quad m_1\text{(old) becomes } m_1 = m_2 = m_3 \text{ (new)} \\ &\cdots \\ &\text{for particle N:} \quad m_N\text{(old) becomes } m_{3N-2} = m_{3N-1} = m_{3N} \text{ (new)} \end{aligned} \tag{A.8}$$

This notation allows us to write the equation (A.6) in the form

$$\sum_{\nu=1}^{3N} \left(P_\nu - m_\nu \ddot{x}_\nu\right)\delta x_\nu = 0 \tag{A.9}$$

As already said the system of N particles subject to k holonomic constraint has $n = 3N - k$ degrees of freedom and thus the coordinates $x_1, \ldots, x_{3N}$ are not independent (due to constraints). Hence we define n independent parameters $q_1, q_2, \ldots, q_n$ which determine the system position and call them the generalized coordinates. Each Cartesian coordinate x_ν can now be expressed in terms of generalized coordinates

$$x_\nu = g_\nu(q_1, \ldots, q_n; t) \tag{A.10}$$

For stationary systems there is no explicit dependence on t. The virtual displacement δx_ν is now

$$\delta x_\nu = \sum_{i=1}^{n} \frac{\partial x_\nu}{\partial q_i} \delta q_i = \sum_{i=1}^{n} a_{\nu i} \delta q_i \tag{A.11}$$

Substituting (A.11) into (A.9)

$$\sum_{\nu=1}^{3N} \left(P_\nu - m_\nu \ddot{x}_\nu\right) \sum_{i=1}^{n} a_{\nu i} \delta q_i = 0 \tag{A.12}$$

i.e.

$$\sum_{\nu=1}^{3N} P_\nu \sum_{i=1}^{n} a_{\nu i} \delta q_i - \sum_{\nu=1}^{3N} m_\nu \ddot{x}_\nu \sum_{i=1}^{n} a_{\nu i} \delta q_i = 0 \tag{A.13}$$

or

$$\sum_{i=1}^{n} \sum_{\nu=1}^{3N} m_\nu \ddot{x}_\nu a_{\nu i} \delta q_i = \sum_{i=1}^{n} \sum_{\nu=1}^{3N} P_\nu a_{\nu i} \delta q_i \tag{A.14}$$

The right-hand side of the equation (A.14) represents the virtual work A of all active forces acting on the system. If we introduce the notation

$$Q_i = \sum_{\nu=1}^{3N} P_\nu a_{\nu i}, \qquad i = 1,\dots,n \tag{A.15}$$

the virtual work acquires the form

$$\delta A = \sum_{i=1}^{n} Q_i \delta q_i \tag{A.16}$$

Q_i is called the generalized force corresponding to the generalized coordinate q_i. One convenient definition of generalized force Q_i sais that it is the coefficient of the virtual displacement ∂q_i in the expression for the virtual work. This definition allows a simple determination of generalized forces, as shown in 2.3.

Let us now transform the left-hand side of the equation (A.14). First, we find the time derivatives of the expression (A.10):

$$\dot{x}_\nu = \sum_{i=1}^{n} \frac{\partial x_\nu}{\partial q_i} \dot{q}_i + \frac{\partial x_\nu}{\partial t} = \sum_{i=1}^{n} a_{\nu i}\dot{q}_i + a_\nu \tag{A.17}$$

and

$$\ddot{x}_\nu = \sum_{i=1}^{n} a_{\nu i}\ddot{q}_i + \sum_{i=1}^{n} \sum_{j=1}^{n} \frac{\partial a_{\nu i}}{\partial q_j} \dot{q}_i\dot{q}_j + \sum_{j=1}^{n} \frac{\partial a_\nu}{\partial q_j} \dot{q}_j + \sum_{i=1}^{n} \frac{\partial a_{\nu i}}{\partial t} \dot{q}_i + \frac{\partial a_\nu}{\partial t} \tag{A.18}$$

From (A.18) it follows that $a_{\nu i}$ can be expressed in the form

$$a_{\nu i} = \frac{\partial \ddot{x}_\nu}{\partial \ddot{q}_i}, \qquad \nu = 1,\dots,3N \tag{A.19}$$

By using (A.19) the left-hand side of the equation (A.14) becomes:

$$\sum_{i=1}^{n} \sum_{\nu=1}^{3N} m_\nu \ddot{x}_\nu a_{\nu i} \delta q_i = \sum_{i=1}^{n} \sum_{\nu=1}^{3N} m_\nu \ddot{x}_\nu \frac{\partial \ddot{x}_\nu}{\partial \ddot{q}_i} \delta q_i \tag{A.20}$$

We now introduce the function

$$S = \frac{1}{2} \sum_{\nu=1}^{3N} m_\nu \ddot{x}_\nu^2 \tag{A.21}$$

and call it the energy of acceleration. The form of the function is analogous to that of kinetic energy, but the accelerations stand instead of velocities; thus the name for the function S. It is evident that

$$\sum_{\nu=1}^{3N} m_\nu \ddot{x}_\nu \frac{\partial \ddot{x}_\nu}{\partial \ddot{q}_i} = \frac{\partial S}{\partial \ddot{q}_i}, \qquad i = 1,\dots,n \tag{A.22}$$

Now, by using (A.20) and (A.21) to transform the left-hand side of the equation (A.14), and (A.15), (A.16) to transform the right-hand side of the same equation, we obtain

$$\sum_{i=1}^{n} \frac{\partial S}{\partial \ddot{q}_i} \delta q_i = \sum_{i=1}^{n} Q_i \delta q_i \tag{A.23}$$

or

$$\sum_{i=1}^{n} \left(\frac{\partial S}{\partial \ddot{q}_i} - Q_i\right) \delta q_i = 0 \tag{A.24}$$

Since the virtual displacements δq_i are independent, it follows that

$$\frac{\partial S}{\partial \ddot{q}_i} = Q_i, \qquad i = 1,\dots,n \tag{A.25}$$

The expressions (A.25) represent the Appel's equations. They form the set of second-order differential equations. The problem of forming the dynamic model (differential equations) now reduces to the formation of the expression for acceleration energy S in terms of generalized coordinates.

Let us now consider the application of Appel's equations to linked systems. From 2.3. it follows that it is enough to find the expression for acceleration energy of a rigid body since it is then possible to find the function S of the whole chain by making the sum over all bodies of the chain. S is expressed in terms of generalized coordinates $q_1,\dots,q_n$ by using the recurrent expressions for accelerations.

Let us briefly discuss the acceleration energy of a rigid body. We start from the definition expression which considers a rigid body as a system of particles

$$S = \frac{1}{2} \sum_{\ell=1}^{N} m_\ell \left(\frac{d^2\vec{r}_\ell}{dt}\right)^2 \tag{A.26}$$

If we introduce the center of mass C with the position vector $\vec{r}_c$, then

$$\vec{r}_\ell = \vec{r}_c + \vec{r}\,'_\ell \tag{A.27}$$

where $\vec{r}\,'_\ell$ defines the position of particle m with respect to the center of mass. Now

$$S = \frac{1}{2}\sum_{\ell=1}^{N} m_\ell\left(\frac{d^2\vec{r}_c}{dt^2} + \frac{d^2\vec{r}\,'_\ell}{dt^2}\right)^2 =$$

$$= \frac{1}{2} M\left(\frac{d^2\vec{r}_c}{dt^2}\right)^2 + \frac{1}{2}\sum_{\ell=1}^{N} m_\ell\left(\frac{d^2\vec{r}\,'_\ell}{dt^2}\right)^2 + \frac{d^2\vec{r}_c}{dt^2}\sum_{\ell=1}^{N} m_\ell \frac{d^2\vec{r}\,'_\ell}{dt^2}, \qquad (A.28)$$

where M is the total mass of the system i.e. the mass of rigid body

$$M = \sum_{\ell=1}^{N} m_\ell \qquad (A.29)$$

The last term on the right-hand side of the expression (A.28) equals zero since

$$\sum_{\ell=1}^{N} m \frac{d^2\vec{r}\,'_\ell}{dt^2} = \frac{d^2}{dt^2}\sum_{\ell=1}^{N} m_\ell \vec{r}\,'_\ell \qquad (A.30)$$

and for the center of mass it holds

$$\sum_{\ell=1}^{N} m_\ell \vec{r}\,'_\ell = 0 \qquad (A.31)$$

Thus, from (A.28) it follows

$$S = \frac{1}{2} M\vec{w}_c^{\,2} + \frac{1}{2}\sum_{\ell=1}^{N} m_\ell \vec{w}\,'^{2}_\ell \qquad (A.32)$$

where $\vec{w}_c$ is the acceleration of the center of mass and $\vec{w}\,'_\ell$ is the acceleration of particle m_ℓ in its relative motion with respect to center of mass. This theorem is analogous to the one holding for kinetic energy, but accelerations here stand instead of velocities. It has been shown that for a rigid body the second term on the right-hand side of the expression (A.32) can be transformed and the acceleration energy written in the form

$$G = \frac{1}{2} m\tilde{\vec{w}}_c^{\,2} + \frac{1}{2}\tilde{\vec{\varepsilon}}\,\tilde{J}\,\tilde{\vec{\varepsilon}} - [2(\tilde{J}\tilde{\vec{\omega}})\times\tilde{\vec{\omega}}]\tilde{\vec{\varepsilon}} \qquad (A.33)$$

where $\vec{\varepsilon}$ is the angular acceleration of the body, $\tilde{J}$ is the inertia tensor and the tilde over a letter indicates that the corresponding variable (vector or tensor) is expressed with respect to a coordinate system fixed to the body. Notice that $\tilde{\vec{w}}_c^{\,2} = \vec{w}_c^{\,2}$.

As already said, the acceleration energy for a chain of rigid bodies is obtained by summation over all segments. Thus, in order to express the acceleration energy in terms of generalized coordinates it is enough to express the accelerations $\vec{w}_c$ and $\vec{\varepsilon}$ of each segment in terms of generalized coordinates. This is not a problem since there exist recurrent expressions for accelerations of linked segments (see (2.3.14)-(2.3.18)). In 2.3 it is explained how Appel's equations are used to form the dynamic equations of open kinematic chains.

Let us now discuss why Appel's equations are considered to be suitable for application to linked bodies systems. Any type of equations describing the dynamics of a system necessarily leads towards second-order differential equations (often called the dynamic model). Thus the first and the second differentiation must appear somewhere in the model forming procedure. For instance, with Lagrange equations

$$\frac{d}{dt}\frac{\partial T}{\partial \dot{q}} - \frac{\partial T}{\partial q} = Q \tag{A.34}$$

the first differentiation is included in kinetic energy T (because of velocities) but the second differentiation $\frac{d}{dt}$ still explicitly appears. With Appel's equations (A.25) there is no explicit differentiation but both differentiations are included in acceleration energy S. Thus, these two differentiations are solved by using the recurrent expressions for accelerations of segments ((2.3.14)-(2.3.18)) when forming the expression for S. So, we conclude that such an approach allows a better adaptation of model forming procedure to the system considered.

Chapter 3: Closed Chain Dynamics

3.1. Introduction

In Chapter 2 we derived an algorithm which solved the complete dynamics of manipulation robots. Let us note that the algorithm considered a manipulator as an open kinematic chain. This fact restricts the set of manipulation tasks which can be analyzed. Let us consider a peg-in-the--hole assembly task (Fig. 3.1). In such a task the manipulator changes its configuration. In the phase of transferring the object it is an open kinematic chain (Fig. 3.1a) and in the phase of insertion a closed chain since its last segment is connected to the ground by means of a joint permitting one translation and one rotation (Fig. 3.1b)

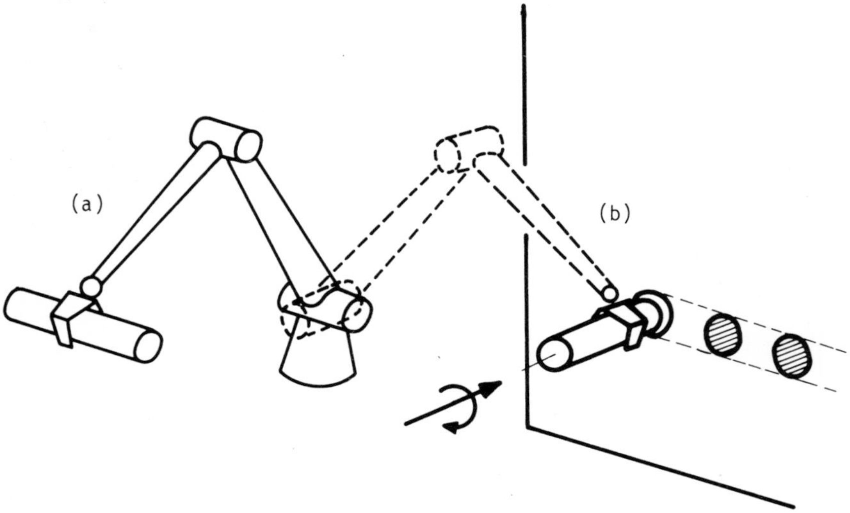

Fig. 3.1. Open and closed chain in an assembly manipulation task

There is another reason for studying closed chains. There exist some special manipulator configurations which contain closed chains (Fig. 3.2). Thus, in order to solve their dynamics in any manipulation task one has to solve closed chains.

Recently, certain efforts have been made towards solving the closed chain dynamics. Paragraph 3.2. provides a short review of these results. What is common to most of these approaches to closed chains is their generality. The methods are derived so as to cover a general kinematic scheme containing arbitrary open or closed chains. This generality, although scientifically justifiable, becomes a disadvantage in the application. In addition, some important cases of closed chains in practice are not taken into account. These methods concentrate on complex mechanisms and do not consider the interaction of the mechanism with its environment. The type of interaction depends on the manipulation task, but this interaction does result in certain boundary conditions which should be satisfied by manipulator gripper. All different types of interaction (boundary conditions) are not covered by the methods discussed. On the other hand, some methods are not sufficiently automatéd and require many jobs to be done analytically by hand. Here, we propose a method which is oriented to practical problems.

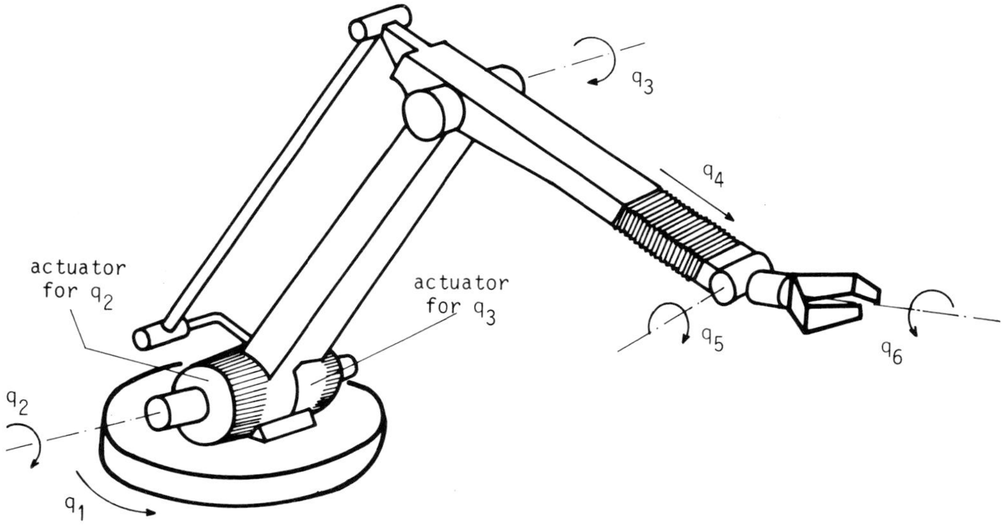

Fig. 3.2. Mechanism containing a closed chain

We consider two cases of closed chains interesting for practice. The first case appears in some special robot mechanisms which are designed to contain closed chains (e.g. manipulator of Fig. 3.2 or ASEA robot). The second case appears in assembly manipulation for instance, or to be more general, this case refers to any manipulation task in which the gripper is not allowed to more freely but its motion is subject to some constraints resulting from the interaction with the environment. An illustration is given in Fig. 3.1. This case also covers the problems of bilateral manipulation.

Let us also mention the third case of closed chains. These are problems of biped gait in double support phase or multilegged walking mechines, but they will not be discussed here.

Paragraph 3.3 discusses the first case and 3.4 the second. These two paragraphs develop the theory which is necessary for the solution of concrete problems. Paragraph 3.6 is devoted to the actual problems of constrained gripper motions.

Paragraph 3.4 discusses the theoretical and practical aspects of constrained gripper motion. The dynamic models of such robot motions are derived. These models can be used to obtain the solutions either to the direct or to the inverse dynamic problem. These solutions assume that the constrains imposed on gripper motion are satisfied at any time instant. Thus the initial conditions also satisfy the constraints. Let us consider the peg-in-the-hole assembly task (Fig. 3.1) in which an open chain becomes closed when the peg enters the hole. The models derived in 3.4 are valid when the peg has already entered the hole. Thus, the problem of insertion still remains to be solved. It usually involves the impact which ensures the initial conditions which satisfy the insertion constraints. Paragraph 3.5 is devoted to the solution of this problem.

3.2. Review of Previous Results

This paragraph will briefly describe the main results achieved in closed chain dynamics.

The method [1] represents a general theory for formulating dynamic equations for robots of arbitrary configuration. The method is based upon d'Alembert's principle and uses the concepts of graph theory for defining the mechanism configuration. Matrix formulation is combined with a symbolic vector-tensor notation in order to obtain a compact form of dynamic model. Systems with tree structure and with closed kinematic chains are distinguished. The method is elaborated up to the program package which forms the nonlinear dynamic model as well as the linearized model of any robot configuration [2, 3]. The method is not suitable for closed chains appearing with constrained gripper motions.

Another general methodology is presented in [4]. It considers a system consisting of arbitrary rigid bodies interconnected by joints permitting one rotation or one translation. There are no constraints upon the kinematic scheme, thus arbitrary closed chains can be contained. The method is based on Gauss' principle of minimal compulsion (in german - "Zwang"). The compulsion function is expressed in terms of external accelerations. The position of bodies with respect to the external coordinate system is defined via 4×4 transformation matrices and, accordingly, the external accelerations are defined via the second derivatives of these matrices. An internal coordinate is defined for each joint to define the relative position of two connected bodies. Thus, the connection relations, if written in the form of accelerations, involve both the external and the internal accelerations. For the solution of simulation problem, the minimization of compulsion function with respect to accelerations has to be performed. This minimization is subject to constraints of the equality type resulting from joint connections. Penalty functions method is suggested for solving this problem. Certain difficulties appear with this method: first, one has to take care of the relations between external and internal accelerations, and, second, the 4×4 matrices defining the position and orientation of a body depend on six independent parameters and this dependence must be taken into account. These problems are not discussed in [4]. There is no algorithmic realization of the method and no discussion on numerical aspects of minimization. The methodology is discussed in detail for the case of open chains only. This theory is shortly described in [8].

Another method using the Gauss' principle is presented in [5, 6]. The general scheme including arbitrary closed chains is considered. The method uses vector notation instead of 4×4 matrices. The compulsion function and the constraints due to joint connections are expressed in terms of internal accelerations. For the solution of constrained minimization problem the conjugate gradient projection method is proposed. A detailed explaination including numerical examples is given.

The problem of interconnected multibody system with friction is discussed in [7]. Newton-Euler's equations are used. The method is rather analytical. Several simple examples are solved.

3.3. Mechanisms Containing a Kinematic Parallelogram

General discussion and definitions. We consider a special mechanism of manipulation robot which contains a closed chain. Two examples are shown in Fig. 3.3a,b. In general, the first (a) has a spherical kinematic scheme, and the second (b) an arthropoid scheme. But, these schemes are more complex since they contain closed chains. Let us note that a closed chain appears because of a hydraulic cylinder (marked by (*) in Fig. 3.3) which is connected to two segments. We also note that the cylinder which drives a joint S_i is connected to segments "i-1" and "i". We shall not regard these schemes as robots with closed chains

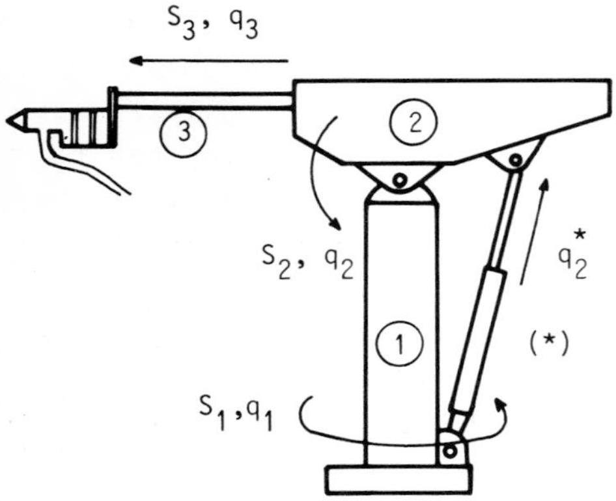

(a) Spray-painting robot UMS-3B

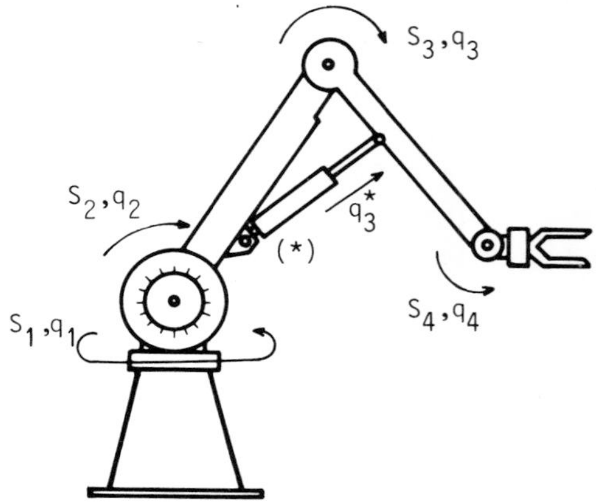

(b) Milacrone robot

Fig. 3.3. Two mechanisms containing closed chains

since there exists a satisfactory approximative approach which allows us to consider these schemes as open chains. The inertial forces of the cylinder are added to the corresponding segment (to segment 1 in case (a), and segment 2 in case (b)). The linear coordinates q_i^* are introduced instead of q_i. The nonlinear dependences between q_i, $\dot{q}_i$, $\ddot{q}_i$, and q_i^*, $\dot{q}_i^*$, $\ddot{q}_i^*$ are solved. We also solve the nonlinear dependence between the joint driving torque P_i and the cylinder force P_i^* [9]. The solution to all these dependences is included in the dynamic analysis algorithm in the form of a special subroutine.

Let us consider another example of a mechanism containing a closed chain. It is the mechanism shown in Fig. 3.2. For this scheme which contains a kinematic parallelogram there is no simple approximative solution. The mechanism has to be considered as a closed chain. The kinematic scheme of such a mechanism is given in Fig. 3.4. Let us define

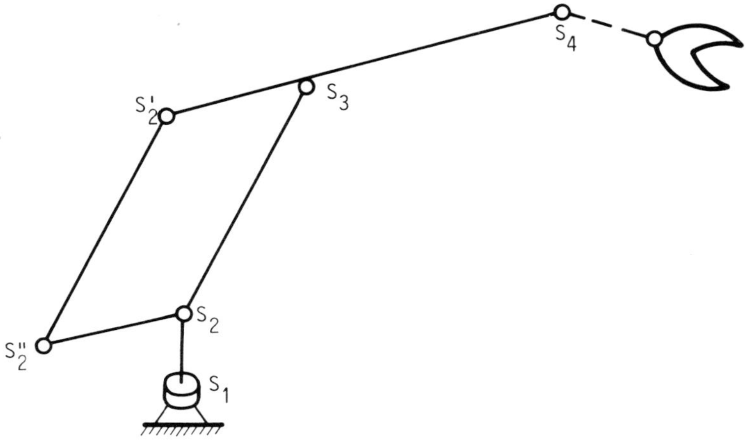

Fig. 3.4. Mechanism with a kinematic parallelogram

the direct branch (solid line in Fig. 3.5). Such a simplified mechanism has n degrees of freedom (d.o.f.) and we define n generalized coordinates $q_1,\ldots,q_n$. It is an open chain and we may use the method from Para. 2.3 to form the dynamic model. In fact we find the matrices W^d, V^d which determine the acceleration energy (2.3.35) and the left hand side of Appel's equations (see eq. (2.3.36)). The upper index "d" indicates that the direct branch is consider only.

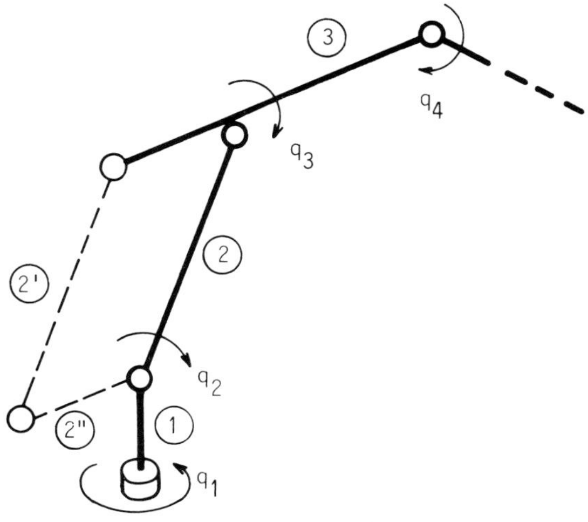

Fig. 3.5. Direct branch

Let us now add the segments 2' and 2" which close the chain (Figs. 3.5 and 3.6). First, we conclude that the number of d.o.f. does not change. Second, we note that there is no drive in the joint S_3 but there are two independent drives in S_2 (in fact S_2 represent two independent joints). One of these drives (P_2) acts between the segments 1 and 2 and the other (P_3) acts between 1 and 2". In this way P_3 acts between 1 and 3 but via 2" and 2' (Fig. 3.6).

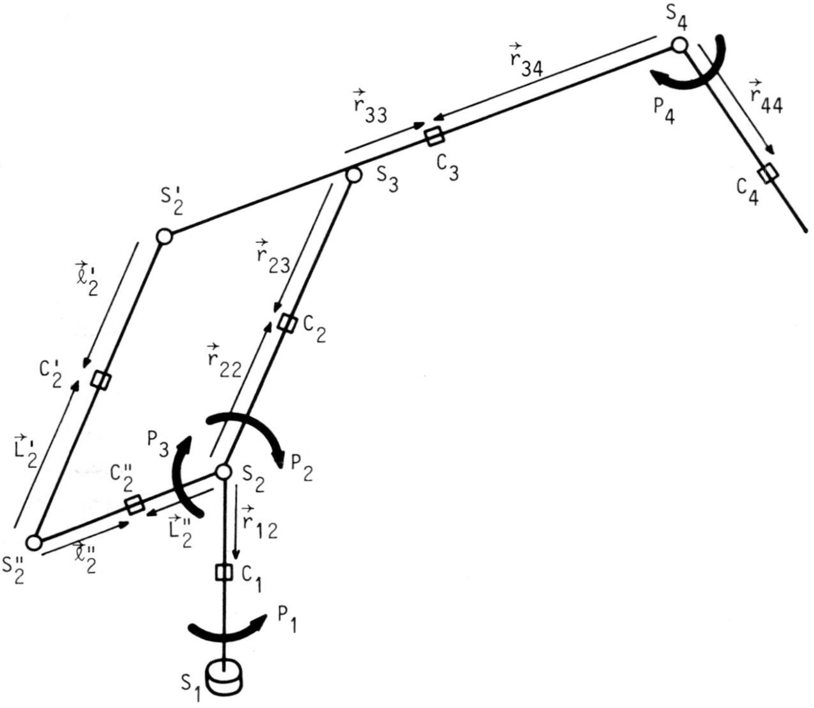

Fig. 3.6. Mechanism geometry and driving torques

<u>Acceleration energy</u>. Now, let us find the components of acceleration energy corresponding to the segments 2" and 2', i.e. let us find the way of computing the corresponding matrices $W_{2"}$, $V_{2"}$ and $W_{2'}$, $V_{2'}$. We first note that for transition matrices it holds

$$A_{2'} = A_2 \qquad A_{2"} = A_3 \tag{3.3.1}$$

$$A_{2",1} = A_{3,1} = A_{3,2}A_{2,1}; \qquad A_{2',2"} = A_{2,3} \tag{3.3.2}$$

It is clear from (3.3.2) that these matrices can be simply computed

$$\delta A_3 = \left[P_3+y_3^d\right]\delta q_3+m_{2''}\vec{g}\delta\vec{r}_{2''}+m_{2'}\vec{g}\delta\vec{r}_{2'} \tag{3.3.19}$$

We conclude that $\vec{e}_2 = \vec{e}_3$ and thus

$$\delta\vec{r}_{2''} = \vec{e}_3\times\vec{L}_{2''}\delta q_3,\ \delta\vec{r}_{2'} = \vec{e}_3\times\left(\vec{L}_{2''}-\vec{\ell}_{2''}\right)\delta q_3 \tag{3.3.20}$$

The generalized force is

$$Q_3 = P_3 + y_3 \tag{3.3.21}$$

$$y_3 = y_3^d + \underbrace{m_{2''}\vec{g}\left(\vec{e}_3\times\vec{L}_{2''}\right)+m_{2'}\vec{g}\left(\vec{e}_3\times\left(\vec{L}_{2''}-\vec{\ell}_{2''}\right)\right)}_{y_3^a} \tag{3.3.22}$$

where y_3^d is the component holding for the direct chain which can be computed from (2.3.57).

For the joints q_4, $q_5,\ldots,q_n$ the forces are

$$Q_j = Q_j^d = P_j + y_j^d; \qquad j=4,\ldots,n \tag{3.3.23}$$

where y_j^d is computed by (2.3.57).

Formation of dynamic model. Combining the left hand and the right hend side of Appel's equations one obtains

$$W\ddot{q} + V^T = MP + Y \tag{3.3.24}$$

where

$$M = \begin{bmatrix} 1 & & & & & \\ & 1 & 1 & & 0 & \\ & & 1 & & & \\ & & & 1 & & \\ & 0 & & & \ddots & \\ & & & & & 1 \end{bmatrix} \tag{3.3.25}$$

Introducing

$$U = Y - V^T \tag{3.3.26}$$

the model (3.3.24) obtains the form

$$W\ddot{q} = MP + U \tag{3.3.27}$$

The matrices W and V are calculated by (3.3.14). The composite vector Y is

$$Y = \left[y_1^d;\ y_2^d + y_2^a;\ y_3^d + y_3^a;\ y_4^d;\ldots;y_n^d\right]^T \tag{3.3.28}$$

where y_2^a, y_3^a are as defined in (3.3.18), (3.3.22).

Now, the sheme of Fig. 2.12 is modified (Fig. 3.9).

In order to avoid the matrix M in dynamic model (3.3.27) we can subtract the third equation of this system from the second one, thus obtaining

$$W^*\ddot{q} = P + U^* \tag{3.3.29}$$

where

$$\left.\begin{aligned} W_{ij}^* &= W_{ij}, & U_i^* &= U_i, \quad i \neq 2 \\ W_{2j}^* &= W_{2j} - W_{3j}, & U_2^* &= U_2 - U_3 \end{aligned}\right\} \tag{3.3.30}$$

If the segments 2" and 2' are considerably smaller than the segment 2, an approximative model can be formed. We may use the model (3.3.27) with the matrix W calculated for the open chain configuration. This open chain is obtained by taking the direct branch and adding the inertial effects of segment 2' to the segment 2. In this case the segment 2" is neglected.

Let us discuss one modification of the mechanism considered (Fig. 3.10). It is the mechanism of ASEA robot which contains a kinematic paralelogram, too. But, instead of driving torques P_2, P_3 (Fig. 3.6), the linear drives of mechanism P_2^*, P_3^* are applied (Fig. 3.10). Nonlinear relationships exist between the linear and the rotational drives.

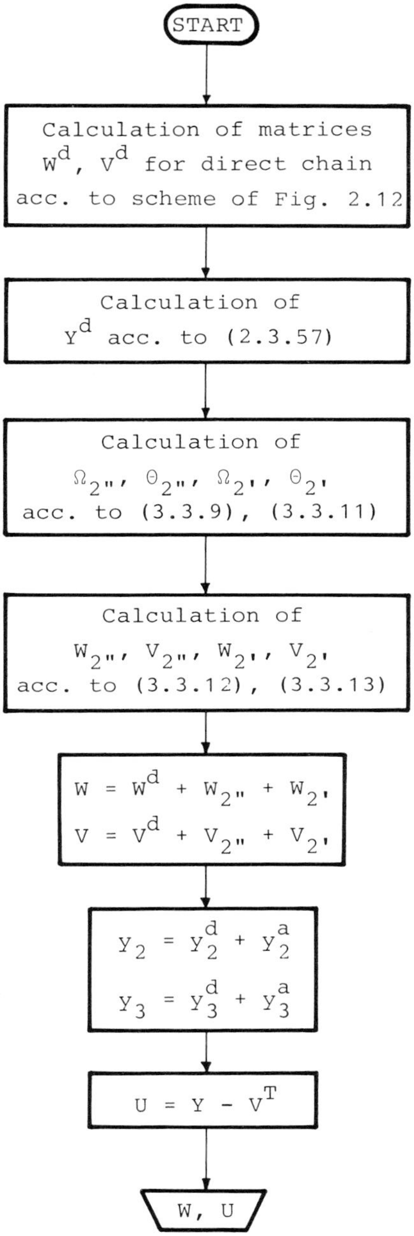

Fig. 3.9. Model-forming procedure for a manipulator with a kinematic parallelogram

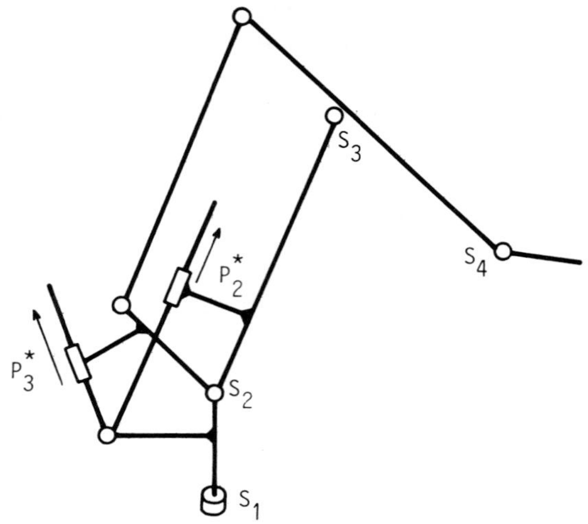

Fig. 3.10. Modified mechanism with a parallelogram

3.4. Manipulators with Constraints on Gripper Motion

This paragraph develops the theory dealing with manipulators having constraints imposed on gripper motion. Each constraint restricts to a certain extent the possibility for gripper motion. In this way the number of d.o.f. of the gripper is reduced. In order to avoid a purely theoretical discussion, we consider only these types of constraints which appear in practical problems. But, there are no obstacles to expanding the theory to cover any type of constraints.

We are mainly interested in manipulators having five or six d.o.f. since they are needed for the solution of manipulation tasks considered. Thus, $n \in \{5, 6\}$ holds. Further, we assume that there is no singularity with the manipulators and the tasks considered.

First, we give an extension of the theory explained in Ch. 2. Then, an example of a stationary constraint is considered (Para. 3.4.2). We locate the problems appearing when a stationary constraint is substituted by a nonstationary one. Friction effects apper to be of main importance. Finally, we elaborate several cases of constraints, those which are interesting for practice (Para. 3.4.3. - 3.4.12).

3.4.1. Theory extension

Let us consider a manipulator as an open chain of n segments, as considered in Para. 3.2. Let us apply an additional external force $\vec{F}_A$ acting on gripper at point A and an additional external moment $\vec{M}_A$ (Fig. 3.11).

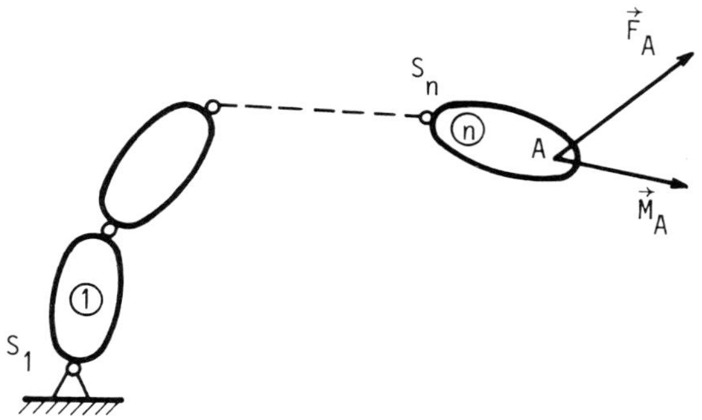

Fig. 3.11. Manipulator with additional force and moment acting on gripper

Now, the dynamic model of such a mechanism will slightly differ from (2.3.2) since $\vec{F}_A$, $\vec{M}_A$ must be included in the generalized forces. One obtains

$$W\ddot{q} = P + U + D_F F_A + D_M M_A \tag{3.4.1}$$

where F_A, M_A, are 3×1 matrices corresponding to $\vec{F}_A$, $\vec{M}_A$. D_F and D_M are obtained from virtual displacement method. The additional component (Q_{iFM}) of generalized force Q_i due to $\vec{F}_A$, $\vec{M}_A$ is

$$Q_{iFM} = \begin{cases} \left(\vec{e}_i \times \vec{r}_A^{\,i}\right)\vec{F}_A + \vec{e}_i \vec{M}_A, & s_i = 0 \\ \vec{e}_i \vec{F}_A, & s_i = 1 \end{cases} \tag{3.4.2}$$

If we introduce

$$D_{F\,(n\times 3)} = \begin{bmatrix} d_{F1}^T \\ \vdots \\ d_{Fn}^T \end{bmatrix}, \qquad \vec{d}_{Fi} = \begin{cases} \vec{e}_i \times \vec{r}_A^{\,i}, & s_i = 0 \\ \vec{e}_i, & s_i = 1 \end{cases} \tag{3.4.3}$$

with

$$\vec{r}_A^{\,i} = \overrightarrow{S_iA} = \sum_{k=i}^{n-1}\left(\vec{r}\,'_{kk}-\vec{r}_{k,k+1}\right) + \vec{r}\,'_{nn} + \vec{p} \tag{3.4.4}$$

and

$$D_{M\,(n\times 3)} = \begin{bmatrix} d_{M1}^T \\ \vdots \\ d_{Mn}^T \end{bmatrix}, \qquad \vec{d}_{Mi} = \begin{cases} \vec{e}_i, & s_i=0 \\ 0, & s_i=1 \end{cases} \tag{3.4.5}$$

then we obtain the form (3.4.1).

This extension was necessary since the constraints produce reaction force and reaction moment acting on gripper. These reactions will be determined so that the constraints are satisfied.

3.4.2. Surface-type constraint

Let us consider an n-d.o.f. manipulator and impose the constraint that the point A of the gripper cannot move freely but is forced to be on a given surface (Fig. 3.12). The immobile surface is defined by

$$h(x, y, z) = 0 \tag{3.4.6}$$

An iterative approach to this problem is given in [8]. Here, we derive another, noniterative one.

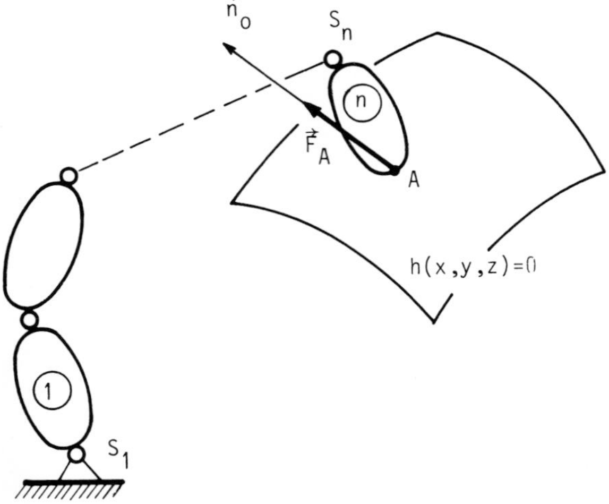

Fig. 3.12. Mechanism with the gripper subject to a surface-type constraint

$\ddot{q}(n\times 1)$.

The constraints produce reactions which are introduced in the dynamic model obtaining (3.4.1), i.e.,

$$W\ddot{q} = P + U + \begin{bmatrix} D_F D_M \end{bmatrix} \begin{bmatrix} F_A \\ M_A \end{bmatrix} \tag{3.4.27}$$

or

$$W\ddot{q} = P + U + DR_A \tag{3.4.28}$$

where the matrices D and R_A are

$$D_{(n\times 6)} = \begin{bmatrix} D_{F\ (n\times 3)} \mid D_{M\ (n\times 3)} \end{bmatrix}, \quad R_{A\ (6\times 1)} = \begin{bmatrix} F_{A\ (3\times 1)} \\ \hline M_{A\ (3\times 1)} \end{bmatrix} \tag{3.4.29}$$

Depending on the constraint imposed and the manipulator configuration (n) there exist some conditions which should be satisfied by the six--component reaction R_A. There are $6-(n-n_r)$ scalar conditions which can be expressed in matrix form:

$$ER_A = 0 \tag{3.4.30}$$

where E is a matrix of dimensions $(6-n+n_r)\times 6$.

Now, the equations (3.4.26), (3.4.28), and (3.4.30) define the complete mathematical model of closed chain configuration.

We first discuss the problem of nominal dynamics calculation. One has to prescribe the manipulation task in terms of gripper motion. Since the manipulator gripper motion is constrained, we prescribe the task in terms of independent parameters X_r. Thus, we prescribe $\ddot{X}_r$. The generalized accelerations $\ddot{q}$ are now computed from (3.4.26) by means of the $n\times n$ Jacobian matrix (J).

The nominal dynamics (prescribed manipulation task) understands that forces which we want to produce are also given. Thus F_A, M_A and, accordingly, R_A are known. These values must be prescribed so that they satisfy (3.4.30). This problem can be avoided by introducing the free components of reactions, as will be shown later in the text. Now, the necessary drives P can be solved from (3.4.28).

Let us discuss the simulation problem. The drives P are prescribed and the motion and the reactions are unknown. Substituting $\ddot{q}$ from (3.4.26) into (3.4.28) it follows that

$$WJ^{-1}J_r\ddot{X}_r = P + U - WJ^{-1}(A_r - A) + DR_A \tag{3.4.31}$$

If we write (3.4.31) and (3.4.30) together, we obtain

$$\left[\begin{array}{c|c} WJ^{-1}J_r & -D \\ \hline 0 & E \end{array}\right] \left[\begin{array}{c} \ddot{X}_r \\ \hline R_A \end{array}\right] = \left[\begin{array}{c} P \\ \hline 0 \end{array}\right] + \left[\begin{array}{c} U - WJ^{-1}(A_r - A) \\ \hline 0 \end{array}\right] \tag{3.4.32}$$

where the dimensions are:

$$\left[\begin{array}{c|c} (n\times n_r) & (n\times 6) \\ \hline ((6-n+n_r)\times n_r) & ((6-n+n_r)\times 1) \end{array}\right] \left[\begin{array}{c} (n_r\times 1) \\ \hline (6\times 1) \end{array}\right] = \left[\begin{array}{c} (n\times 1) \\ \hline ((6-n+n_r)\times 1) \end{array}\right] +$$

$$\left[\begin{array}{c} (n\times 1) \\ \hline ((6-n+n_r)\times 1) \end{array}\right] .$$

Thus, (3.4.32) represents a system of n_r+6 equations which can be solved for n_r+6 unknowns $(\ddot{X}_r, R_A)$. Now, by using (3.4.25) and (2.4.94) we may obtain $\ddot{X}_q$ and $\ddot{q}$. This allows us to form an iterative simulation procedure which finally gives the manipulator motion. Reactions are also obtained.

The dimensionality of system (3.4.32) can be reduced if we note that the six-component reaction R_A depends on $n-n_r$ independent components. Let us introduce the reduced reaction vector R_{Ar} of dimension $(n-n_r)\times 1$ consisting of these independent components. Now, R_A can be expressed as

$$R_A = GR_{Ar} \tag{3.4.33}$$

where G is a matrix of dimensions $(6\times(n-n_r))$.

Substituting (3.4.33) into (3.4.31) it follows that

$$WJ^{-1}J_r\ddot{X}_r = P + U - WJ^{-1}(A_r - A) + DGR_{Ar} \tag{3.4.34}$$

or

$$\left[WJ^{-1}J_r \;\vdots\; -DG \right] \left[\begin{array}{c} \ddot{X}_r \\ \hdashline R_{Ar} \end{array} \right] = P + U - WJ^{-1}(A_r - A) \tag{3.4.35}$$

where the dimensions are

$$\left[(n\times n_r) \;\vdots\; (n\times(n-n_r)) \right] \left[\begin{array}{c} (n_r\times 1) \\ \hdashline (n-n_r)\times 1 \end{array} \right] = (n\times 1) .$$

Thus, (3.4.35) represents a system of n equations which can be solved for n unknowns ($\ddot{X}_r$, R_{Ar}). This approach using independent components is suitable for nominal dynamics as well since we can prescribe all the independent components without taking care of (3.4.30).

In the following paragraphs we apply this theory to several concrete types of constraints.

3.4.4. Gripper moving along a surface

We consider again a manipulator with the gripper which can not move freely but its point A is forced to move along a given surface. This time, the independent parameters approach is applied.

Let us define the relative position of the gripper point A with respect to a surface by means of two parameters u_1 and u_2 (Fig. 3.14). This leads us to the parametric form of the moving surface (nonstationary constraint):

$$\begin{aligned} x &= f_x(u_1, u_2, t) \\ y &= f_y(u_1, u_2, t) \\ z &= f_z(u_1, u_2, t) \end{aligned} \tag{3.4.36}$$

Now $n_r = n-1$ and the reduced position vector is

$$X_r = \begin{cases} \left[u_1 \;\; u_2 \;\; \theta \;\; \varphi \right]^T, & \text{for } n=5 \\ \left[u_1 \;\; u_2 \;\; \theta \;\; \varphi \;\; \psi \right]^T, & \text{for } n=6 \end{cases} \tag{3.4.37}$$

i.e., $u_3 = \Theta$, $u_4 = \varphi$, $u_5 = \psi$

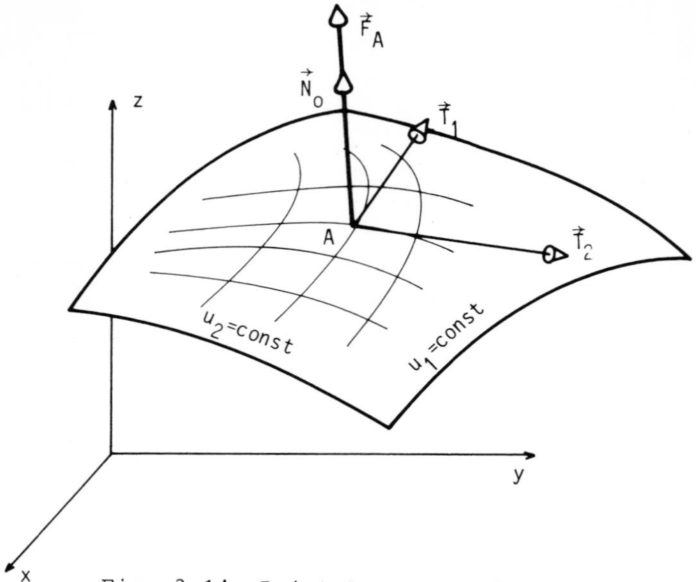

Fig. 3.14. Point A on a moving surface

The first and the second derivative of (3.4.36) give

$$\begin{aligned}
\dot{x} &= \frac{\partial f_x}{\partial u_1}\dot{u}_1 + \frac{\partial f_x}{\partial u_2}\dot{u}_2 + \frac{\partial f_x}{\partial t} \\
\dot{y} &= \frac{\partial f_y}{\partial u_1}\dot{u}_1 + \frac{\partial f_y}{\partial u_2}\dot{u}_2 + \frac{\partial f_y}{\partial t} \qquad (3.4.38)\\
\dot{z} &= \frac{\partial f_z}{\partial u_1}\dot{u}_1 + \frac{\partial f_z}{\partial u_2}\dot{u}_2 + \frac{\partial f_z}{\partial t}
\end{aligned}$$

and

$$\ddot{x} = \underbrace{\frac{\partial^2 f_x}{\partial u_1^2}\dot{u}_1^2 + 2\frac{\partial^2 f_x}{\partial u_1 \partial u_2}\dot{u}_1\dot{u}_2 + \frac{\partial^2 f_x}{\partial u_2^2}\dot{u}_2^2 + 2\frac{\partial^2 f_x}{\partial u_1 \partial t}\dot{u}_1 + 2\frac{\partial^2 f_x}{\partial u_2 \partial t}\dot{u}_2 + \frac{\partial^2 f_x}{\partial t^2}}_{\alpha_x} +$$

$$+ \frac{\partial f_x}{\partial u_1}\ddot{u}_1 + \frac{\partial f_x}{\partial u_2}\ddot{u}_2$$

$$\ddot{y} = \underbrace{\frac{\partial^2 f_y}{\partial u_1^2}\dot{u}_1^2 + 2\frac{\partial^2 f_y}{\partial u_1 \partial u_2}\dot{u}_1\dot{u}_2 + \frac{\partial^2 f_y}{\partial u_2^2}\dot{u}_2^2 + 2\frac{\partial^2 f_y}{\partial u_1 \partial t}\dot{u}_1 + 2\frac{\partial^2 f_y}{\partial u_2 \partial t}\dot{u}_2 + \frac{\partial^2 f_y}{\partial t^2}}_{\alpha_y} +$$

$$+ \frac{\partial f_y}{\partial u_1}\ddot{u}_1 + \frac{\partial f_y}{\partial u_2}\ddot{u}_2$$

$$\ddot{z} = \underbrace{\frac{\partial^2 f_z}{\partial u_1^2}\dot{u}_1^2 + 2\frac{\partial^2 f_z}{\partial u_1 \partial u_2}\dot{u}_1\dot{u}_2 + \frac{\partial^2 f_z}{\partial u_2^2}\dot{u}_2^2 + 2\frac{\partial^2 f_z}{\partial u_1 \partial t}\dot{u}_1 + 2\frac{\partial^2 f_z}{\partial u_2 \partial t}\dot{u}_2 + \frac{\partial^2 f_z}{\partial t^2}}_{\alpha_z} +$$

$$+ \frac{\partial f_z}{\partial u_1}\ddot{u}_1 + \frac{\partial f_z}{\partial u_2}\ddot{u}_2 \qquad (3.4.39)$$

The three equations from (3.4.39) can be written together in the matrix form:

$$\begin{bmatrix} \ddot{x}_A \\ \ddot{y}_A \\ \ddot{z}_A \end{bmatrix} = \begin{bmatrix} \frac{\partial f_x}{\partial u_1} & \frac{\partial f_x}{\partial u_2} \\ \frac{\partial f_y}{\partial u_1} & \frac{\partial f_y}{\partial u_2} \\ \frac{\partial f_z}{\partial u_1} & \frac{\partial f_z}{\partial u_2} \end{bmatrix} \begin{matrix} \ddot{u}_1 \\ \\ \ddot{u}_2 \end{matrix} + \begin{bmatrix} \alpha_x \\ \alpha_y \\ \alpha_z \end{bmatrix} \qquad (3.4.40)$$

When we use this acceleration form of the constraint we assume that initial conditions satisfy the position and the velocity form, i.e., that (3.4.36) and (3.4.38) are satisfied at the initial time instant.

Now, the Jacobian form (3.4.25) can be obtained. The reduced Jacobian and the reduced adjoint matrix are:

$$J_r = \left[\begin{array}{cc|c} \frac{\partial f_x}{\partial u_1} & \frac{\partial f_x}{\partial u_2} & \\ \frac{\partial f_y}{\partial u_1} & \frac{\partial f_y}{\partial u_2} & O_{(3\times(n-3))} \\ \frac{\partial f_z}{\partial u_1} & \frac{\partial f_z}{\partial u_2} & \\ \hline \multicolumn{2}{c|}{O_{((n-3)\times 2)}} & I_{((n-3)\times(n-3))} \end{array}\right], \qquad A_r = \left[\begin{array}{c} \alpha_x \\ \alpha_y \\ \alpha_z \\ \hline O_{((n-3)\times 1)} \end{array}\right]$$

$$(3.4.41)$$

where α_x, α_y, α_z are as defined in eq. (3.4.39) and I is a unit matrix of the corresponding dimension.

The Jacobian J and the adjoint matrix A appearing in (3.4.26) and (3.4.31) as well as in dynamic models (3.4.32) and (3.4.35) are defined in Para. 2.4.

Let us consider the reactions. It is clear that there exists a reaction force $\vec{F}_A$ perpendicular to the surface and the reaction moment equals zero ($\vec{M}_A=0$) (Fig. 3.14). If we define two tangents

$$\vec{T}_1 = \left\{\frac{\partial f_x}{\partial u_1}, \frac{\partial f_y}{\partial u_1}, \frac{\partial f_z}{\partial u_1}\right\}$$
$$\vec{T}_2 = \left\{\frac{\partial f_x}{\partial u_2}, \frac{\partial f_y}{\partial u_2}, \frac{\partial f_z}{\partial u_2}\right\} \tag{3.4.42}$$

having the unit vectors

$$\vec{T}_{o1} = \vec{T}_1/|\vec{T}_1|, \qquad \vec{T}_{o2} = \vec{T}_2/|\vec{T}_2| \tag{3.4.43}$$

then it holds that $\vec{F}_A \perp \vec{T}_{o1}$ and $\vec{F}_A \perp \vec{T}_{o2}$ or

$$\vec{T}_{o1}\cdot\vec{F}_A = 0, \qquad \vec{T}_{o2}\cdot\vec{F}_A = 0 \tag{3.4.44}$$

The conditions (3.4.44) together with the condition $\vec{M}_A=0$ can be written in the form (3.4.30) where

$$E = \left[\begin{array}{c|c} T_{o1}^T & O_{(1\times 3)} \\ \hline T_{o2}^T & O_{(1\times 3)} \\ \hline O_{(3\times 3)} & I_{(3\times 3)} \end{array}\right]_{(5\times 6)} \tag{3.4.45}$$

Now, all the elements of the dynamic model (3.4.32) are determined and the model can be solved for $n_r + 6$ unknowns: $\ddot{X}_r$, R_A.

The concept of independent reaction components can also be applied. The unit vector $\vec{N}_o$ perpendicular to the surface can be obtained as

$$\vec{N}_o = \vec{T}_{o1}\times\vec{T}_{o2} \tag{3.4.46}$$

Now it holds that $\vec{F}_A || \vec{N}_o$ and, accordingly,

$$\vec{F}_A = S\,\vec{N}_o \tag{3.4.47}$$

where

$$S = |\vec{F}_A| \tag{3.4.48}$$

is the independent component.

Now, the reduced reaction vector is

$$R_{Ar} = [S]_{(1\times 1)} \tag{3.4.49}$$

since $n-n_r=1$. The matrix G which transforms the reaction components (eq. (3.4.33)) is

$$G = \left[N_o^T \,\vdots\, O_{(1\times 3)} \right]^T \tag{3.4.50}$$

and has the dimensions (6×1).

Now, all the elements of the dynamic model (3.4.35) are determined and the model can be solved for n unknowns $\ddot{X}_r$, S

Friction effects. Let us introduce the friction force

$$\vec{F}_f = -\mu S \vec{v}_{Aro} \tag{3.4.51}$$

where μ is friction coefficient and $\vec{v}_{Aro}$ is the unit vector

$$\vec{v}_{Aro} = \vec{v}_{Ar} / |\vec{v}_{Ar}| \tag{3.4.52}$$

and $\vec{v}_{Ar}$ is the relative velocity of gripper point A with respect to the surface. This relative velocity can be expressed by the difference

$$\vec{v}_{Ar} = \vec{v}_A - \vec{v}_s \tag{3.4.53}$$

where $\vec{v}_A$ is the velocity of point A and $\vec{v}_s$ is the velocity of the corresponding surface point. Since $\vec{v}_A$ can be obtained from (3.4.38) and $\vec{v}_s = \left\{\frac{\partial f_x}{\partial t}, \frac{\partial f_y}{\partial t}, \frac{\partial f_z}{\partial t}\right\}$, one obtains

$$v_{Ar} = \begin{bmatrix} \frac{\partial f_x}{\partial u_1} & \frac{\partial f_x}{\partial u_2} \\ \frac{\partial f_y}{\partial u_1} & \frac{\partial f_y}{\partial u_2} \\ \frac{\partial f_z}{\partial u_1} & \frac{\partial f_z}{\partial u_2} \end{bmatrix} \begin{bmatrix} \dot{u}_1 \\ \dot{u}_2 \end{bmatrix} \tag{3.4.54}$$

The friction force produces an additional component of generalized forces and thus the model (3.4.34) becomes

$$WJ^{-1}J_r\ddot{X}_r = P + U - WJ^{-1}(A_r - A) + D(G+G')S \tag{3.4.55}$$

where

$$G' = \left[-\mu v_{Aro}^T \;\vdots\; O_{(1\times 3)}\right]^T \tag{3.4.56}$$

Finally the dynamic model (3.4.35) becomes

$$\left[WJ^{-1}J_r \;\vdots\; -D(G+G')\right]\begin{bmatrix}\ddot{X}_r\\ \hline S\end{bmatrix} = P + U - WJ^{-1}(A_r - A) \tag{3.4.57}$$

and it can be solved for n unknowns $(\ddot{X}_r, S)$.

Jamming. The discussion presented holds when the joint A slides over the surface. If a manipulator starts from a resting position, it may happen that the driving torques applied can not move the point A over the surface because of friction. Something very similar happens if at one time instant the relative velocity becomes zero and we want to continue the relative motion. The case when the driving torques applied cannot produce the relative motion because of friction is called jamming. In that case there is no sliding of point A over the surface. The friction force is less than μS and is sonsidered as an unknown force. Let us find this force.

If we introduce the unknown friction force F_f' then the model (3.4.34) becomes

$$WJ^{-1}J_r\ddot{X}_r = P + U - WJ^{-1}(A_r - A) + D\,G\,S + D_F F_f'$$

where G is defined by (3.4.50). The friction has two independent components along two tangential unit vectors. Thus

$$\vec{F}_f' = F_{f1}'\vec{T}_{o1} + F_{f2}'\vec{T}_{o2}$$

or in matrix form

$$F_f' = \left[T_{o1} \;\vdots\; T_{o2}\right]\begin{bmatrix}F_{f1}'\\ F_{f2}'\end{bmatrix} = G''\begin{bmatrix}F_{f1}'\\ F_{f2}'\end{bmatrix}$$

Now the model obtains the form

$$WJ^{-1}J_r\ddot{X}_r = P + U - WJ^{-1}(A_r - A) + D\,G\,S + D_F G'' \begin{bmatrix} F'_{f1} \\ F'_{f2} \end{bmatrix}$$

Since in the case of jamming it is $\ddot{u}_1 = 0$, $\ddot{u}_2 = 0$ the model obtained can be solved for n unknowns: $\ddot{\theta}$, $\ddot{\varphi}$, $\ddot{\psi}$, S, F_{f1}, F_{f2} if n=6 or $\ddot{\theta}$, $\ddot{\varphi}$, S, F_{f1}, F_{f2} if n=5. The friction force has the absolute value

$$|\vec{F}'_f| = \sqrt{F'^2_{f1} + F'^2_{f2}}$$

This absolute value is tested against maximal possible value μS. If $|\vec{F}'_f| < \mu S$ then there is jamming of point A and the value of $|\vec{F}'_f|$ is correct. If we obtain $|\vec{F}'_f| > \mu S$, this value is not correct since sliding will start. In this case there is no jamming.

Let it be noted that we are talking about jamming of point A only. u_1 and u_2 are stopped but the gripper still has n_r-2 d.o.f. (θ, φ, ψ for n=6 or ψ, φ for n=5). Point A behaves as if it were fixed on a surface (see Para. 3.4.6) or some kind of rolling can appear (see. Para. 3.4.12).

3.4.5. Gripper moving along a line

We consider a manipulator with a gripper subject to a line-type constraint. It means that the gripper cannot move freely but its point A is forced to move along a given line. The discussion of this problem will be similar to the discussion of surface-type constraint (Para. 3.4.4), and therefore it will not be as detailed.

Let us define the relative position of the gripper point A with respect to the line by means of a parameter u (Fig. 3.15).

The parameter equations of a moving line (nonstationary constraint) can be written in the form

$$x = f_x(u, t)$$

$$y = f_y(u, t) \tag{3.4.58}$$

$$z = f_z(u, t)$$

Now, $n_r = n - 2$ and the reduced position vector is

$$X_r = \begin{cases} [u\ \theta\ \varphi]^T, & \text{for } n=5 \\ [u\ \theta\ \varphi\ \psi]^T, & \text{for } n=6 \end{cases} \tag{3.4.59}$$

i.e. $u_2 = \theta$, $u_3 = \varphi$, $u_4 = \psi$.

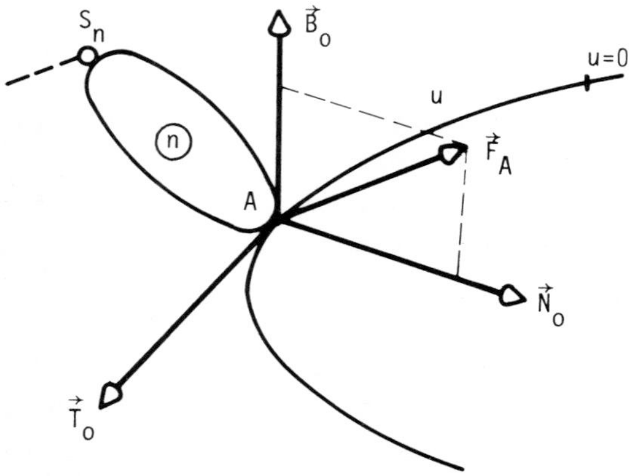

Fig. 3.15. Point A on a moving line

The first and the second derivative of (3.4.58) give

$$\begin{aligned} \dot{x} &= \frac{\partial f_x}{\partial u}\dot{u} + \frac{\partial f_x}{\partial t} \\ \dot{y} &= \frac{\partial f_y}{\partial u}\dot{u} + \frac{\partial f_y}{\partial t} \\ \dot{z} &= \frac{\partial f_z}{\partial u}\dot{u} + \frac{\partial f_z}{\partial t} \end{aligned} \tag{3.4.60}$$

and

$$\begin{aligned} \ddot{x} &= \underbrace{\frac{\partial^2 f_x}{\partial u^2}\dot{u}^2 + 2\frac{\partial^2 f_x}{\partial u \partial t}\dot{u} + \frac{\partial^2 f_x}{\partial t^2}}_{\alpha_x} + \frac{\partial f_x}{\partial u}\ddot{u} \\ \ddot{y} &= \underbrace{\frac{\partial^2 f_y}{\partial u^2}\dot{u}^2 + 2\frac{\partial^2 f_y}{\partial u \partial t}\dot{u} + \frac{\partial^2 f_y}{\partial t^2}}_{\alpha_y} + \frac{\partial f_y}{\partial u}\ddot{u} \end{aligned} \tag{3.4.61}$$

$$\ddot{z} = \underbrace{\frac{\partial^2 f_z}{\partial u^2}\dot{u}^2 + 2\frac{\partial^2 f_z}{\partial u \partial t}\dot{u} + \frac{\partial^2 f_z}{\partial t^2}}_{\alpha_z} + \frac{\partial f_z}{\partial u}\ddot{u} \qquad (3.4.61)$$

The equations (3.4.61) can be written in the matrix form

$$\begin{bmatrix} \ddot{x}_A \\ \ddot{y}_A \\ \ddot{z}_A \end{bmatrix} = \begin{bmatrix} \frac{\partial f_x}{\partial u} \\ \frac{\partial f_y}{\partial u} \\ \frac{\partial f_z}{\partial u} \end{bmatrix} [\ddot{u}] + \begin{bmatrix} \alpha_x \\ \alpha_y \\ \alpha_z \end{bmatrix} \qquad (3.4.62)$$

This acceleration form of the constraint requires that (3.4.58) and (3.4.60) be satisfied by the initial state.

Now, the Jacobian form (3.4.25) can be obtained. The reduced Jacobian and the reduced adjoint matrix are:

$$J_r = \left[\begin{array}{c|c} \frac{\partial f_x}{\partial u} & \\ \frac{\partial f_y}{\partial u} & O_{(3\times(n-3))} \\ \frac{\partial f_z}{\partial u} & \\ \hline O_{(n-3)\times 1} & I_{((n-3)\times(n-3))} \end{array}\right], \quad A_r = \left[\begin{array}{c} \alpha_x \\ \alpha_y \\ \alpha_z \\ \hline O_{((n-3)\times 1)} \end{array}\right] \qquad (3.4.63)$$

The reaction force $\vec{F}_A$ is perpendicular to the line and, accordingly, to the tangent $\vec{T}$ as well. The tangent is

$$\vec{T} = \left\{\frac{\partial f_x}{\partial u}, \frac{\partial f_y}{\partial u}, \frac{\partial f_z}{\partial u}\right\} \qquad (3.4.64)$$

with the unit vector

$$\vec{T}_o = \vec{T}/|\vec{T}| \qquad (3.4.65)$$

Now, we have the conditions:

$$\vec{M}_A = 0, \quad \vec{T}_o\vec{F}_A = 0 \qquad (3.4.66)$$

which can be written in the form (3.4.30) where

$$E = \left[\begin{array}{c|c} T_o^T & O_{(1\times 3)} \\ \hline O_{(3\times 3)} & I_{(3\times 3)} \end{array}\right]_{(4\times 6)} \tag{3.4.67}$$

Now, all the elements of dynamic model (3.4.32) are determined and the model can be solved.

If the concept of independent reaction component is to be appied then we define two unit vectors $\vec{N}_o$ and $\vec{B}_o$ perpendicular to each other and each of them perpendicular to the line.

These vectors are

$$\vec{B}_o = \vec{B}/|\vec{B}|, \quad \vec{B} = \left\{\frac{\partial f_x}{\partial u}, \frac{\partial f_y}{\partial u}, \frac{\partial f_z}{\partial u}\right\} \times \left\{\frac{\partial^2 f_x}{\partial u^2}, \frac{\partial^2 f_y}{\partial u^2}, \frac{\partial^2 f_z}{\partial u^2}\right\} \tag{3.4.68}$$

$$\vec{N}_o = \vec{B}_o \times \vec{T}_o \tag{3.4.69}$$

and the force $\vec{F}_A$ can be expressed in terms of two independent scalar components S_N, S_B:

$$\vec{F}_A = S_N \vec{N}_o + S_B \vec{B}_o \tag{3.4.70}$$

The reduced reaction vector is

$$R_{Ar} = \left[S_N \; S_B\right]^T \tag{3.4.71}$$

The matrix G which transforms the reaction components (eq. (3.4.33)) is

$$G = \left[\begin{array}{c|c} N_{o\,(3\times 1)} & B_{o\,(3\times 1)} \\ \hline O_{(3\times 1)} & O_{(3\times 1)} \end{array}\right]_{(6\times 2)} \tag{3.4.72}$$

Now, all the elements of the dynamic model (3.4.35) are determined and the model can be solved.

Friction effects. Let us introduce the friction force

$$\vec{F}_f = -\mu S \vec{v}_{Aro} \tag{3.4.73}$$

where

$$S = |\vec{F}_A| = \sqrt{S_N^2 + S_B^2} \tag{3.4.74}$$

and

$$\vec{v}_{Aro} = \vec{v}_{Ar}/|\vec{v}_{Ar}| \tag{3.4.75}$$

$$\vec{v}_{Ar} = \left[\frac{\partial f_x}{\partial u} \quad \frac{\partial f_y}{\partial u} \quad \frac{\partial f_z}{\partial u}\right]^T \dot{u}$$

The friction force produces an additional component of generalized forces and thus the model (3.4.34) becomes:

$$WJ^{-1}J_r\ddot{X}_r = P + U - WJ^{-1}(A_r-A)+D\,G\,R_{Ar} + D_F F_f$$

or

$$WJ^{-1}J_r\ddot{X}_r = P + U - WJ^{-1}(A_r-A)+D_F(N_oS_A+B_oS_B-\mu v_{Aro}\sqrt{S_N^2+S_B^2}) \tag{3.4.76}$$

where D_F is as defined in (3.4.3).

The equation (3.4.76) defines the dynamic model in the presence of friction. However, the system is not linear with respect to S_N, S_B. If the nominal dynamics has to be calculated, this nonlinearity is not important since S_A, S_B are given and the system is solved for P. But, in a simulation problem the system has to be solved for $\ddot{X}_r$, S_A, S_B and the nonlinearity makes the problem rather complicated.

3.4.6. Spherical joint constraint

We consider a manipulator having the gripper connected to the ground, or to an object which moves according to a given law, by means of a spherical joint (Fig. 3.16).

The constraint can be expressed by the requirement that the point A moves according to

$$x_A = f_x(t)$$

$$y_A = f_y(t) \tag{3.4.77}$$

$$z_A = f_z(t)$$

(nonstationary constraint).

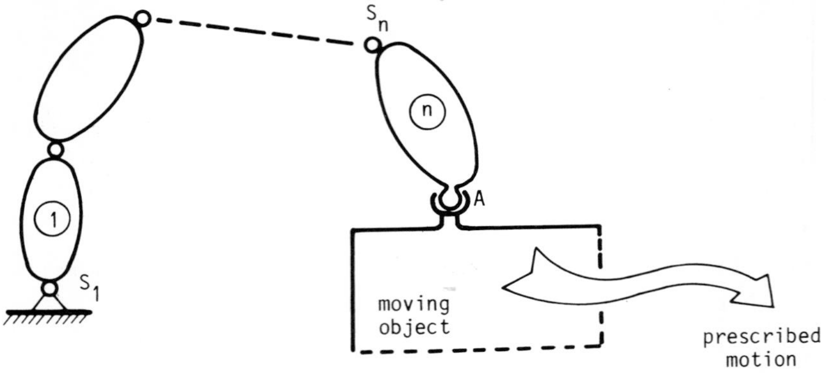

Fig. 3.16. Gripper with spherical joint constraint

The velocity and the acceleration forms of this constraint are

$$v_A = \left[\dot{f}_x(t) \quad \dot{f}_y(t) \quad \dot{f}_z(t)\right]^T \tag{3.4.78}$$

and

$$w_A = \left[\ddot{f}_x(t) \quad \ddot{f}_y(t) \quad \ddot{f}_z(t)\right]^T \tag{3.4.79}$$

Now, $n_r=n-3$ and there are only three free parameters forming the reduced position vector

$$X_r = \begin{cases} [\theta \ \ \varphi]^T, & \text{for } n=5 \\ [\theta \ \ \varphi \ \ \psi]^T, & \text{for } n=6 \end{cases} \tag{3.4.80}$$

i.e., $u_1=\theta$, $u_2=\varphi$, $u_3=\psi$.

The Jacobian form (3.4.25) can be obtained if the reduced Jacobian and the reduced adjoint matrix are defined as

$$J_r = \begin{bmatrix} O_{(3\times(n-3))} \\ \hline I_{((n-3)\times(n-3))} \end{bmatrix}, \qquad A_r = \begin{bmatrix} \ddot{f}_x(t) \\ \ddot{f}_y(t) \\ \ddot{f}_z(t) \\ \hline O_{((n-3)\times 1)} \end{bmatrix} \tag{3.4.81}$$

The reaction force F_A has three independent components and the reaction moment equals zero ($\vec{M}_A=0$). The form (3.4.30) can be used if

$$E = \begin{bmatrix} O_{(3\times3)} & I_{(3\times3)} \end{bmatrix}_{(3\times6)} \qquad (3.4.82)$$

Now, all the elements of dynamic model (3.4.32) are determined and the model can be solved.

If we wish to use the form (3.4.33) then

$$R_{Ar} = F_{A\ (3\times1)}, \qquad G = \begin{bmatrix} I_{(3\times3)} \\ \hline O_{(3\times3)} \end{bmatrix} \qquad (3.4.83)$$

and the dynamic model (3.4.35) can be used.

3.4.7. Two d.o.f. joint constraint

We consider a manipulator having the gripper connected to the ground (or to an object which moves according to a given law) by means of a two d.o.f. joint. Let the joint permit a relative translation along the unit vector $\vec{h}$ and a relative rotation around $\vec{h}$ (Fig. 3.17). Let it be stressed that when we talk about the gripper we understand the whole segment "n" of the chain.

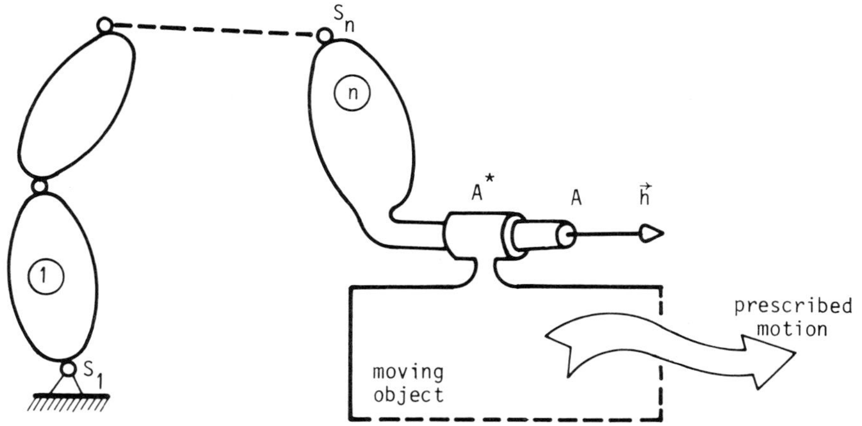

Fig. 3.17. Gripper subject to a two d.o.f. joint constraint

Let us first discuss the prescribed motion of the object to which the gripper is connected (nonstationary constraint). Let us define the linear motion of the object (i.e., of its point A^* shown in Figs. 3.17 and 3.18) by

$$x_A^* = f_1(t), \quad y_A^* = f_2(t), \quad z_A^* = f_3(t) \tag{3.4.84}$$

and the angular motion of the object by

$$\theta^* = f_4(t), \quad \varphi^* = f_5(t), \quad \psi^* = f_6(t) \tag{3.4.85}$$

(see Fig. 3.18). The angles θ^*, φ^*, ψ^* are defined for the moving object but in a way analogous to the definition of gripper angles θ, φ, ψ in Para. 2.4.

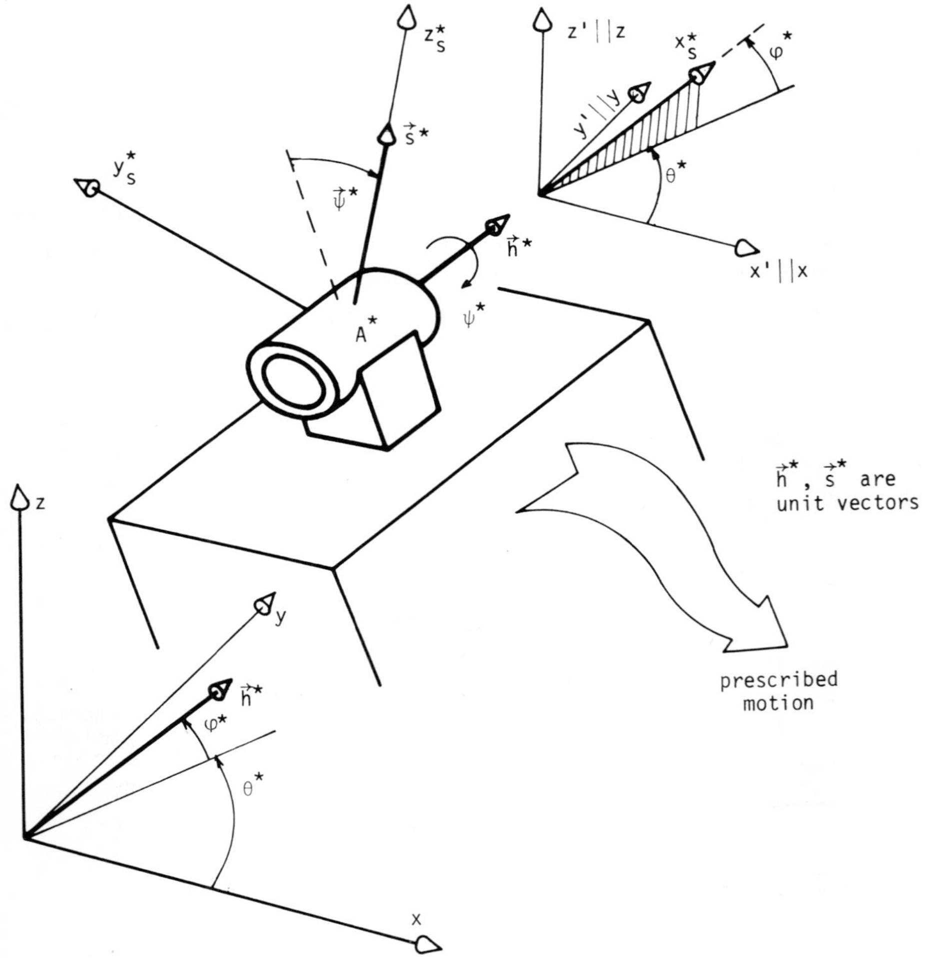

Fig. 3.18. Definition of object motion

Let us define an orientation coordinate system $O_s^* x_s^* y_s^* z_s^*$ corresponding to the moving object, in a way analogous to the definition of gripper orientation system ($O_s x_s y_s z_s$ in Para. 2.4). The origin O_s^* coincides with A^*. Axis x_s^* is along $\vec{h}^*$, z_s^* is along $\vec{s}^*$, and y_s^* along $\vec{\ell}^* = \vec{s}^* \times \vec{h}^*$.

If we connect the gripper to the moving object, as shown in Fig. 3.17, then $\vec{h}$ (on the gripper) has to coincide with $\vec{h}^*$ (on the object) and accordingly x_s^* coincides with x_s. This means that

$$\theta = \theta^*, \qquad \varphi = \varphi^* \tag{3.4.86}$$

We note that the relative translation occurs along x_s^*. Let u_1 be the coordinate of gripper point A with respect to axis x_s^*. Since the y_s^* and z_s^* coordinates of point A equal zero it holds that

$$u_1 = x_{sA}^* = \overline{O_s^* A} = \overline{A^* A} \tag{3.4.87}$$

(see Fig. 3.19). Thus u_1 defines the relative translation.

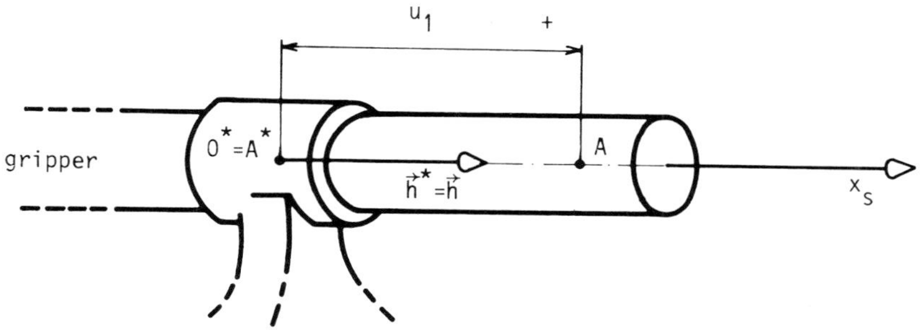

Fig. 3.19. Relative translation

Relative rotation is performed around $\vec{h}^*$ (note $\vec{h}^* = \vec{h}$) and is defined by

$$u_2 = \psi - \psi^* \tag{3.4.88}$$

If a six d.o.f. manipulator (n=6) is considered, u_2 is a free parameter since ψ can be changed as we wish (ψ^* is given by (3.4.85)). But, with a five d.o.f. manipulator (n=5) the angle ψ is not free but it follows from x_A, y_A, z_A, θ, φ, and hence, u_2 is not a free parameter.

Let us now define the reduced position vector determining the position relative to the joint constraint. Since $n_r = n-4$ it follows that

$$X_s = \begin{cases} [u_1], & \text{for } n=5 \\ [u_1 \ u_2]^T, & \text{for } n=6 \end{cases} \tag{3.4.89}$$

The position of gripper point A can be expressed in the form

$$\vec{r}_A = \vec{r}_{A*} + \overrightarrow{A^*A} \tag{3.4.90}$$

and $\overrightarrow{A^*A}$ can be expressed in the object orientation system as

$$\overrightarrow{A^*A} = \{u_1, \ 0, \ 0\} \tag{3.4.91}$$

Thus, we obtain the matrix relation

$$r_A = r_{A*} + A_s^*[u_1 \ 0 \ 0]^T \tag{3.4.92}$$

where A_s^* is the transition matrix of the object orientation system. A_s^* is obtained in a may analogous to that for gripper orientation system (Para. 2.4). The difference is in that angles θ^*, φ^*, ψ^* appear instead of θ, φ, ψ. The derivatives $\dot{A}_s^*$, $\ddot{A}_s^*$ can also be obtained by analogy.

The first and the second derivative of (3.4.92) give

$$v_A = v_{A*} + \dot{A}_s^*[u_1 \ 0 \ 0]^T + A_s^*[\dot{u}_1 \ 0 \ 0]^T \tag{3.4.93}$$

and

$$w_A = w_{A*} + \ddot{A}_s^*[u_1 \ 0 \ 0]^T + 2\dot{A}_s^*[\dot{u}_1 \ 0 \ 0]^T + A_s^*[\ddot{u}_1 \ 0 \ 0]^T \tag{3.4.94}$$

From (2.4.21) if follows that

$$\dot{A}_s^* = F_{\theta*}\dot{\theta}^* + F_{\varphi*}\dot{\varphi}^* + F_{\psi*}\dot{\psi}^*$$

and

$$\ddot{A}_s^* = F_{\theta*}\ddot{\theta}^* + F_{\varphi*}\ddot{\varphi}^* + F_{\psi*}\ddot{\psi}^* + G^* \tag{3.4.95}$$

where $F_{\theta*}$, $F_{\varphi*}$, $F_{\psi*}$ and G^* are defined by (2.4.21) but with θ^*, φ^*, ψ^* instead of θ, φ, ψ.

Keeping in mind the structure of matrices $A_{\theta*}$, $A_{\varphi*}$, $A_{\psi*}$ (see eq.

(2.4.16)) it is clear that the acceleration w_A given by (3.4.94) will not depend on ψ^* or its derivatives $\dot{\psi}^*$, $\ddot{\psi}^*$. The relation (3.4.94) can be transformed in to

$$w_A = \underbrace{w_{A^*} + 2\dot{A}_s^*[\dot{u}_1 \ \ 0 \ \ 0]^T + f_{\theta^*1}\ddot{\theta}^* + f_{\varphi^*1}\ddot{\varphi}^*}_{\alpha_w} + a_{s1}^*\ddot{u}_1 \tag{3.4.96}$$

where f_{θ^*1}, f_{φ^*1}, a_{s1}^* are the first columns of the matrices F_{θ^*}, F_{φ^*}, A_s^*, respectively. w_A^* is, determined by (3.4.84), i.e.,

$$w_A = \begin{bmatrix} \ddot{f}_1(t) & \ddot{f}_2(t) & \ddot{f}_3(t) \end{bmatrix}^T \tag{3.4.97}$$

and θ^*, φ^* and their derivatives are given by (3.4.85), i.e.,

$$\dot{\theta}^* = \dot{f}_4(t), \quad \dot{\varphi}^* = \dot{f}_5(t), \quad \ddot{\theta}^* = \ddot{f}_4(t), \quad \ddot{\varphi}^* = \ddot{f}_5(t)$$

Keeping in mind (3.4.86) and (3.4.96) we may write the Jacobian form (3.4.25) where the reduced Jacobian and the reduced acjoint matrix are

$$J_r = \left[\begin{array}{c|c} a^*_{s1\,(3\times1)} & O_{(3\times(n-5))} \\ \hline O_{(1\times1)} & O_{(1\times(n-5))} \\ \hline O_{(1\times1)} & O_{(1\times(n-5))} \\ \hline O_{((n-5)\times1)} & I_{((n-5)\times(n-5))} \end{array}\right], \qquad A_r = \left[\begin{array}{c} \alpha_{w\,(3\times1)} \\ \hline \ddot{f}_{4\,(1\times1)} \\ \hline \ddot{f}_{5\,(1\times1)} \\ \hline \ddot{f}_{6\,((n-5)\times1)} \end{array}\right] \tag{3.4.98}$$

The joint constraint imposed produces a reaction force $\vec{F}_A$ and the reaction moment $\vec{M}_A$. The following conditions hold: $\vec{F}_A \perp \vec{h}^*$ and $\vec{M}_A \perp \vec{h}^*$. Hence, $\vec{h}^*\vec{F}_A = 0$ and $\vec{h}^*\vec{M}_A = 0$ or, in the matrix form:

$$h^{*T}F_A = 0, \qquad h^{*T}M_A = 0 \tag{3.4.99}$$

These conditions can be written in the form (3.4.30) where

$$E = \left[\begin{array}{c|c} h^T & O_{(1\times3)} \\ \hline O_{(1\times3)} & h^T \end{array}\right]_{(2\times6)} \tag{3.4.100}$$

Now, all the elements of dynamic model (3.4.32) are determined and the model can be solved.

If the concept of independent reaction components is to be applied, we introduce four independent reaction components. Since $\vec{F}_A \perp \vec{h}$, $\vec{M}_A \perp \vec{h}$ and, accordingly, $\vec{F}_A \perp x^*_s$, $\vec{M}_A \perp \vec{x}^*_s$, we express these reactions in terms of independent components along the unit vectors $\vec{s}^*$ and $\vec{\ell}^* = \vec{s}^* \times \vec{h}^*$:

$$\vec{F}_A = S_1 \vec{s}^* + S_2 \vec{\ell}^*, \qquad \vec{M}_A = S_3 \vec{s}^* + S_4 \vec{\ell}^* \tag{3.4.101}$$

The reduced reaction vector is

$$R_{Ar} = \begin{bmatrix} S_1 & S_2 & S_3 & S_4 \end{bmatrix}^T \tag{3.4.102}$$

The matrix G which transforms the reaction components (eq. (3.4.33) is

$$G = \left[\begin{array}{c|c|c|c} s^*_{(3\times 1)} & \ell^*_{(3\times 1)} & O_{(3\times 1)} & O_{(3\times 1)} \\ \hline O_{(3\times 1)} & O_{(3\times 1)} & s^*_{(3\times 1)} & \ell^*_{(3\times 1)} \end{array} \right]_{(6\times 4)} \tag{3.4.103}$$

Now, all the elements of the dynamic model (3.4.35) are determined and the model can be solved.

Friction effects. We remember that the relative motion of the gripper with respect to the joint connection is defined by $u_1 = \overline{A^*A}$ and $u_2 = \psi - \psi^*$.

Let us find the friction force. We substitute the reaction moment $\vec{M}_A$ by two forces $\vec{F}_{M1}$, $\vec{F}_{M2}$, and the reaction force $\vec{F}_A$ by two forces $\vec{F}_{F1}$, $\vec{F}_{F2}$ (Fig. 3.20). The lengths ℓ_1 and ℓ_2 are shown in Fig. 3.21a,b and they are defined as

$$\ell_1 = u_1, \qquad \ell_2 = L$$

if $u_1 + L < L^*$ (Fig. 3.21b) or

$$\ell_1 = u_1, \qquad \ell_2 = L^* - u_1$$

if $u_1 + L > L^*$ (Fig. 3.21a).

Now:

$$\vec{F}_{F1} = \frac{\ell_2}{\ell_1+\ell_2}\vec{F}_A, \quad \vec{F}_{F2} = \frac{\ell_1}{\ell_1+\ell_2}\vec{F}_A \tag{3.4.104}$$

and

$$\vec{F}_{M1} = \frac{\vec{h}\times\vec{M}_A}{\ell_1+\ell_2}, \quad \vec{F}_{M2} = -\frac{\vec{h}\times\vec{M}_A}{\ell_1+\ell_2} \tag{3.4.105}$$

The total forces are

$$\vec{F}_1 = \vec{F}_{F1} + \vec{F}_{M1}, \quad \vec{F}_2 = \vec{F}_{F2} + \vec{F}_{M2} \tag{3.4.106}$$

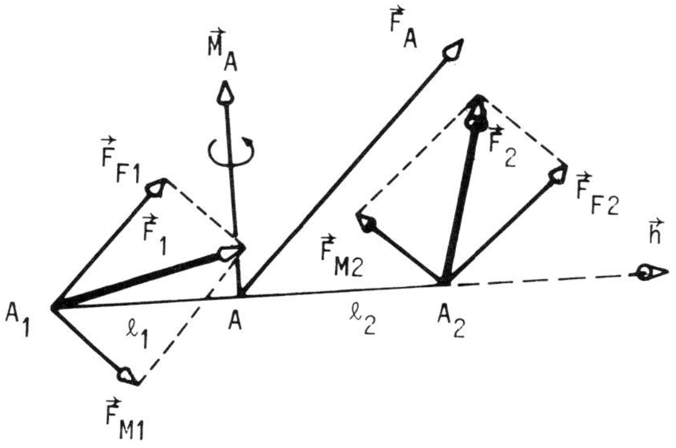

Fig. 3.20. Transformation of reactions

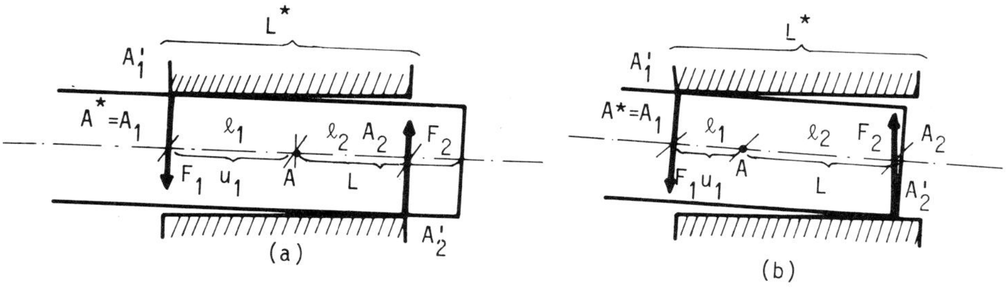

Fig. 3.21. Decomposition of reactions

These forces determine the position of gripper points A'_1 and A'_2 which are in connection with the cylindrical constraint (Figs. 3.21, 3.22). The friction forces $\vec{F}_{f1}$ and $\vec{F}_{f2}$ act in these points.

The absolute values of these friction forces ($\vec{F}_{f1}$ and $\vec{F}_{f2}$) are $\mu|\vec{F}_1|$

and $\mu|\vec{F}_2|$. The forces act in the directions opposite to the relative velocities of points A_1', A_2'. Each velocity has two components: longitudinal (along $\vec{h}$) and tangential (along $\vec{T}$). Thus, the relative velocities of A_1' and A_2' with respect to cylindrical constraint are

$$\vec{v}_{1r} = \dot{u}_1\vec{h} + R\dot{u}_2\vec{T}_1; \qquad \vec{v}_{2r} = \dot{u}_1\vec{h} + R\dot{u}_2\vec{T}_2 \tag{3.4.107}$$

where R is the radius of the cylinder, and $\vec{T}_1$, $\vec{T}_2$ are tangential unit vectors in points A_1', A_2' (Fig. 3.22). The vectors $\vec{T}_1$, $\vec{T}_2$ can be found as

$$\vec{T}_1 = (\vec{F}_1\times\vec{h})/|\vec{F}_1|, \qquad \vec{T}_2 = (\vec{F}_2\times\vec{h})/|\vec{F}_2|$$

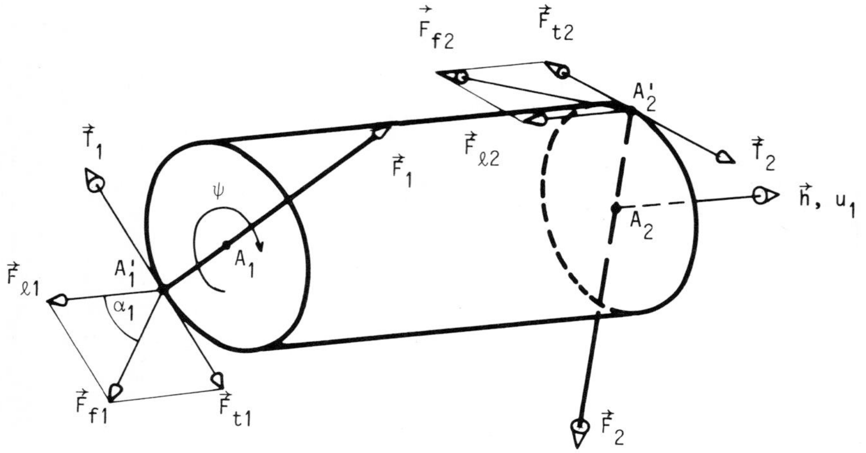

Fig. 3.22. Determination of friction forces

Now, the friction forces are

$$\vec{F}_{f1} = -\mu|\vec{F}_1|\vec{v}_{1ro}, \qquad \vec{F}_{f2} = -\mu|\vec{F}_2|\vec{v}_{2ro} \tag{3.4.108}$$

where $\vec{v}_{1ro}$ and $\vec{v}_{2ro}$ are unit vectors:

$$\vec{v}_{1ro} = \vec{v}_{1r}/|\vec{v}_{1r}|, \qquad \vec{v}_{2ro} = \vec{v}_{2r}/|\vec{v}_{2r}|$$

Each friction force has a longitudinal ($\vec{F}_\ell$) and a tangential component ($\vec{F}_t$):

$$\vec{F}_{\ell 1} = (\vec{F}_{f1}\cdot\vec{h})\vec{h} = -\mu|\vec{F}_1|\cos\alpha\ \vec{h} = -\mu|\vec{F}_1|\frac{\dot{u}_1}{\sqrt{\dot{u}_1^2+R^2\dot{u}_2^2}}\ \vec{h}$$

$$\vec{F}_{\ell 2} = (\vec{F}_{f2}\cdot\vec{h})\vec{h} = -\mu|\vec{F}_2|\cos\alpha\ \vec{h} = -\mu|\vec{F}_2|\frac{\dot{u}_1}{\sqrt{\dot{u}_1^2+R^2\dot{u}_2^2}}\vec{h}$$

$$\vec{F}_{t1} = (\vec{F}_{f1}\cdot\vec{T}_1)\vec{T}_1 = -\mu|\vec{F}_1|\sin\alpha\ \vec{T}_1 = -\mu|\vec{F}_1|\frac{R\dot{u}_2}{\sqrt{\dot{u}_1^2+R^2\dot{u}_2^2}}\vec{T}_1$$

$$\vec{F}_{t2} = (\vec{F}_{f2}\cdot\vec{T}_2)\vec{T}_2 = -\mu|\vec{F}_2|\sin\alpha\ \vec{T}_2 = -\mu|\vec{F}_2|\frac{R\dot{u}_2}{\sqrt{\dot{u}_1^2+R^2\dot{u}_2^2}}\vec{T}_2$$

These forces can be arranged in a different way to find the total longitudinal friction force:

$$\vec{F}_{\ell} = \vec{F}_{\ell 1} + \vec{F}_{\ell 2} = -\mu(|\vec{F}_1|+|\vec{F}_2|)\cos\alpha\ \vec{h} \qquad (3.4.109)$$

and the total friction moment (scalar value) around $\vec{h}$:

$$M_f = (|\vec{F}_{t1}|+|\vec{F}_{t2}|)R = -\mu(|\vec{F}_1|+|\vec{F}_2|)R\ \sin\alpha \qquad (3.4.110)$$

The friction forces $\vec{F}_{f1}$, $\vec{F}_{f2}$ produce additional components of generalized forces and thus the model (3.4.34) becomes

$$WJ^{-1}J_r\ddot{X}_r = P + U - WJ^{-1}(A_r-A) + D\,G\,R_{Ar} + D_F(F_{f1}+F_{f2}) \qquad (3.4.111)$$

where D_F is defined by (3.4.3).

The equation (3.4.111) defines the dynamic model in the presence of friction. However, the system is not linear with respect to the independent reaction components $R_{Ar} = \begin{bmatrix} S_1 & S_2 & S_3 & S_4\end{bmatrix}^T$ since the friction forces depend nonlinearly on these components. If the nominal dynamics has to be calculated, this nonlinearity causes no additional problems since these independent components (i.e., the reactions F_A, M_A) are given and the system is solved for P. But, in a simulation problem the system (3.4.111) has to be solved for $\ddot{X}_r$, S_1, S_2, S_3, S_4 and the presence of nonlinearity makes the problem rather complicated.

This discussion on friction holds in the case when there is sliding in points A_1' and A_2'. But, if friction is strong enough then jamming can appear and the discussion does not apply. Let us be more precise. If friction is strong enough then there will be no sliding in points A_1' and A_2'. In such case rolling or jamming will appear. Also, there can be a rolling with sliding. These problems are rather complex and they will not be considered here.

3.4.8. Rotational joint constraint

We consider a manipulator having the gripper connected to the ground, or to an object which moves according to a given lew, by means of a joint permitting one rotation only. Let $\vec{h}$ be the unit vector of rotation axis (Fig. 3.23).

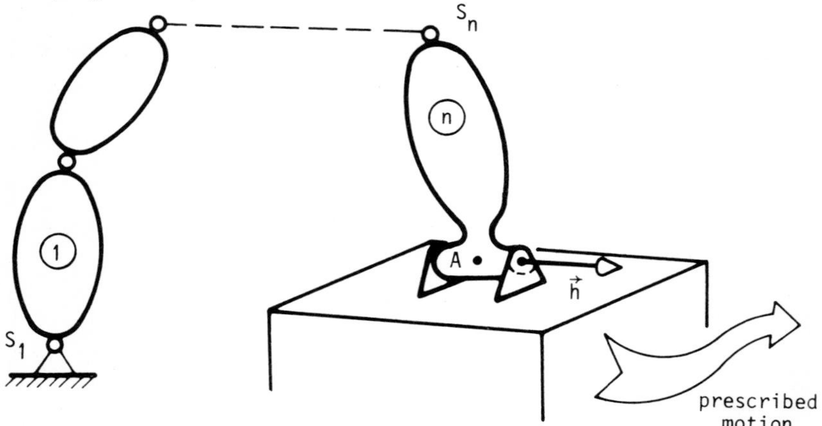

Fig. 3.23. Gripper subject to a rotational joint constraint

Let us first discuss the prescribed motion of the object to which the gripper is connected. Let this motion be defined by

$$x_A = f_1(t), \qquad y_A = f_2(t), \qquad z_A = f_3(t) \tag{3.4.113}$$

$$\theta^* = f_4(t), \qquad \varphi^* = f_5(t), \qquad \psi^* = f_6(t) \tag{3.4.114}$$

as was done in the previous Para. 3.4.7. Note that $A = A^*$ here, and, accordingly, $x_A = x_A^*$, $y_A = y_A^*$, $z_A = z_A^*$. Thus we use A only. The joint connection also produces: $\theta = \theta^*$ and $\varphi = \varphi^*$.

The parameter

$$u = \psi - \psi^* \tag{3.4.115}$$

defines the relative position of the gripper with respect to the moving object. If a six d.o.f. manipulator is considered, u is a free parameter and $n_r = n - 5 = 1$. With a five d.o.f. manipulator the angle ψ is not free but it follows from x_A, y_A, z_A, θ, ψ and, accordingly, u is not free. In that case: $n_r = n - 5 = 0$. Since there are no free parameters with five d.o.f. manipulators, this problem is interesting for six d.o.f. manipulators only.

Let us define the reduced position vector:

$$X_r = \begin{cases} \text{not existing,} & \text{for } n = 5 \\ [u], & \text{for } n = 6 \end{cases} \tag{3.4.116}$$

The Jacobian form (3.4.25) can be obtained if the reduced Jacobian and the reduced adjoint matrix are defined as

$$J_r = \begin{bmatrix} O_{(5\times(n-5))} \\ \hline I_{((n-5)\times(n-5))} \end{bmatrix}, \quad A_r = \left[\ddot{f}_1 \ \ddot{f}_2 \ \ddot{f}_3 \ \ddot{f}_4 \ \ddot{f}_5 \,\middle|\, \ddot{f}_6 \ {}_{(n-5)\times 1}\right]^T \tag{3.4.117}$$

For n=5 this form directly reduces to (2.4.13), (2.4.14) since there are no independent parameters.

The joint constraint produces a reaction moment $\vec{M}_A$ satisfying $\vec{M}_A \perp \vec{h}$, i.e.,

$$\vec{M}_A \cdot \vec{h} = 0 \tag{3.4.118}$$

and a reaction force $\vec{F}_A$ which need not be perpendicular to $\vec{h}$. Thus, the condition imposed on reactions can be expressed by (3.4.30) where

$$E = \begin{bmatrix} O_{(1\times 3)} & h^T \end{bmatrix}_{(1\times 6)} \tag{3.4.119}$$

Now, all the elements of dynamic model (3.4.32) are determined and the model can be solved.

Friction is treated in a way similar to the discussion in the previous paragraph, but here, there is no longitudinal motion and, accordingly, no longitudinal friction components.

3.4.9. Linear joint constraint

We consider a manipulator having the gripper connected to the ground, or to an object which moves according to a given law, by means of a joint permitting one translation only. Let $\vec{h}$ be the unit vector of translation axis (Fig. 3.24).

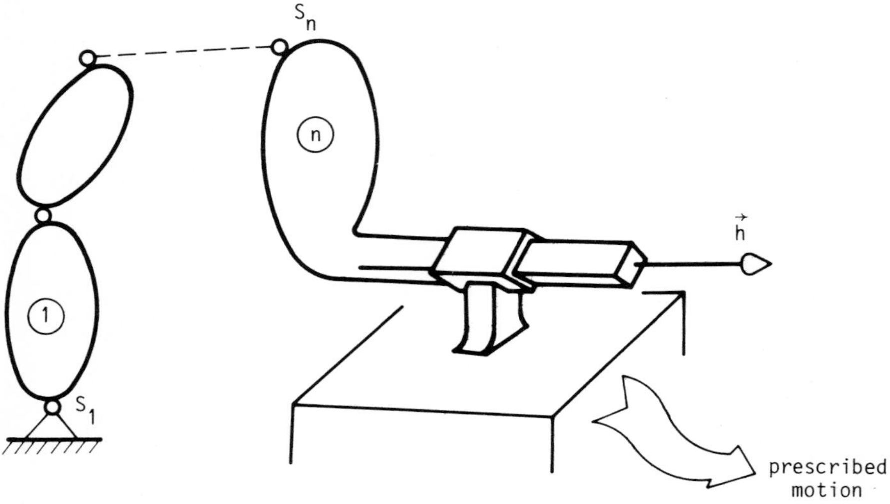

Fig. 3.24. Gripper subject to a linear joint constraint

Let us first discuss the prescribed motion of the object to which the gripper is connected (nonstationary constraint). Let us define this motion in the same way as it was done in Para. 3.4.7. Thus, the linear motion is defined by

$$x_A^* = f_1(t), \quad y_A^* = f_2(t), \quad z_A^* = f_3(t) \tag{3.4.120}$$

and angular motion by

$$\theta^* = f_4(t), \quad \varphi^* = f_5(t), \quad \psi^* = f_6(t) \tag{3.4.121}$$

If we connect the gripper to the moving object (Fig. 3.24), then $\vec{h}$ (on the gripper) has to coincide with $\vec{h}^*$ (on the object) and $\vec{s}$ has to coincide with $\vec{s}^*$ (since relative rotation around $\vec{h}$ is not possible). Thus

$$\theta = \theta^*, \quad \varphi = \varphi^*, \quad \psi = \psi^* \tag{3.4.122}$$

Introducing orientation coordinate systems for the gripper and for the object (as was done in Para. 3.4.7), we find that the relative translation can be defined by

$$u = x^*_{sA} = \overline{O^*_s A} = \overline{A^* A} \tag{3.4.123}$$

(see Fig. 3.19).

If a six d.o.f. manipulator (n=6) is considered, u is a free parameter. But, with a five d.o.f. manipulator (n=5) this parameter is not free since such a manipulator cannot solve the position along with the total orientation. Hence, for five d.o.f. manipulators this problem makes almost no sense.

Since $n_r = n - 5$ we introduce the reduced position vector as

$$X_r = \begin{cases} \text{not existing, for } n=5 \\ [u], \qquad \text{for } n=6 \end{cases} \tag{3.4.124}$$

Repeating the transformations (3.4.90) to (3.4.98) (from Para. 3.4.7), we obtain the acceleration

$$w_A = a^*_{s1}\ddot{u} + \alpha_w \tag{3.4.125}$$

By differentiation, (3.4.121) gives

$$\dot{\theta}^* = \dot{f}_4(t), \quad \dot{\varphi}^* = f_5(t), \quad \dot{\psi}^* = \dot{f}_6(t) \tag{3.4.126}$$

$$\ddot{\theta}^* = \ddot{f}_4(t), \quad \ddot{\varphi}^* = \ddot{f}_5(t), \quad \ddot{\psi}^* = \ddot{f}_6(t) \tag{3.4.127}$$

The Jacobian form (3.4.25) can be obtained if the reduced Jacobian and the reduced adjoint matrix are defined as

$$J_r = \begin{bmatrix} a^*_{s1} \ (3\times 1) \\ \text{-----------} \\ 0_{(3\times 1)} \end{bmatrix} ; \qquad A_r = \begin{bmatrix} \alpha_w^T & \ddot{f}_4 & \ddot{f}_5 & \ddot{f}_6 \end{bmatrix}^T \tag{3.4.128}$$

The joint constraint produces a reaction force $\vec{F}_A$ satisfying $\vec{F}_A \perp \vec{h}$, i.e.,

$$\vec{F}_A \cdot \vec{h} = 0 \tag{3.4.129}$$

and a reaction moment $\vec{M}_A$ which need not be perpendicular to $\vec{h}$. Thus, the condition imposed on reactions can be expressed by (3.4.30) where

$$E = \left[h^T \quad 0_{(1\times3)}\right]_{(1\times6)} \tag{3.4.130}$$

Now, all the elements of dynamic model (3.4.32) are determined and the model can be solved.

If the concept of independent reaction components has to be applied, we introduce five independent reaction components. Since $\vec{F}_A \perp \vec{h}$ and, accordingly, $\vec{F}_A \perp x_s^*$ we express it in terms of independent components along the unit vectors $\vec{s}$ and $\vec{\ell} = \vec{s}\times\vec{h}$:

$$\vec{F}_A = S_1\vec{s} + S_2\vec{\ell} \tag{3.4.131a}$$

The reaction moment has three independent components:

$$\vec{M}_A = S_3\vec{h} + S_4\vec{s} + S_5\vec{\ell} \tag{3.4.131b}$$

The reduced reaction vector is

$$R_{Ar} = \left[S_1 \; S_2 \; S_3 \; S_4 \; S_5\right] \tag{3.4.132}$$

The matrix G which transforms the reaction components (eq. (3.4.33)) is

$$G = \left[\begin{array}{c|c|c|c|c} s & \ell & 0_{(3\times1)} & 0_{(3\times1)} & 0_{(3\times1)} \\ \hline 0_{(3\times1)} & 0_{(3\times1)} & h & s & \ell \end{array}\right]_{(6\times5)} \tag{3.4.133}$$

Now all the elements of the dynamic model (3.4.35) are determined and the model can be solved.

3.4.10. Constraint permitting no relative motion

We consider a manipulator having the gripper connected to the ground (or to an object which moves according to a given law) in such a way that no relative motion is possible (Fig. 3.25).

In this case the constraint can be expressed in the form.

$$x_A = f_1(t), \quad y_A = f_2(t), \quad z_A = f_3(t) \tag{3.4.136a}$$

$$\theta = f_4(t), \quad \varphi = f_5(t), \quad \psi = f_6(t) \tag{3.4.136b}$$

This problem makes sense for six d.o.f. manipulators only. However, even in that case there are no free parameters since $n_r = n-6=6-6=0$.

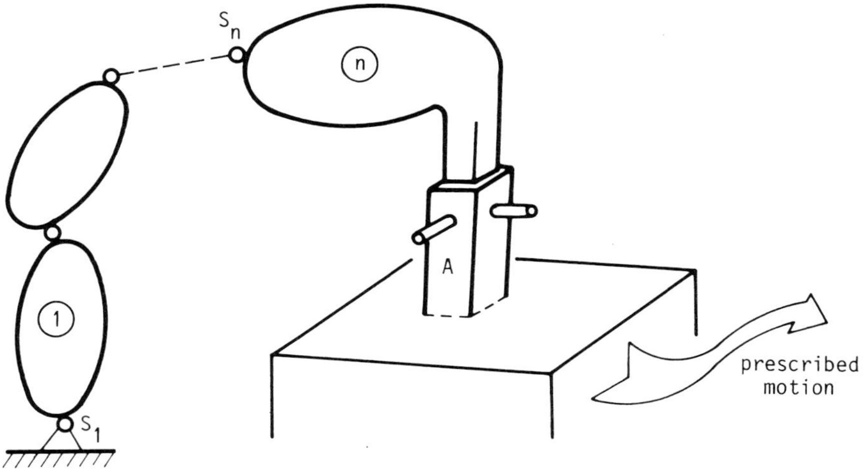

Fig. 3.25. Constraint permitting no relative motion

The second derivative of (3.4.136a,b) gives

$$\left[\ddot{x}_A \; \ddot{y}_A \; \ddot{z}_A \; \ddot{\theta} \; \ddot{\varphi} \; \ddot{\psi}\right]^T = \left[\ddot{f}_1 \; \ddot{f}_2 \; \ddot{f}_3 \; \ddot{f}_4 \; \ddot{f}_5 \; \ddot{f}_6\right]^T$$

If we wish to write this relation in the form (3.4.25), it is clear that the reduced position vector and the reduced Jacobian do not exist. The reduced adjoint matrix is

$$A_r = \left[\ddot{f}_1 \; \cdots \; \ddot{f}_6\right]^T_{(6\times 1)} \tag{3.4.137}$$

The reaction force and reaction moment have three independent components each. It means that there are no conditions for reactions. If we consider the projections on external Cartesian system as independent components, then the relation (3.4.30) is unnecessary, i.e., it does not exist.

Now, in a simulation problem, we do not use (3.4.32). We use (3.4.26) to solve $\ddot{q}$ and then (3.4.28) to solve reaction R_A.

3.4.11. Bilateral manipulation

We consider two manipulators connected to each other by means of their grippers. In previous paragraphs (3.4.2 - 3.4.10) we considered closed chains obtained by connecting the last segment (gripper) of an open chain to the ground or to an object which moves according to a given law. Here we consider a closed chain formed of two open chains by connecting their last segments (grippers) to each other. The connection between the two grippers can be of any type considered in Para. 3.4.4 - 3.4.10. All the theory from these paragraphs can easily be extended by considering another gripper instead of a moving object.

From all different types of connections we choose the three cases which are most interesting for practice.

Two d.o.f. joint connection. We consider two manipulators having the grippers connected to each other by means of a joint permitting one relative translation (along the unit vector $\vec{h}$) and one relative rotation (around $\vec{h}$) (Fig. 3.26)

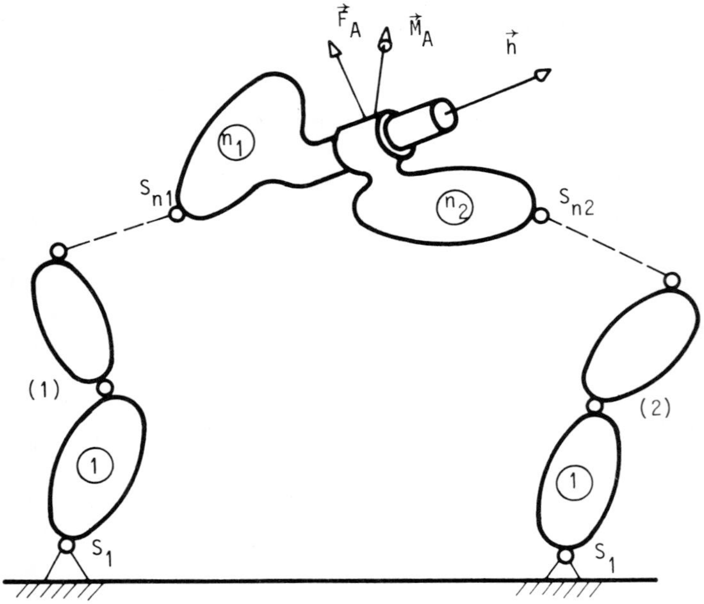

Fig. 3.26. Two d.o.f. joint connection

Let the position vectors of the two manipulators (1) and (2) be

$$X_{g1} = \left[x_{A1}\ y_{A1}\ z_{A1}\ \theta_1\ \varphi_1\ \psi_1\right]^T, \quad X_{g2} = \left[x_{A2}\ y_{A2}\ z_{A2}\ \theta_2\ \varphi_2\ \psi_2\right]^T \tag{3.4.138}$$

If one of the manipulators (or both) has five d.o.f., the corresponding angle ψ does not appear in the position vector. We restrict our consideration to manipulators having five or six d.o.f. For simplicity we assume $n_1 = n_2 = n$.

Let us apply the theory derived in Para. 3.4.7 but here the gripper of manipulator (2) plays the role of moving object. The motion of this gripper (2) is defined by

$$x_{A2}(t), \quad y_{A2}(t), \quad z_{A2}(t), \quad \theta_2(t), \quad \varphi_2(t), \quad \psi_2(t) \tag{3.4.139}$$

i.e., $X_{g2}(t)$. Now the kinematics and the dynamics of manipulator (2) are described by

$$\ddot{X}_{g2} = J_2\ddot{q}_2 + A_2 \tag{3.4.140}$$

$$W_2\ddot{q}_2 = P_2 + U_2 - D_2R_A \tag{3.4.141}$$

The sign "-" of the term D_2R_A is due to action and reaction (R_A is acting on manipulator (1) and $-R_A$ on manipulator (2)).

Let us define the relative translation of the grippers by

$$u_1 = \overline{A_2A_1}. \tag{3.4.142}$$

It is the relation (3.4.87) but with A_1 instead of A and A_2 instead of A^*. The relative rotation is defined by

$$u_2 = \psi_1 - \psi_2 \tag{3.4.143}$$

It is the relation (3.4.88) but with ψ_1 instead of ψ and ψ_2 instead of ψ^*.

The relation (3.4.94) becomes here

$$w_{A1} = w_{A2} + \ddot{A}_s^{(2)}[u_1\ 0\ 0]^T + 2\dot{A}_s^{(2)}[\dot{u}_1\ 0\ 0]^T + A_s^{(2)}[\ddot{u}_1\ 0\ 0]^T \tag{3.4.144}$$

and (3.4.95) becomes

$$\ddot{A}_s^{(2)} = F_\theta^{(2)}\ddot{\theta}_2 + F_\varphi^{(2)}\ddot{\varphi}_2 + F_\psi^{(2)}\ddot{\psi}_2 + G^{(2)} \tag{3.4.145}$$

Substituting (3.4.145) into (3.4.144) one obtains

$$w_{A1} = w_{A2} + \underbrace{2\dot{A}_s^{(2)}[\dot{u}_1 \;\; 0 \;\; 0]^T}_{\beta_w} + f_{\theta 1}^{(2)}\ddot{\theta}_2 + f_{\varphi 1}^{(2)}\ddot{\varphi}_2 + a_{s1}^{(2)}\ddot{u}_1 \tag{3.4.146}$$

where $f_{\theta 1}^{(2)}$, $f_{\varphi 1}^{(2)}$, $a_{s1}^{(2)}$ are the first columns of the matrices $F_\theta^{(2)}$, $F_\varphi^{(2)}$, $A_s^{(2)}$, respectively. When the connection between the two grippers is made, then

$$\theta_1 = \theta_2, \quad \varphi_1 = \varphi_2 \tag{3.4.147}$$

and accordingly

$$\ddot{\theta}_1 = \ddot{\theta}_2, \quad \ddot{\varphi}_1 = \ddot{\varphi}_2 \tag{3.4.148}$$

Now, the vector $\ddot{X}_{g1}$ can be expressed as

$$\underbrace{\left[\begin{array}{c} w_{A1}\ (3\times1) \\ \hline \ddot{\theta}_1\ (1\times1) \\ \hline \ddot{\varphi}_1\ (1\times1) \\ \hline \ddot{\psi}_{1\,((n-5)\times1)} \end{array}\right]_{(n\times1)}}_{\ddot{X}_{g1}\ (n\times1)} = \underbrace{\left[\begin{array}{c:c:c:c} I_{(3\times3)} & f_{\theta 1}^{(2)}\ (3\times1) & f_{\varphi 1}^{(2)}\ (3\times1) & O_{(3\times(n-5))} \\ \hdashline O_{(1\times3)} & I_{(1\times1)} & O_{(1\times1)} & O_{(1\times(n-5))} \\ \hdashline O_{(1\times3)} & O_{(1\times1)} & I_{(1\times1)} & O_{(1\times(n-5))} \\ \hdashline O_{((n-5)\times3)} & O_{((n-5)\times1)} & O_{((n-5)\times1)} & I_{((n-5)\times(n-5))} \end{array}\right]}_{J_{1,2}\ (n\times n)} \cdot$$

$$\cdot \underbrace{\left[\begin{array}{c} w_{A2}\ (3\times1) \\ \hline \ddot{\theta}_2\ (1\times1) \\ \hline \ddot{\varphi}_2\ (1\times1) \\ \hline \ddot{\psi}_{2\,((n-5)\times1)} \end{array}\right]}_{\ddot{X}_{g2}\ (n\times1)} + \underbrace{\left[\begin{array}{c:c} a_{s1}^{(2)}\ (3\times1) & O_{(3\times(n-5))} \\ \hdashline O_{(1\times1)} & O_{(1\times(n-5))} \\ \hdashline O_{(1\times1)} & O_{(1\times(n-5))} \\ \hdashline O_{((n-5)\times1)} & I_{((n-5)\times(n-5))} \end{array}\right]}_{J'_{r(n\times(n-4))}} \underbrace{\left[\begin{array}{c} \ddot{u}_1\ (1\times1) \\ \hline \ddot{u}_{2\,((n-5)\times1)} \end{array}\right]}_{\ddot{X}_{r1\,((n-4)\times1)}} +$$

$$+ \underbrace{\begin{bmatrix} \beta_w \ (3\times 1) \\ \hdashline O_{(1\times 1)} \\ \hdashline O_{(1\times 1)} \\ \hdashline O_{((n-5)\times 1)} \end{bmatrix}}_{A'_r \ (n\times 1)} \tag{3.4.149}$$

or in the form

$$\ddot{X}_{g1} = J_{1,2}\ddot{X}_{g2} + J'_r\ddot{X}_{r1} + A'_r \tag{3.4.150}$$

Let us note that, if the position of manipulator (2) is given by position vector X_{g2} of dimension $(n_2\times 1)$, the position of manipulator (1) can be given in terms of the reduced position vector

$$X_{r1} = \begin{cases} [u_1], & \text{for } n=5 \\ [u_1 \ u_2]^T, & \text{for } n=6 \end{cases} \tag{3.4.151}$$

of dimension $(n_1-4)\times 1$.

In this way we have n_2+n_1-4 free and independent parameters defining the position of the whole closed system. These are X_{g2} and X_{r1}.

Kinematics and dynamics of the manipulator (1) are described by

$$\ddot{X}_{g1} = J_1\ddot{q}_1 + A_1 \tag{3.4.152}$$

$$W_1\ddot{q}_1 = P_1 + U_1 + D_1R_A \tag{3.4.153}$$

Substituting (3.4.150) into (3.4.152) one obtains

$$J_{1,2}\ddot{X}_{g2} + J'_r\ddot{X}_{r1} + A'_r = J_1\ddot{q}_1 + A_1 \tag{3.4.154}$$

and substituting $\ddot{q}$ form (3.4.154) into (3.4.153):

$$W_1J_1^{-1}J_{1,2}\ddot{X}_{g2} + W_1J_1^{-1}J'_r\ddot{X}_{r1} = P_1 + U_1 - W_1J_1^{-1}(A'_r-A_1) + D_1R_A \tag{3.4.155}$$

Reaction vector R_A contains two components: force F_A and moment M_A. Since $\vec{F}_A \perp \vec{h}$ and $\vec{M}_A \perp \vec{h}$ it follows that

$$h^T F_A = 0, \quad h^T M_A = 0 \tag{3.4.156}$$

This can be written in the form

$$E R_A = 0 \tag{3.4.157}$$

where

$$E = \left[\begin{array}{c|c} h^T & O_{(1\times 3)} \\ \hline O_{(1\times 3)} & h^T \end{array}\right], \quad R_A = \left[F_A^T \quad M_A^T\right]^T \tag{3.4.158}$$

Combining the relations (3.4.140) and (3.4.141) one obtains

$$W_2 J_2^{-1} \ddot{X}_{g2} = P_2 + U_2 - W_2 J_2^{-1} A_2 - D_2 R_A \tag{3.4.159}$$

Now, (3.4.155), (3.4.159) and (3.4.157) can be written together in the form

$$\left[\begin{array}{c|c|c} W_1 J_1^{-1} J_r' & W_1 J_1^{-1} J_{1,2} & -D_1 \\ \hline O & W_2 J_2^{-1} & +D_2 \\ \hline O & O & E \end{array}\right] \left[\begin{array}{c} \ddot{X}_{r1} \\ \hline \ddot{X}_{g2} \\ \hline R_A \end{array}\right] = \left[\begin{array}{c} P_1 \\ \hline P_2 \\ \hline 0 \end{array}\right] + \left[\begin{array}{c} U_1 - W_1 J_1^{-1}(A_r' - A_1) \\ \hline U_2 - W_2 J_2^{-1} A_2 \\ \hline 0 \end{array}\right]$$

(3.4.160)

In the case of nominal dynamics calculation, the accelerations $\ddot{X}_{r1}$, $\ddot{X}_{g2}$ and reaciton R_A are given. The system (3.4.160) is solved for P_1, P_2.

In the case of simulation problem, the drives P_1, P_2 are known (3.4.160) represents a system of n_1+n_2+2 equations which should be solved for $(n_1-4)+n_2-6 = n_1+n_2+2$ unknowns ($\ddot{X}_{r1}$, $\ddot{X}_{g2}$, R_A).

The concept of independent reaction components can also be applied. The reaction vector is

$$R_A = G R_{Ar} \tag{3.4.161}$$

where R_{Ar} is the reduced reaction vector

$$R_{Ar} = \left[s_1 \; s_2 \; s_3 \; s_4 \right] \tag{3.4.162}$$

containing four independent components along the unit vectors $\vec{s}$ and $\vec{\ell} = \vec{s}\times\vec{h}$. Matrix G is

$$G = \left[\begin{array}{c|c|c|c} s_{(3\times 1)} & \ell_{(3\times 1)} & O_{(3\times 1)} & O_{(3\times 1)} \\ \hline O_{(3\times 1)} & O_{(3\times 1)} & s_{(3\times 1)} & \ell_{(3\times 1)} \end{array} \right]_{(6\times 4)} \tag{3.4.163}$$

Now, (3.4.155) and (3.4.159) become

$$W_1J_1^{-1}J_{1,2}\ddot{X}_{g2} + W_1J_1^{-1}J_r'X_{r1} = P_1 + U_1 - W_1J_1^{-1}(A_r'-A_1) + D_1G\,R_{Ar} \tag{3.4.164}$$

and

$$W_2J_2^{-1}\ddot{X}_{g2} = P_2 + U_2 - W_2J_2^{-1}A_2 - D_2G\,R_{Ar} \tag{3.4.165}$$

These two equations can be written together giving:

$$\left[\begin{array}{c|c|c} W_1J_1^{-1}J_r' & W_1J_1^{-1}J_{1,2} & -D_1G \\ \hline O & W_2J_2^{-1} & +D_2G \end{array} \right] \left[\begin{array}{c} \ddot{X}_{r1} \\ \hline \ddot{X}_{g2} \\ \hline R_{Ar} \end{array} \right] = \left[\begin{array}{c} P_1 \\ \hline P_2 \end{array} \right] +$$

$$+ \left[\begin{array}{c} U_1 - W_1J_1^{-1}(A_r'-A_r) \\ \hline U_2 - W_2J_2^{-1}A_2 \end{array} \right] \tag{3.4.166}$$

This system is simpler than (3.4.160) since it represents a set of n_1+n_2 equations which should be solved for $n_1-4+n_2+4 = n_1+n_2$ unknowns $(\ddot{X}_{r1}, \ddot{X}_{g2}, R_{Ar})$.

The friction effects are taken into account in the same way as was done in Para. 3.4.7.

Linear joint connection. We consider two manipulators having the grippers connected to each other by a joint permitting one relative translation only (Fig. 3.27). Let $\vec{h}$ be the unit vector of this translation.

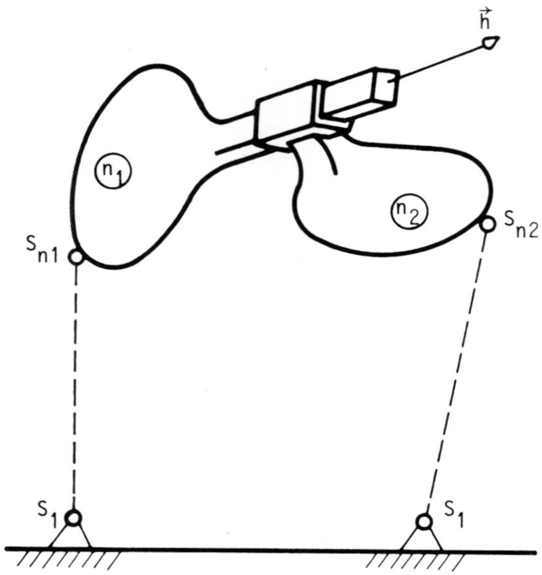

Fig. 3.27. Grippers connected by means of a linear joint

For the solution of this manipulation task it is necessary that at least one of the manipulators has six d.o.f. Let it be manipulator (1). Thus $n_1=6$ and n_2 can be equal to five or six. For simplicity, we assume $n_1=n_2=n=6$.

We apply the theory explained in Para. 3.4.9 combined with the transformations just described.

The position of manipulator (2) is defined by the position vector

$$X_{g2} = \begin{bmatrix} x_{A2} & y_{A2} & z_{A2} & \theta_2 & \varphi_2 & \psi_2 \end{bmatrix}, \tag{3.4.167}$$

The joint connection results in:

$$\theta_1 = \theta_2, \quad \varphi_1 = \varphi_2, \quad \psi_1 = \psi_2 \tag{3.4.168}$$

For the given vector X_{g2} the position of manipulator (1) can be defined by

$$u = \overline{A_2A_1} \tag{3.4.169}$$

i.e., by the reduced position vector

$$X_{r1} = [u]_{1\times 1} \tag{3.4.170}$$

Analogously to what was done for two d.o.f. joint, we can obtain

$$\underbrace{\begin{bmatrix} w_{A1} \\ \hline \ddot{\theta}_1 \\ \hline \ddot{\varphi}_1 \\ \hline \ddot{\psi}_1 \end{bmatrix}}_{} = \underbrace{\left[\begin{array}{c|c|c|c} I_{(3\times 3)} & f_{\theta 1}^{(2)} & f_{\varphi 1}^{(2)} & 0 \\ \hline 0_{(1\times 3)} & 1 & 0 & 0 \\ \hline 0_{(1\times 3)} & 0 & 1 & 0 \\ \hline 0_{(1\times 3)} & 0 & 0 & 1 \end{array}\right]}_{J_{1,2}} \begin{bmatrix} w_{A2} \\ \hline \ddot{\theta}_2 \\ \hline \ddot{\varphi}_2 \\ \hline \ddot{\psi}_2 \end{bmatrix} + \underbrace{\begin{bmatrix} a_{s1}^{(2)} \\ \hline 0 \\ \hline 0 \\ \hline 0 \end{bmatrix}}_{J'_r} [\ddot{u}] +$$

$$+ \underbrace{\begin{bmatrix} \beta_w \\ \hline 0 \\ \hline 0 \\ \hline 0 \end{bmatrix}}_{A'_r} \tag{3.4.171a}$$

or

$$\ddot{X}_{g1} = J_{1,2}\ddot{X}_{g2} + J'_r\ddot{X}_{r1} + A'_r \tag{3.4.171b}$$

The joint connection produces a reaction force $\vec{F}_A$ perpendicular to $\vec{h}$ and $\vec{M}_A$ which need not be perpendicular to $\vec{h}$. Thus

$$h^T F_A = 0 \tag{3.4.172a}$$

and this condition can be written in the form

$$E\,R_A = 0 \tag{3.4.172b}$$

where

$$E = \left[h^T \;\vdots\; 0_{(1\times 3)}\right]_{(1\times 6)}, \qquad R_A = \left[F_A^T \;\vdots\; M_A^T\right]^T \tag{3.4.173}$$

Now, all the elements of dynamic model (3.4.160) are determined and the model can be solved. The dimension of the model in n_1+n_2+1.

The concept of independent reaction components can also be applied.

Connection permitting no relative motion. We consider two manipulators having the grippers connected to each other by means of a connection permitting no relative motion (Fig. 3.28).

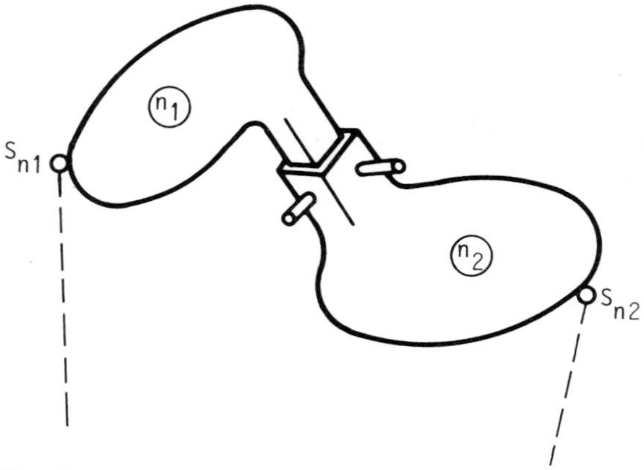

Fig. 3.28. Two grippers connected by means of a connection permitting no relative motion

Let us assume that six d.o.f. manipulators are considered. Then, the position vectors are

$$X_{g1} = \left[x_{A1}\ y_{A1}\ z_{A1}\ \theta_1\ \varphi_1\ \psi_1\right]^T, \quad X_{g2} = \left[x_{A2}\ y_{A2}\ z_{A2}\ \theta_2\ \varphi_2\ \psi_2\right]^T \tag{3.4.174}$$

The connection results in relation

$$X_{g1} = X_{g2} = X_g \tag{3.4.175}$$

and, accordingly,

$$\ddot{X}_{g1} = \ddot{X}_{g2} = \ddot{X}_g$$

The reaction force $\vec{F}_A$ and reaction moment $\vec{M}_A$ have three independent components each.

Now, the dynamic model can be written in the form

3.5. Impact Problems

Paragraph 3.4. discussed the theoretical aspects of constrained gripper motions. The dynamic models of such robot motions were derived. In the discussion of model solution it was assumed that the initial state satisfies the constraints imposed. Let us consider two time intervals T_1 and T_2. In the interval T_1 the gripper is not subject to any constraint. It is a free motion when the robot is not in contact with the constraint. If a peg-in-the-hole assembly problem is considered, the interval T_1 represents the phase in which the peg is moved towards the hole (Fig. 3.1a). We can calculate the nominal dynamics in this time interval in such a way that the peg comes into the hole without any collision, i.e., that the terminal state of interval T_1 satisfies the constraint. In the interval T_2 the peg is inserted into the hole. The terminal state of T_1 becomes the initial state for T_2. The constraint is imposed and the chain is closed. In the nominal dynamics calculation we prescribe this insertion and the reactions (for instance we prescribe reactions to equal zero).

But, if a perturbed motion is considered (simulation problem), the peg will come to the hole but the terminal state of T_2 will not satisfy the constraint. The impact will happen and it will ensure the initial state (for T_2) which satisfies the hole constraint.

With manipulation systems the problems of impact were studied in [11] only. Here, we make a further development of robot impact theory. Let it be noted that all the impacts which will be considered are completely nonelastic.

3.5.1. General methodology

Let us consider a manipulator with n d.o.f. If there is no constraint on gripper motion, then the dynamics can be described by the matrix model

$$W\ddot{q} = P + U \qquad (3.5.1)$$

and the kinematics by the second order Jacobian form

$$\ddot{X}_g = J\ddot{q} + A \qquad (3.5.2)$$

where X_g is the position vector (see Ch. 2). If we substitute $\ddot{q}$ from (3.5.2) into (3.5.1), the dynamic model obtains the form:

$$WJ^{-1}\ddot{X}_g = P + U - WJ^{-1}A \tag{3.5.3}$$

Let T_1 be the time interval of this nonconstrained motion.

Let us now consider the collision of manipulator gripper with some given constraint and let this impact last for Δt.

Let $(q', \dot{q}')$, or equivalently $(X_g', \dot{X}_g')$, be the mechanism state just before the impact, i.e., the initial state for the impact time interval Δt. It is the terminal state of nonconstrained motion and is obtained by integrating the model (3.5.3) over the time interval of nonconstrained motion (interval T_1). We introduce the reaction force $\vec{F}_A$ and reaction moment $\vec{M}_A$ which form the reaction vector

$$R_A = \left[F_A^T \; M_A^T\right]^T \quad (6\times 1) \tag{3.5.4}$$

The dynamic model of manipulator which is in contact with a constraint is obtained by introducing reactions in the model (3.5.3), i.e.,

$$WJ^{-1}\ddot{X}_g = P + U - WJ^{-1}A + D\,R_A \tag{3.5.5}$$

where D is defined by (3.4.29), (3.4.3). This model holds for the interval of impact (Δt) as well as for the interval T_2 of further constrained motion. We are interested now in the interval Δt. If we integrate (3.5.5) over the interval Δt and if $\Delta t \to 0$, then

$$WJ^{-1}\Delta\dot{X}_g = (P+U-WJ^{-1}A)\Delta t + DR_A\Delta t \tag{3.5.6}$$

or

$$WJ^{-1}(\dot{X}_g'' - \dot{X}_g') = DR_A\Delta t \tag{3.5.7}$$

where X_g'' is the position vector at the end of impact time interval. $DR_A\Delta t$ does not equal zero although $\Delta t \to 0$. This is due to the fact that $R_A \to \infty$ for $\Delta t \to 0$.

The velocity $\dot{X}_g''$ can be expressed in terms of free parameters:

$$\dot{X}''_g = J_r \dot{X}''_r \tag{3.5.8}$$

where X''_r is the reduced position vector and J_r is the reduced Jacobian.

The reaction vector satisfies the condition (3.4.30), i.e.,

$$ER_A = 0 \tag{3.5.9}$$

where the matrix E depends on the constraint imposed.

If we substitute (3.5.8) into (3.5.7) and combine the resulting equation with (3.5.9), we obtain

$$\begin{bmatrix} WJ^{-1}J_r & -D \\ O & E \end{bmatrix} \begin{bmatrix} \dot{X}''_r \\ R_A \Delta t \end{bmatrix} = \begin{bmatrix} XJ^{-1}\dot{X}'_g \\ O \end{bmatrix} \tag{3.5.10}$$

This model represents a set of n_r+6 equations which should be solved for n_r+6 unknowns: $\dot{X}''_r$, $R_A\Delta t$. Let it be noted that R_A and Δt are not considered separately but the product $R_A\Delta t$ represent one variable (the reaction force momentum).

The initial state for the constrained motion (in the interval T_2) understands $(X''_r, \dot{X}''_r)$. The position X''_r follows directly from terminal position X'_g (since position can not be changed in $\Delta t \to 0$), and the velocity $\dot{X}''_r$ is obtained from (3.5.10).

If we wish to apply the concept of independent reaction components, we introduce (3.4.33) into (3.5.7) to obtain:

$$WJ^{-1}(\dot{X}''_g - \dot{X}'_g) = DGR_{Ar}\Delta t \tag{3.5.11}$$

If we introduce (3.5.8) into (3.5.11), the following form can be obtained

$$\begin{bmatrix} WJ^{-1}J_r & -DG \end{bmatrix} \begin{bmatrix} \dot{X}''_r \\ R_{Ar}\Delta t \end{bmatrix} = WJ^{-1}\dot{X}'_g \tag{3.5.12}$$

This is a system of n equations with n unknowns: $\dot{X}''_r$, $R_{Ar}\Delta t$.

Generalization. Let us consider a more general case. In the time interval T_1 a manipulator moves subject to some constraints. X_{r1} is the corresponding reduced position vector. Let the terminal state be $(X'_{r1}, \dot{X}'_{r1})$. Suppose that a collision with an additional constraint happens at the end of T_1 and that the impact lasts for Δt. Let X_{r2} be the reduced position vector corresponding to this new set of constraints, and $(X''_{r2}, \dot{X}''_{r2})$ defines the terminal state of Δt, i.e., the initial conditions for the interval T_2 in which the manipulator motion is subject to this new set of constraints. Since

$$\dot{X}'_g = J_{r1}\dot{X}'_{r1}, \qquad \dot{X}''_g = J_{r2}\dot{X}''_{r2} \tag{3.5.13}$$

the equation (3.5.7) becomes

$$WJ^{-1}(J_{r2}\dot{X}''_{r2} - J_{r1}\dot{X}'_{r1}) = DR_A\Delta t \tag{3.5.14}$$

If this relation is written together with (3.5.9), the following model is obtained:

$$\left[\begin{array}{c|c} WJ^{-1}J_{r2} & -D \\ \hline O & E \end{array}\right] \left[\begin{array}{c} \dot{X}''_{r2} \\ \hline R_A\Delta t \end{array}\right] = \left[\begin{array}{c} WJ^{-1}J_{r1}\dot{X}'_{r1} \\ \hline O \end{array}\right] \tag{3.5.15}$$

The initial state for T_2 is $(X''_{r2}, \dot{X}''_{r2})$. X''_{r2} follows directly from X'_{r1} and $\dot{X}''_{r2}$ is obtained from (3.5.15).

If the concept of independent reaction components is used, then the model (3.5.14) has the form

$$\left[\begin{array}{c|c} WJ^{-1}J_{r2} & -DG \end{array}\right] \left[\begin{array}{c} \dot{X}''_{r2} \\ \hline R_{Ar}\Delta t \end{array}\right] = WJ^{-1}J_{r1}\dot{X}'_{r1} \tag{3.5.16}$$

This impact theory can be applied to all the constraints from Para. 3.4.4 - 3.4.10.

<u>Friction effects</u>. Let us discuss the case of a constraint with friction. The friction force $\vec{F}_f$ follows directly from the reactions $\vec{F}_A, \vec{M}_A$. Since $|\vec{F}|\to\infty$ and $|\vec{M}_A|\to\infty$ when $\Delta t\to 0$, it is clear that $|\vec{F}_f|\to\infty$ too. Hence $\vec{F}_f\Delta t$ does not equal zero even if $\Delta t\to 0$. This component must be expressed in terms of reactions and then included into (3.5.10) in an analogous was as in Para. 3.4.4 - 3.4.10. A more detailed discussion including jamming problem will be given in Para. 3.5.4.

3.5.2. Impact in the case of bilateral manipulation

Let us consider two manipulators. In the interval T_1 they are moving independently. Then, an impact happens which lasts for Δt. After that, in the interval T_2 the manipulators are connected to each other (see Para. 3.4.11) and move in accordance with this constraint. Let $(X'_{g1}, \dot{X}'_{g1})$ and $(X'_{g2}, \dot{X}'_{g2})$ be the states just before the impact, and let $(X''_{g1}, \dot{X}''_{g1})$ and $(X''_{g2}, \dot{X}''_{g2})$ be the states after the impact.

From (3.4.152), (3.4.153) and (3.4.140), (3.4.141) it follows that

$$W_1J_1^{-1}\ddot{X}_{g1} = P_1 + U_1 - W_1J_1^{-1}A_1 + D_1R_A \tag{3.5.17}$$

and

$$W_2J_2^{-1}\ddot{X}_{g2} = P_2 + U_2 - W_2J_2^{-1}A_2 - D_2R_A \tag{3.5.18}$$

The singn "+" of the term D_1R_A in (3.5.17) and the sign "-" of D_2R_A in (3.5.18) are due to action and reaction.

By integrating these models over the interval Δt and assuming that $\Delta t \to 0$, one obtains

$$W_1J_1^{-1}(\dot{X}''_{g1}-\dot{X}'_{g1}) = D_1R_A\Delta t \tag{3.5.19}$$

and

$$W_2J_2^{-1}(\dot{X}''_{g2}-\dot{X}'_{g1}) = -D_2R_A\Delta t \tag{3.5.20}$$

It holds that (see eq. (3.4.150) and apply it to velocities)

$$\dot{X}''_{g1} = J_{1,2}\dot{X}''_{g2} + J'_r\dot{X}''_{r1} \tag{3.5.21}$$

where the Jacobians $J_{1,2}$, J'_r and the reduced speed vector $\dot{X}''_{r1}$ are defined in (3.4.149). (3.5.19) now becomes

$$W_1J_1^{-1}(J_{1,2}\dot{X}''_{g2}+J_r\dot{X}''_{r1}-\dot{X}'_{g1}) = D_1R_A\Delta t \tag{3.5.22}$$

Let us write the equations (3.5.22), (3.5.20), and (3.5.9) together into form

$$\begin{bmatrix} W_1 J_1^{-1} J_r' & W_1 J_1^{-1} J_{1,2} & -D_1 \\ 0 & W_2 J_2^{-1} & D_2 \\ 0 & 0 & E \end{bmatrix} \begin{bmatrix} \dot{X}_{r1}'' \\ \dot{X}_{g2}'' \\ R_A \Delta t \end{bmatrix} = \begin{bmatrix} W_1 J_1^{-1} \dot{X}_{g1}' \\ W_2 J_2^{-1} \dot{X}_{g2}' \\ 0 \end{bmatrix} \tag{3.5.23}$$

This system can be solved for $\dot{X}_{r1}''$, $\dot{X}_{g2}''$, $R_A\Delta t$. The positions X_{g1}'', X_{g2}'' (which follow directly from X_{g1}', X_{g2}'), together with these velocities define the initial state for the interval T_2.

3.5.3. Extension of surface-type constraint

We consider the surface-type constraint explained in Para. 3.4.12 (Figs. 3.29 and 3.30). We restrict our consideration to stationary constraint without friction.

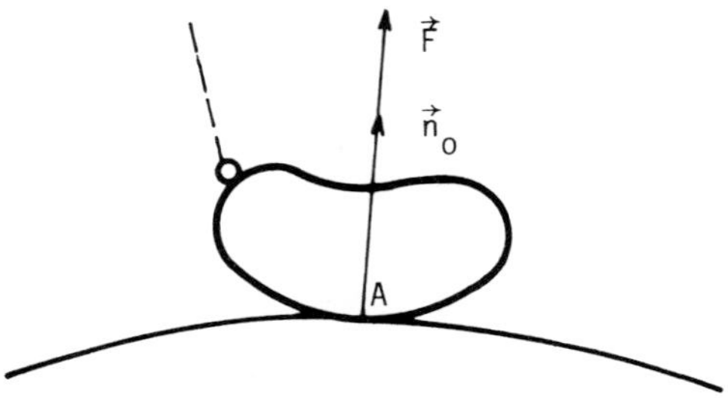

Fig. 3.30a. Surface-type constraint and the reaction force

Let A be the point in which a contact is obtained. $(q', \dot{q}')$ is the state before the impact. During the impact interval Δt the dynamics is described by

$$W\ddot{q} = P + U + D_F F = P + U + D_F n_o R \tag{3.5.24}$$

If we integrate (3.5.24) over Δt and assume $\Delta t \to 0$, we obtain

$$W(\dot{q}''-\dot{q}') = D_F n_o R\Delta t \tag{3.5.25}$$

where $(q'', \dot{q}'')$ is the state after the impact. Note that $q'=q''$. From (3.5.25)

$$\dot{q}'' = \dot{q}' + W^{-1}D_F n_o R\Delta t \tag{3.5.26}$$

The velocity $\vec{v}''$ of point A at the time instant after the impact must satisfy $\vec{v}''\perp\vec{n}_o$ and, accordingly,

$$n_o^T v'' = 0 \tag{3.5.27}$$

This velocity can be expressed in terms of generalized velocity:

$$v'' = J_v \dot{q}'' \tag{3.5.28}$$

In Ch. 2 the n×n Jacobian was considered. It represented the relation between the gripper velocities ($\vec{v}$, $\vec{\omega}$) and the joint velocities ($\dot{q}$). The first three rows of this Jacobian matrix represent $J_{v\ (3\times n)}$ i.e. the relation between $\vec{v}$ and $\dot{q}$. Substituting (3.5.26) into (3.5.28) and then into (3.5.27) we obtain

$$n_o^T J_v \dot{q}' + n_o^T J_v W^{-1} D_F n_o R\Delta t = 0 \tag{3.5.29}$$

wherefrom the reaction momentum can be obtained

$$R\Delta t = -\underbrace{(n_o^T J_v W^{-1} D_F n_o)^{-1}}_{(1\times 1)} \underbrace{n_o^T J_v \dot{q}'}_{(1\times 1)} \tag{3.5.30}$$

Now, $\dot{q}''$ is found from (3.5.26)

$$\dot{q}'' = \dot{q}' - W^{-1}D_F n_o (n_o^T J_v W^{-1} D_F n_o)^{-1} n_o^T J_v \dot{q}' \tag{3.5.31}$$

and the state (q'', $\dot{q}''$) affer the impact is determined.

3.5.4. Jamming problems

We consider the collision with a constraint in the presence of friction. Let us write the model (3.5.12) in the form

$$WJ^{-1}J_r \dot{X}_r'' - DGR_{Ar}\Delta t = WJ^{-1}\dot{X}_g' \tag{3.5.32}$$

If the friction force $\vec{F}_f$ is introduced, the model becomes

$$WJ^{-1}J_r\dot{X}_r'' - DGR_{Ar}\Delta t - D_F F_f \Delta t = WJ^{-1}\dot{X}_g' \tag{3.5.33}$$

It has already been said that $F_f\Delta t$ does not equal zero when $\Delta t \to 0$. This is so because $R_{Ar} \to \infty$ and, accordingly, $F_f \to \infty$. The model (3.5.33) is the basis for the analysis of jamming. We discuss two most interesting cases.

Surface-type constraint. This constraint was discussed in Para. 3.4.4. The impact model (3.5.33) obtains here the form

$$WJ^{-1}J_r\dot{X}_r'' - D(G+G')S\Delta t = WJ^{-1}\dot{X}_g' \tag{3.5.34}$$

where G is defined by (3.4.50), and

$$G' = \left[-\mu v_{Aro}^T \quad 0_{(1\times 3)}\right]^T \tag{3.5.35}$$

S is the absolute value of reaction force. We note that v_{Aro} depends on $\dot{X}_r''$ (see eqs. (3.4.52), (3.4.54)).

Now, the impact model (3.5.34) should be solved for n unknowns $\dot{X}_r''$, $S\Delta t$. Since v_{Aro} appearing in (3.5.35) depends on $\dot{X}_r''$ it is clear that the model is nonlinear with respect to $\dot{X}_r''$.

Let us now discuss the jamming problem. It may happen that the solution to the system gives $\dot{X}_r'' = 0$, which means jamming, but it is not so easy to check whether the point A is jammed or not. We do it in the following way. We consider the solution $\dot{u}_1'' = 0$, $\dot{u}_2'' = 0$ and search for the critical value μ_c of friction coefficient which produces this solution. If the value μ of the real example is $\mu \geqslant \mu_c$, jamming will happen, and if $\mu < \mu_c$, there will be no jamming.

With the surface-type constraint jamming reffers to point A only and the gripper still has n_r-2 d.o.f. (θ, φ for n=5 or θ, φ, ψ for n=6). In fact the gripper will behave as if there were a spherical joint constraint or a constraint discussed in Para. 3.4.12.

Two d.o.f. joint constraint. This is a cylindrical assembly problem and was discussed in Para. 3.4.7. Here, the impact model (3.5.33) obtains the form

$$WJ^{-1}J_r\dot{X}_r'' - DGR_{Ar}\Delta t - D_F(F_{f1}+F_{f2})\Delta t = WJ^{-1}\dot{X}_g' \tag{3.5.36}$$

Calculation of friction is explained in Para. 3.4.7 and we conclude that these forces depend on reactions F_A, R_A (and accordingly on R_{Ar}) and on $\dot{u}_1''$, $\dot{u}_2''$ (and accordingly on $\dot{X}_r''$). The model (3.5.36) should be solved for n unknowns $\dot{X}_r''$, $R_{Ar}\Delta t$. Let it be noted that the model is non-linear since F_{f1}, F_{f2} depends nonlinearly on $\dot{X}_r''$, R_{Ar}.

The solution $\dot{u}_1'' = 0$, $\dot{u}_2'' = 0$ indicates jamming of the gripper inside the cylindrical constraint. In order to check the jamming conditions, we consider the solution $\dot{u}_1'' = 0$, $\dot{u}_2''=0$ and find the critical value μ_c which produces this solution. If the exact value μ is $\mu \geqslant \mu_c$, jamming will happen, and if $\mu < \mu_c$, it will not. Rolling is not considered.

3.6. Practical Cases of Constrained Gripper Motion

This paragraph discusses some problems which appear with practical manipulation tasks and which require the theory developed in the previous paragraphs.

3.6.1. Tasks with surface-type constraints

Task of writing and drawing. Let us consider a manipulation task in which a manipulator has to write or draw on a given surface (Fig. 3.31).

It a 5 d.o.f. manipulator has to write a letter or a sign on a given surface, we have to prescribe the independent parameters $u_1(t)$ and $u_2(t)$ in such a way as to obtain the desired letter. We also prescribe the angles $\theta(t)$ and $\varphi(t)$ so as to obtain the desired orientation of the pencil (Fig. 3.32). In this way the complete reduced position vector $X_r(t)$ (3.4.37) is prescribed and we can solve the nominal dynamics. This calculation of nominal dynamics also includes the time interval T_1 in which the pencil is moving towards the surface. We prescribe this motion so as to avoid impact.

When the perturbed motion is considered, the pencil comes to the surface and the terminal state of T_1 does not satisfy the surface constraint. The impact appears and we solve it by using the theory from Para. 3.5.1 (eq. 3.5.10 or 3.5.12). Thus obtaining the initial conditions for constrained motion we solve this motion by using the theory explained in Para. 3.4.4.

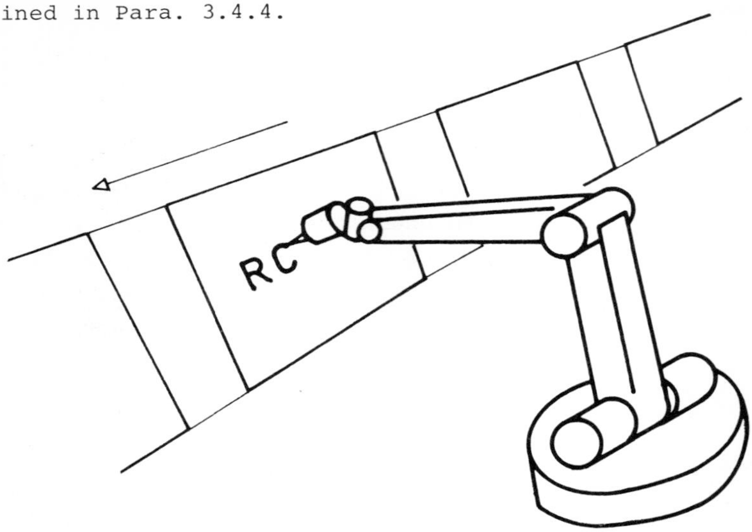

Fig. 3.31. Task of writing

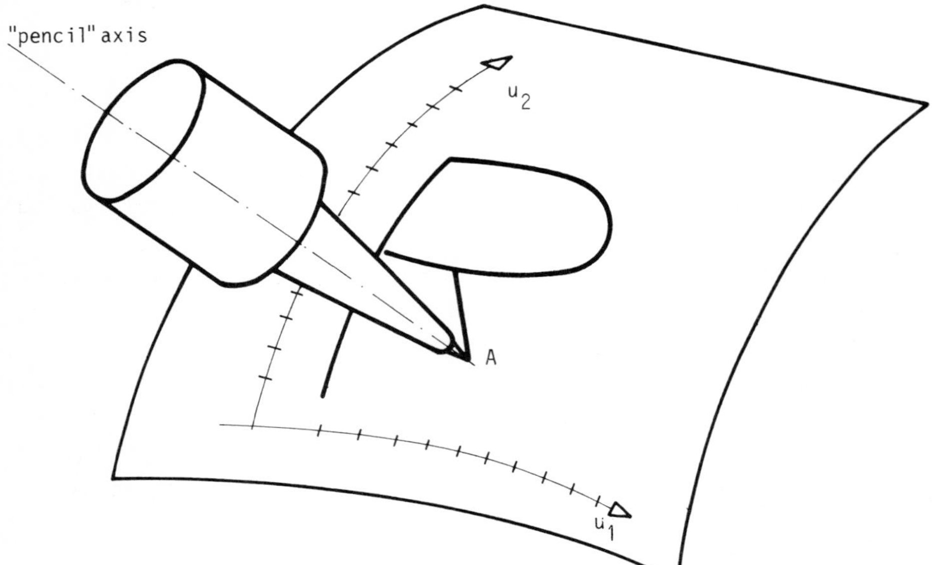

Fig. 3.32. Writing on a surface

Grinding task. Fig. 3.33a,b shows the task of fine grinding. A moving surface results in relative velocity and friction force F_f.

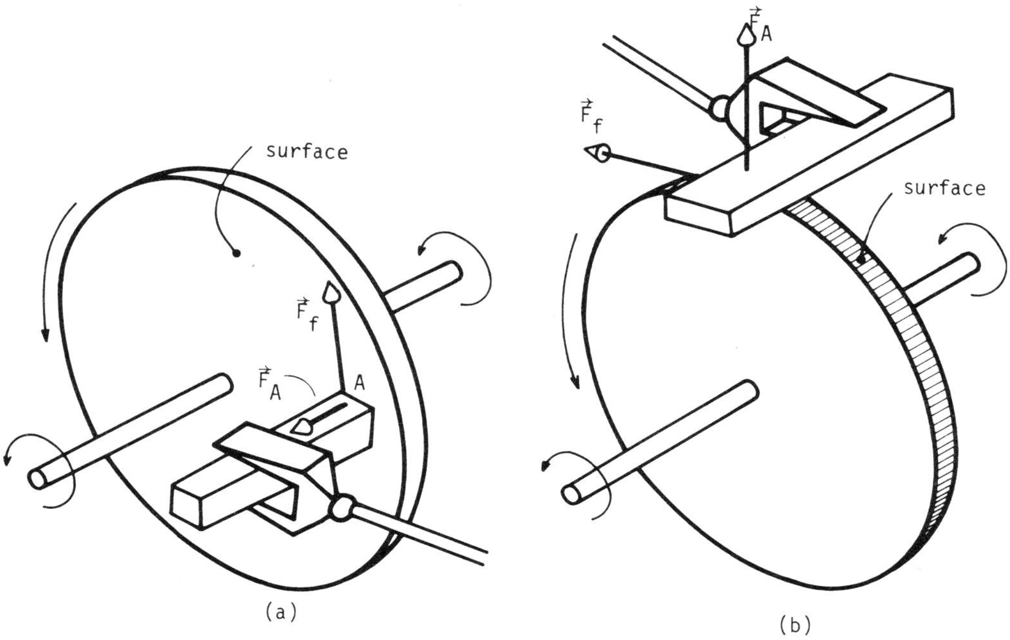

Fig. 3.33. Grinding task

In the case (a) a plane surface is considered. It rotates as shown in the figure. But, in practice this is not an ideal plane and the rotation axis is not exactly perpendicular to the plane. Thus, the motion of the surface is not a simple rotation (Fig. 3.34). The reaction and, accordingly, the friction are not constant and produce vibrations of the working object (and the gripper).

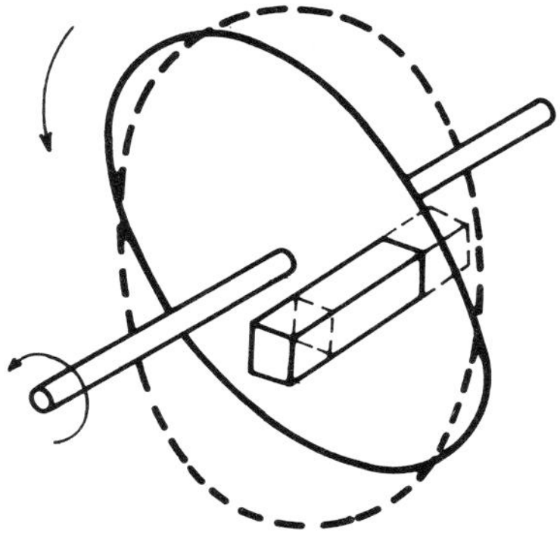

Fig. 3.34. Vibrations

In the case shown in Fig. 3.33b the cylindrical surface is considered. Rotation is not ideal since it is not an ideal cylinder (Fig. 3.35a) and the rotation axis is not in the exact center of the circle.

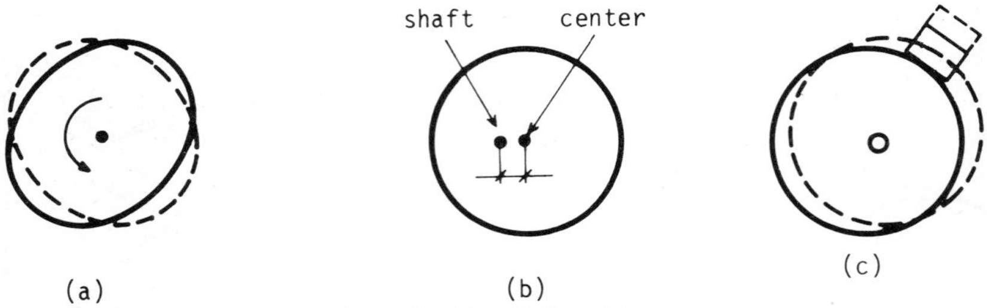

Fig. 3.35. Vibrations

For this reason the reaction is not constant and vibrations of the working object appear. All these effects can be included in the calculation. But, there are some effects which can not be taken into account. These are high frequency vibrations due to grains of grinding wheel. Thus, this calculation of grinding dynamics is approximative. If a polishing task is considered then the discussion performed can also be applied but the problem of high frequency vibrations is not important.

3.6.2. Cylindrical and rectangular assembly tasks

Cylindrical problem. Let us consider a manipulation task in which a cylindrical working object has to be inserted into a cylindrical hole (Fig. 3.36).

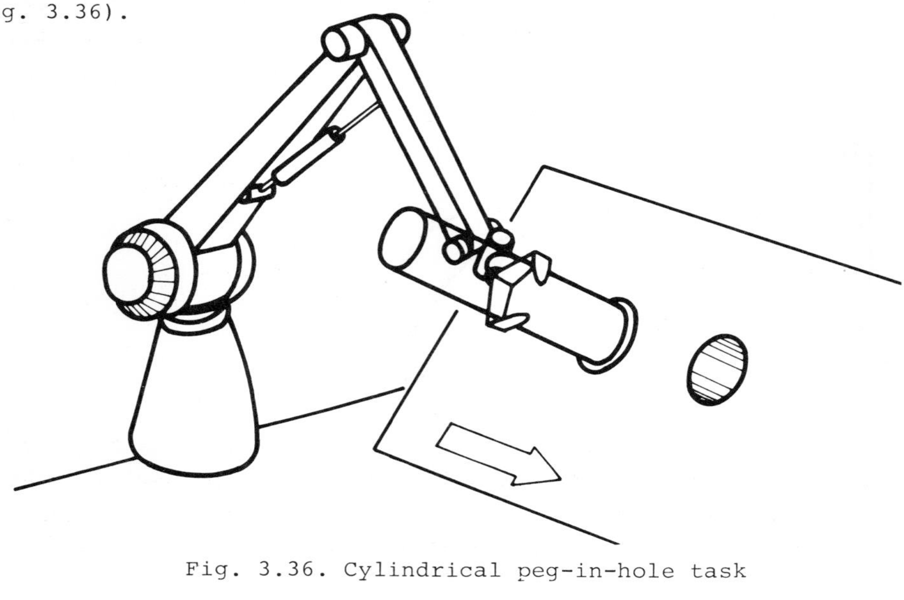

Fig. 3.36. Cylindrical peg-in-hole task

When the nominal dynamics is calculated we prescribe the motion in such a way that no impact happens during the insertion.

Let us now discuss the perturbed motion. In the first time interval T_1 the peg is moved towards the hole. It is a motion without any constraints. At the begining of the insertion an impact happens (Fig. 3.37a,b). The impact is of the type discussed in Para. 3.5.3. The constraint is discussed in 3.4.12. In order to find the point of collision (point A in Figs. 3.30 and 3.37) we have to solve the equations (3.4.177) and (3.4.178) defining the cylindrical hole and the cylindrical peg. While the peg is moving towards the hole there is no real solution to this equation system. The impact happens when this system has one real solution only. Thus, in each iteration the system has to be solved numerically.

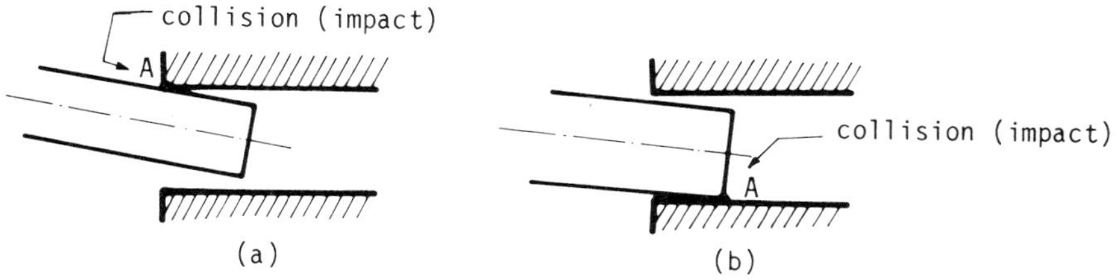

Fig. 3.37. Two cases of collision

In the next time interval (T_2) the motion of the peg is constrained. It is a surface-type constraint, the one discussed in Para. 3.4.12. Then the second impact happens (point A_2 in Fig. 3.38a or A_1 in Fig. 3.38b).

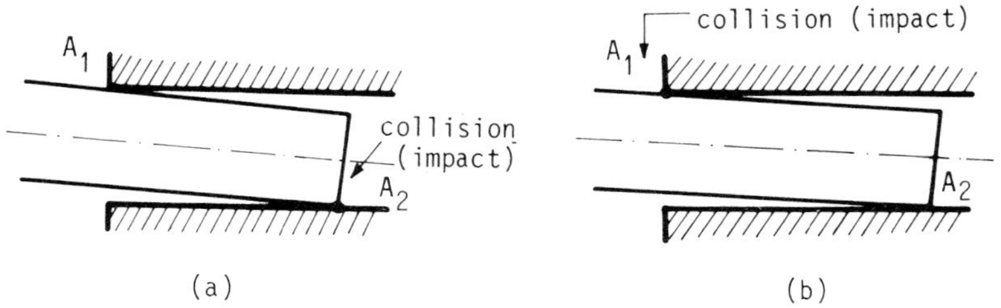

Fig. 3.38. The second impact

After this second impact the motion of the manipulator is subject to two-d.o.f. joint constraint (Para. 3.4.7). However, if there are large perturbances, the motion after the second impact cannot be considered in this way but it is subject to two constraints of surface type (Fig.

3.39a). Cylindrical joint constraint (two d.o.f) can be used when the axis of the peg (*) is close enough to the hole axis (**) (see Fig. 3.39b).

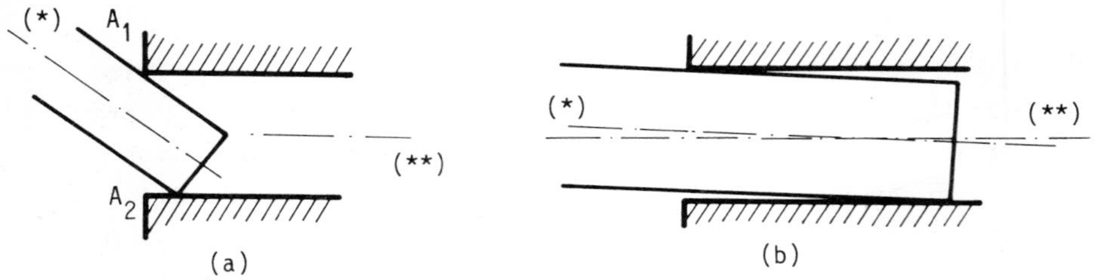

Fig. 3.39. Two constraints of surface type

The problem of friction and jamming was considered in Para. 3.4.7 and 3.5.

Rectangular problem. A rectangular assembly task is shown in Fig. 3.40. Such a manipulation task requires six d.o.f. manipulators since the total orientation of working object is needed.

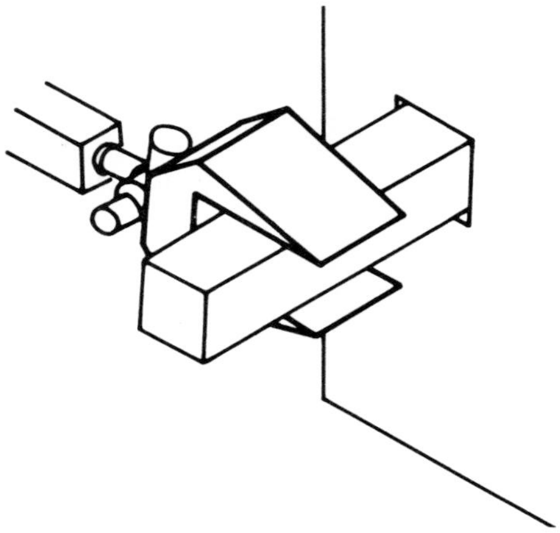

Fig. 3.40. Rectangular assembly task

3.6.3. Constraint permitting no relative motion

We consider the final phase of an assembly task (Fig. 3.41a). In the phase of insertion we face the rectangular assembly problem. But, in this final phase the working object is fixed and no relative motion is possible.

Another example is shown in Fig. 3.41b. When the first two screws are screwed-in the object becomes fixed and no relative motion is possible.

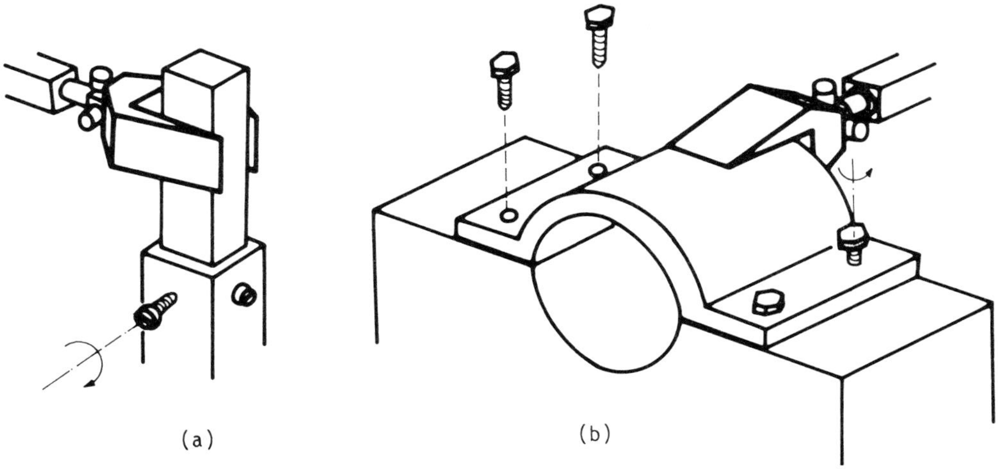

Fig. 3.41. Final phases of assembly tasks

The theory covering these problems is explained in Para. 3.4.10.

3.6.4. Practical problems of bilateral manipulation

Assembly tasks. Two most interesting tasks with bilateral manipulation are cylindrical and rectangular assembly tasks (Fig. 3.42). The whole discussion on cylindrical assembly task given in Para. 3.6.2 can be applied to bilateral manipulation.

Case of no relative motion. The connection of the two grippers can be such that no relative motion is permitted. It happens in a task when one "arm" hands over the working object to the other. Then, there exists a time interval when both arms are in connection with the object (Fig. 3.43).

Something very similar can happen if two manipulators are used together to move a very heavy load.

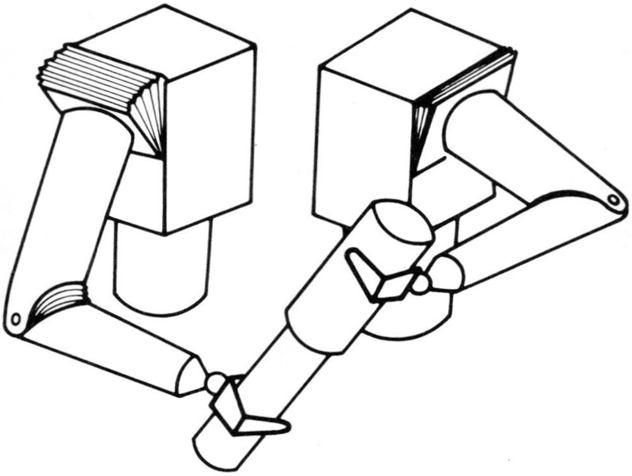

Fig. 3.42. Bilateral manipulator in cylindrical assembly task

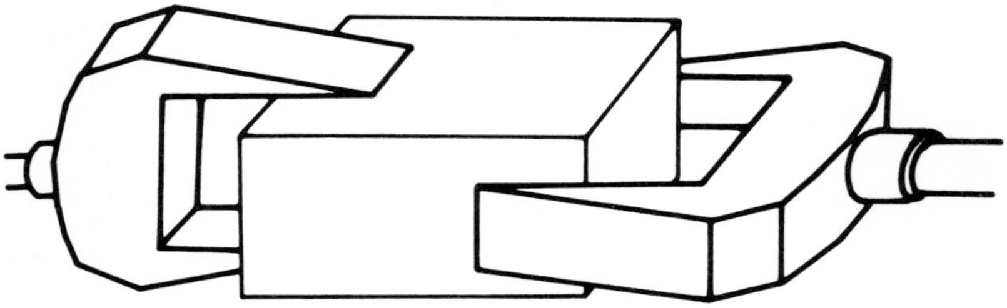

Fig. 3.43. Two grippers connected to a working object

3.7. Examples

Example 1

We consider an arthropoid six d.o.f. manipulator shown in Fig. 3.44. Its data are given in the table in Fig. 3.46. The manipulator has to draw a circle on a moving plane (Figs. 3.44, 3.45). Radius of the circle is R = 0.4m. At the initial time istant the manipulator is in the resting position and at the same time instant the plane begins to move up with the constant acceleration $a = 0.6\ m/s^2$ (Fig. 3.44). During drawing the pencil has to be perpendicular to the plane and produce the force S = 30N upon the plane. The friction coefficient is $\mu=0.3$. The drawing task has to be performed in T=2s with the triangular velocity profile along the circle trajectory.

If we follow the theory explained in Para. 3.4.4, the plane constraint is expressed in the form

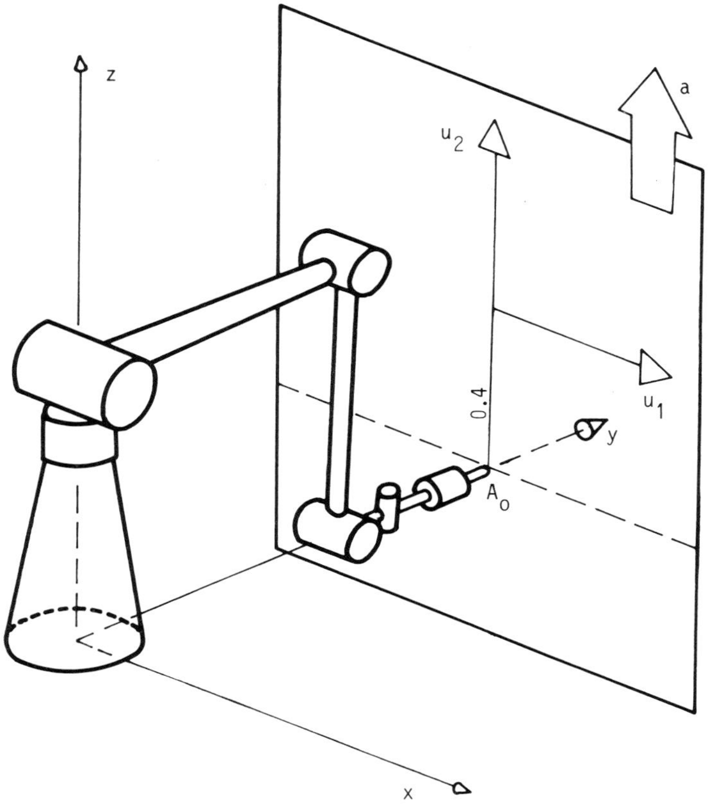

Fig. 3.44. Arthropoid manipulator in a drawing task

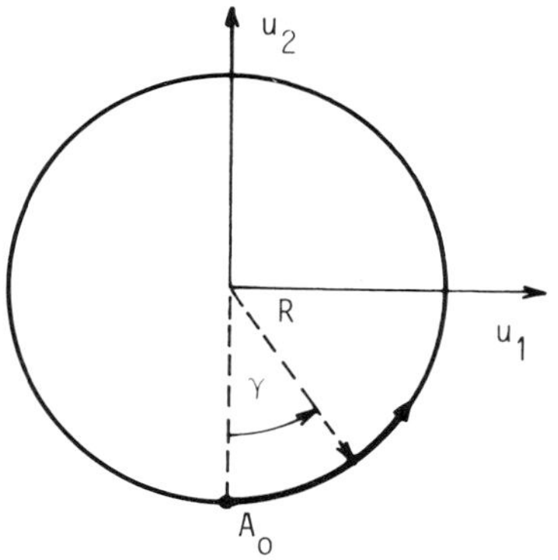

Fig. 3.45. Circle to be drawn

Segment	1	2	3	4	5	6
ℓ_i	0.8	0.8	0.8	0.1	0.1	0.2
m_i	-	5.	5.	1.	1.	2.
I_{xi}	-	0.25	0.25	0.002	0.002	0.01
I_{yi}	-	0.01	0.25	0.002	0.002	0.002
I_{zi}	0.2	0.25	0.01	0.002	0.002	0.01

Fig. 3.46. Manipulator data

$$x = f_x = u_1, \quad y = f_y = 1.2 \quad z = f_z = u_2 + \frac{1}{2}at^2 + 0.4$$

In this example we first illustrate the calculation of nominal dynamics.

Hence, we prescribe the relative motion $u_1(t)$, $u_2(t)$ in such a way as to obtain the circle trajectory and the desired velocity profile. Thus

$$u_1 = R \sin\gamma, \qquad u_2 = R \cos\gamma$$

where

$$\ddot{\gamma} = \begin{cases} +8\pi/T^2, & t<T/2 \\ -8\pi/T^2, & t\geq T/2 \end{cases}$$

The manipulator is driven by CEM-PARVEX D.C. motors: model M17 for the first three joints, and model F9M2 HA for gripper joints. Reduction ratio is N = 100 for all joints except for S_2 where it is N = 150.

The results of nominal dynamics calculation are presented in Figs. 3.47, 3.48, 3.49. Fig. 3.47 shows the relative motion of point A (i.e. the time histories of parameters u_1 and u_2). Fig. 3.48 shows the corre-spondending time histories of the internal coordinates (q). Finally, Fig. 3.49 shows the nominal input control voltages.

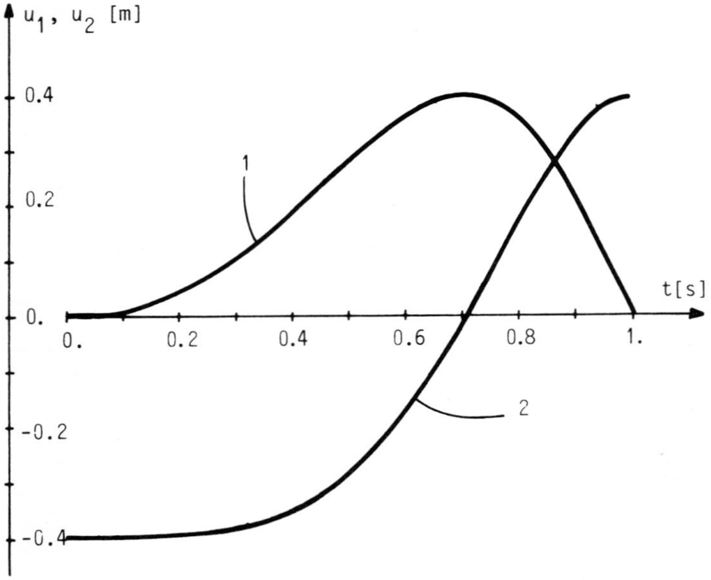

Fig. 3.47. Relative motion with respect to the surface

Now, we present the results of simulation.

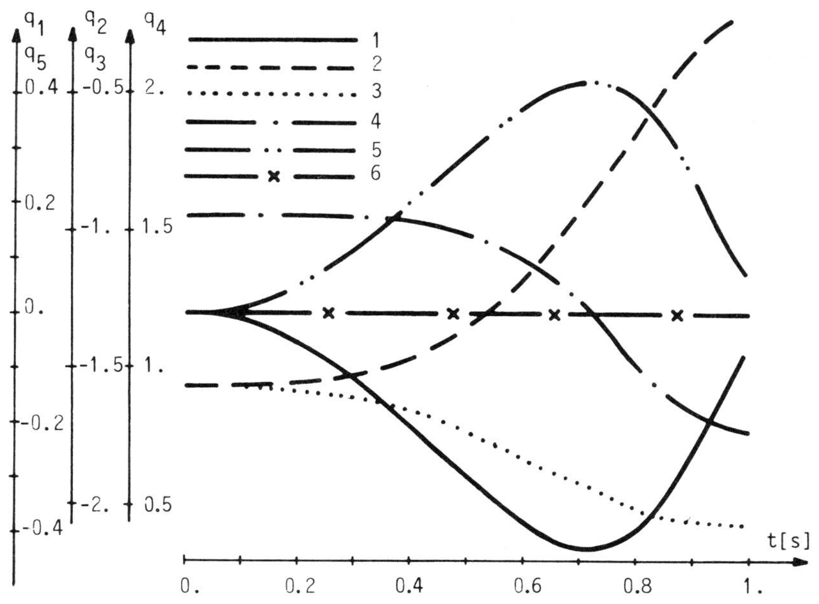

Fig. 3.48. Nominal motion - internal coordinates

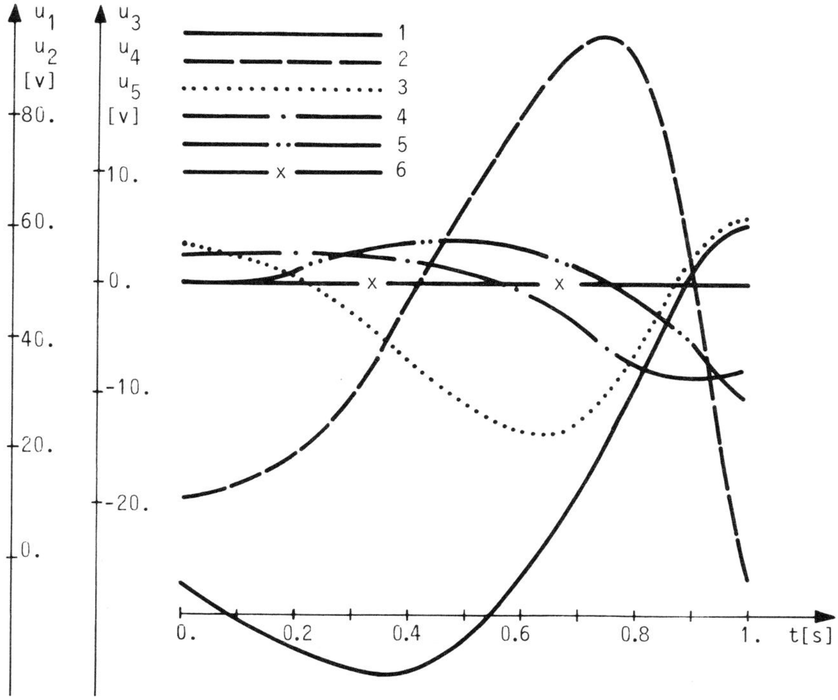

Fig. 3.49. Nominal motion - input control voltages

Let the manipulator start from a perturbed initial position. We first apply the nominal control voltage with no feedback. The sampling period is 10ms and the integration step is 1 ms. Figs. 3.50, 3.51, 3.52 present the results of such simulation. Fig. 3.50 shows the perturbation in initial position and the deviations of internal coordinates from their nominal values. Such perturbed motion produces the error in drawing the circle (Fig. 3.51). Finally, Fig. 3.52 shows the nominal value of surface reaction (i.e. the force produced upon the surface) and its real value obtained by simulation.

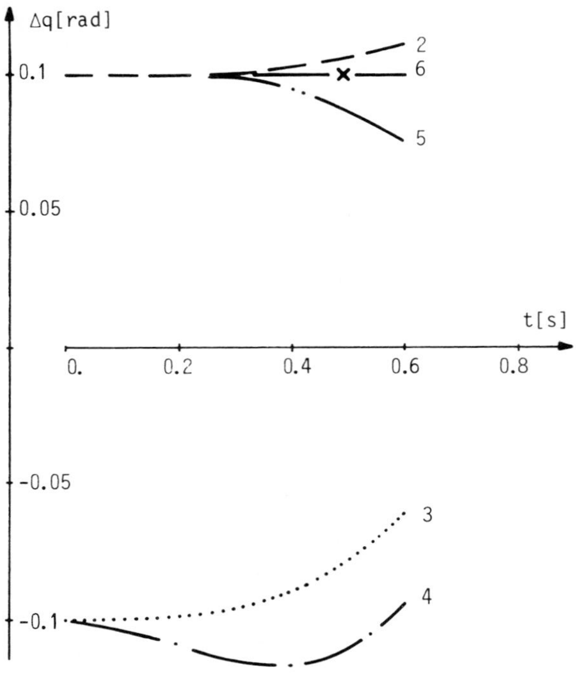

Fig. 3.50. Simulation - deviations from nominal motion

Now, we make a new simulation applying the nominal control and the local control (position feedback and velocity feedback). The results are presented in Figs. 3.53, 3.54. Fig. 3.53 shows the deviations form nominal motion. We see that stability is obtained. Tracking the circle trajectory is presented in Fig. 3.54. Let it be stressed that an efficient control of force S requires a force feedback but it is not the topic of this book.

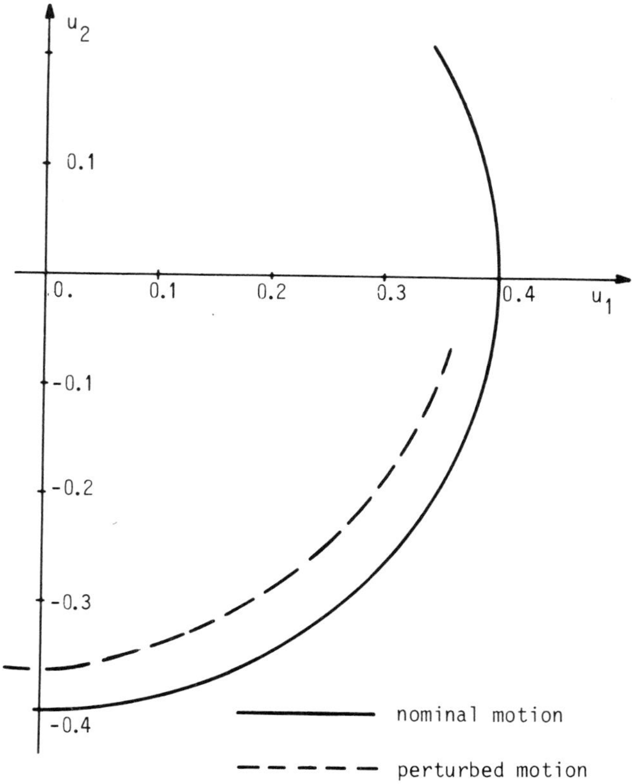

Fig. 3.51. Error in drawing the circle

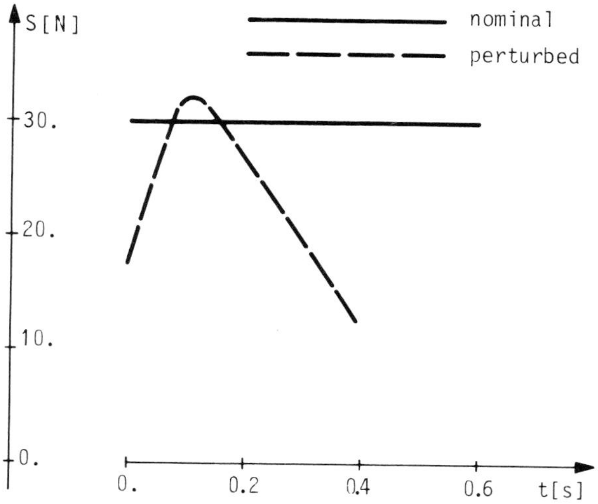

Fig. 3.52. Reaction of the surface (i.e. the force produced upon the surface)

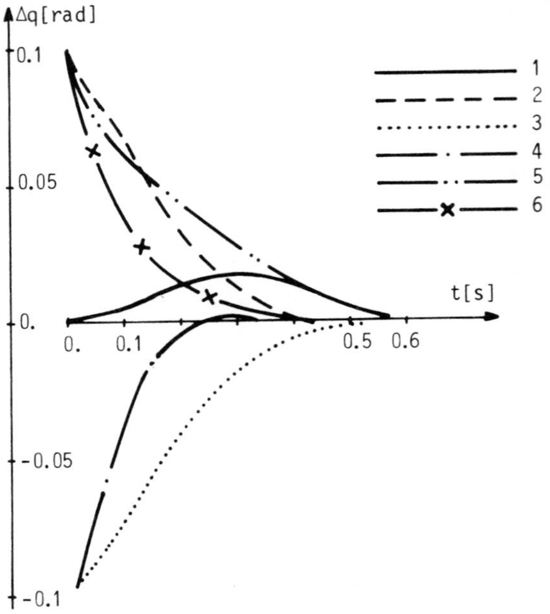

Fig. 3.53. Deviations from nominal motion

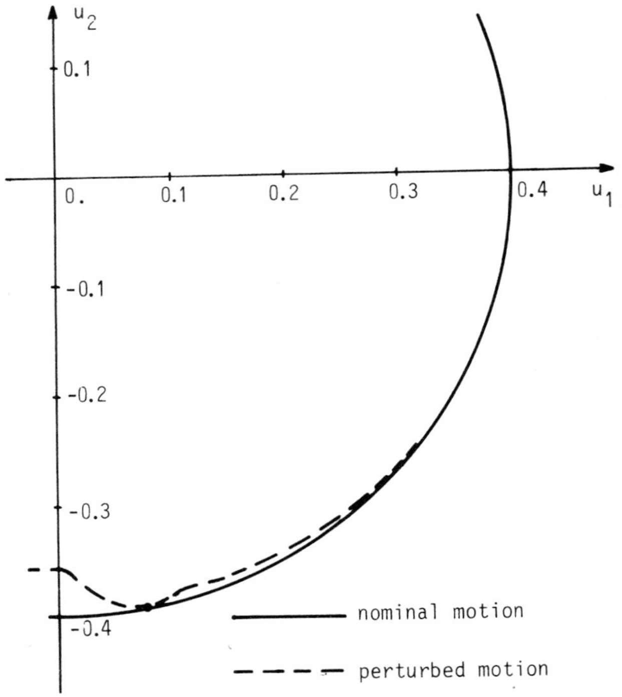

Fig. 3.54. Tracking the circular trajectory

Example 2

We consider again the manipulator from Example 1 (data given in Fig. 3.46). This time the manipulator carries a 5 kg mass working object. The object is cylindrical and should be inserted into a cylindrical hole (Fig. 3.55). The hole is moving with a constant acceleration $a = 2$ m/s^2, thus a nonstationary constraint is considered. It is assumed that there is no friction. The manipulation task consists in moving the object into the hole for 0.3 m in the time interval T = 0.4 s, keeping all the time the initial orientation of the object. The manipulator starts from the resting position shown in Fig. 3.55. Let the relative coordinate u_1 (insertion coordinate) change with triangular velocity profile. Another relative coordinate (u_2) is kept constant.

We consider the motion without impact i.e. we consider the motion when the insertion has already begun. The theory explained in Para. 3.4.7. is used.

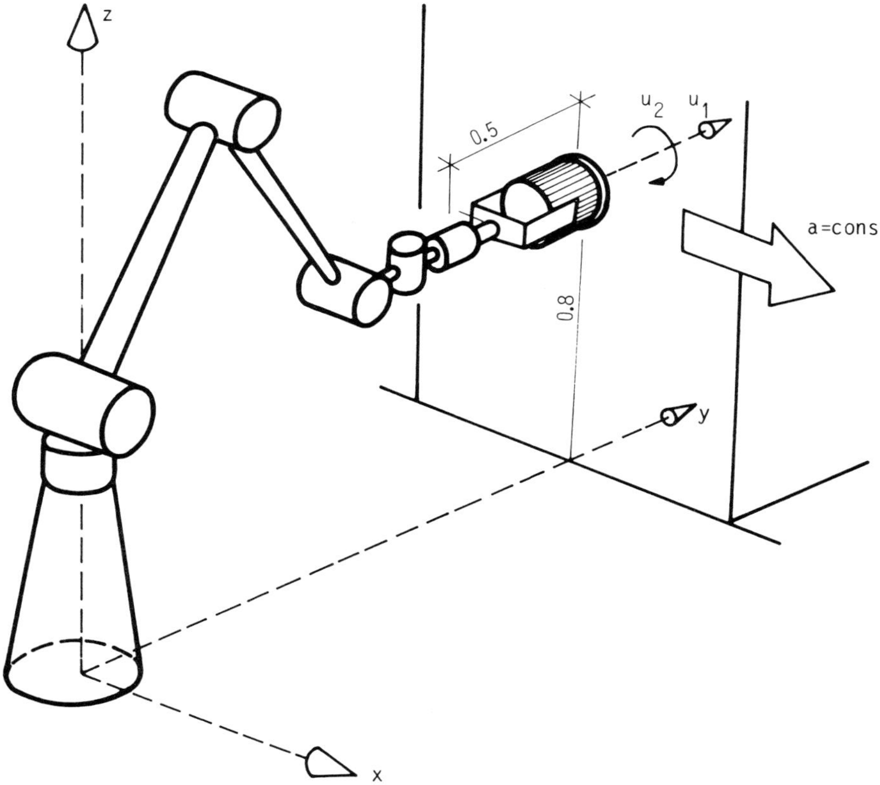

Fig. 3.55. Arthropoid manipulator in a peg-in-hole task

First, we calculate the nominal dynamics, and then we make a simulation. In the case of simulation the manipulator starts from a perturbed initial position but still satisfying the hole constraint. The first simulation is made applying the nominal control voltage with no feedback. The sampling period is 10 ms and the integration step is 1 ms. Fig. 3.56 shows the perturbation in initial position and the deviations of internal coordinates from their nominal values. Fig. 3.57 shows the same but for the case when local feedback is introduced. Finally, Fig. 3.58 shows the insertion i.e. the relative coordinate u_1 (insertion coordinate). There are three cases: nominal, perturbed with no feedback, and perturbed with a local feedback. Let it be noted that the local feedback (position and velocity feedback) can ensure good trajectory tracking but an efficient control of reactions (reaction force and reaction moment of the hole) cannot be achieved without a force feedback.

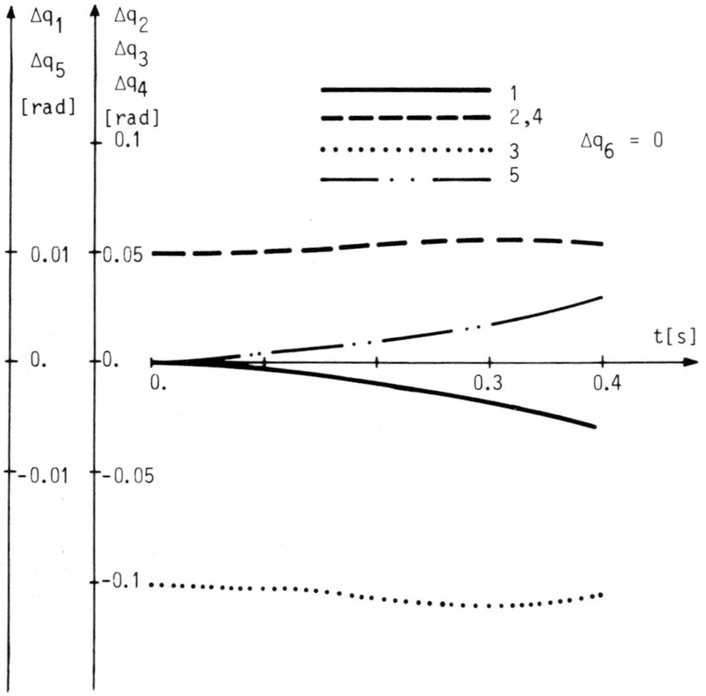

Fig. 3.56. Perturbed motion with no feedback: deviations from nominal motion

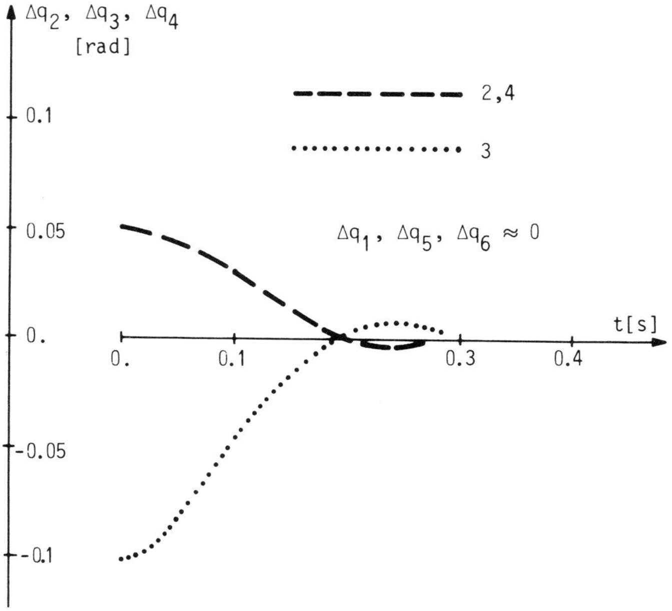

Fig. 3.57. Perturbed motion with local feedback: deviations from nominal motion

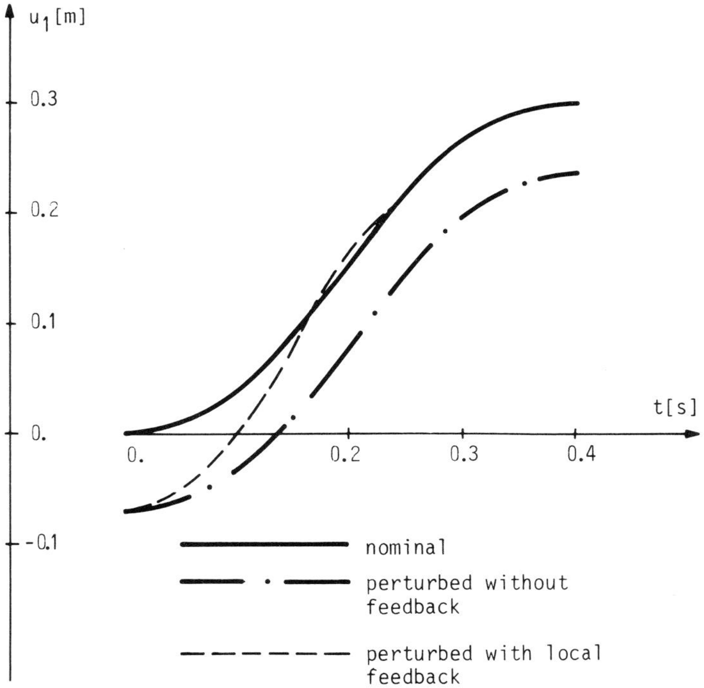

Fig. 3.58. Insertion relative coordinate

References

[1] Wittenburg J., Dynamics of Systems of Rigid Bodies, Stuttgart, series "Leitfäden der angewandten Mathematic und Mechanik (LAMM)", Vol. 33, 1977.

[2] Wittenburg J., "Proceedings of the NATO Advanced Study Institute on Computer Aided Analysis and Optimization of Mechanical System Dynamics", editor: F.J. Haug, Springer Berlin - Heidelberg - New York, 1984.

[3] Wolz U., "MESA VERDE - Preliminary Manual", Institutsbericht 84/1, Institut für Mechanik, Universität Karlsruhe, 1984.

[4] Popov E.P. Vereschagin A.F., Zenkevich S.A., Manipulation Robots: Dynamics and Algorithms, (in Russian), "Nauka", Moscow, 1978.

[5] Lilov L., Loren M., "Dynamic Analysis of Multirigid-Body Systems Based on the Gauss Principle", ZAMMG2, 1982.

[6] Lilov L., "Structure, Kinematics, and Dynamics of Multibody Systems", (in Russian), Advances in Mechanics, No. 1-2, 1983.

[7] Schiehlen W., "Reibungsbehaftete Bindungen in Mehrkorpersustemen", Ingenieur-Archiv 53, Springer-Verlag Berlin - Heidelberg - New York, 1983.

[8] Vukobratović M., Potkonjak V., Scientific Fundamentals of Robotics, Vol. 1: Dynamics of Manipulation Robots, Springer-Verlag, Berlin - Heidelberg - New York, 1982.

[[9] Vukobratović M., Stokić D., Scientific Fundamentals of Robotics, Vol. 2: Control of Manipulation Robots, Springer-Verlag, Berlin - Heidelberg - New York, 1982.

[10] Gottfried B.S., Weisman J., Introduction to Optimization Theory, Prentice-Hall, 1973.

[11] Chumenko V.N., Yuschenko A.S., "Impact Effect upon Manipulation Robot Mechanism", (in Russian), Technical Cybernetics, No. 4, 1981.

Chapter 4: Computer-Aided Design of Manipulation Robots

Introduction

The main idea of this book is to show the possibilities of using computers in the process of manipulator design and to derive the corresponding software. We have decided to study this field for two main reasons. The first reason is the fact that computers offer the possibility of automating the design procedure to a certain extent. The second, and more important, reason lies in our intention to design manipulators on the basis of their exact dynamic characteristics. Any calculation which uses the exact dynamic model is not possible if it has to be done by hand. The dynamic equations are so complex that even the setting of model has to be performed by using a computer. For this purpose we have derived a computer-aided method which sets and solves the exact dynamic model. This method has represented a basis for developing a computer procedure for the complete dynamic analysis of manipulator motion. Chapter 2 and Chapter 3 dealt with this dynamic analysis algorithm which makes possible the calculation of all relevant dynamic characteristics of manipulator work. This calculation can be done in advance i.e. before the device is built. The algorithm also includes the tests of some important dynamic characteristics. We assume that the approach using the exact dynamic models leads towards a more advanced design and devices with better characteristics in work.

An important characteristics is the elastic deformation of manipulator segments. One practical approach is discussed in 2.5.4 and Vol. 7 of this series will probably be devoted to this problem. Each method which calculates the elastic deformations and other characteristics of flexible mechanisms [2-5] can in principle be used in the process of mechanism design (determination of mechanism dimensions only). The only requirement is that the method is general enough to cover different manipulator mechanisms. It should be pointed out that the design of a manipulator includes not only the design of its mechanism but also the choice of actuators, reduces, etc. Some of these methods are more exact and others are approximative. A very important characteristic of a method is the possibility of using a computer. Some of these methods are suitable for a computer and some other are not. There is also a difference from the point of view of computer time consumption. We have chosen the most suitable method satisfying our requirements and incorporated it into the dynamic analysis algorithm. In this way we have obtained the algorithm which calculates all the relevant dynamic characteristics of manipulator work.

The algorithm for dynamic analysis represents a very useful tool for the manipulator design process. It is clear that for a successful design it is necessary to analyze and examine the dynamic characteristics of various configurations in order to find the one best suited to a particular application. The dynamic analysis algorithm described in this book permits a fast analysis of a great number of various manipulator configurations. We see that the algorithm can be a very useful tool but the design in the way described is still very extensive.

The next step is to find a more systematic way of choosing manipulator parameters by using the dynamic analysis algorithm. The fact that the algorithm includes the tests of some important characteristics allows us to define an interactive procedure for computer-aided design of manipulators. This procedure is based on interactive work with the algorithm and is described in paragraph 4.1. [6].

The next step is the choice of optimal manipulator parameters. The results obtained by using the interactive procedure cannot be considered as completely optimal, since no optimality criterion has been used. If we wish to obtain truly optimal parameters we have to define the optimality criteria. These criteria: speed criterion, energy criterion, and the combined one, are defined in 4.2.1. [2,7] Para. 4.2.2. introduces the constraints and thus completes the formulation of optimization problem.

The optimization procedure for the choice of manipulator segments parameters based on energy criterion is described in 4.3. One-parameter optimization and multiparameter optimization are considered.

The working speed criterion is discussed in 4.4. This discussion is rather theoretical but useful for better understanding of optimal reducer choice in one of the methods proposed.

4.5. presents two methods for motor selection: a simple checking method and a systematic procedure based on the complete analysis of power requirements. Two cases are considered: D.C. electromotors and hydraulic actuators [9].

In 4.6. we present the organization of the program package for computer aided design on the basis of dynamical analysis. All possibilities of such a package are explained.

Each of the design methods proposed (interactive design, determination of dimensions, motor choice) is illustrated by examples. Some if the robots appearing in examples are existing models and some of them are models being under design.

4.1. Interactive Procedure for Computer-Aided Design of Manipulators

The dynamic analysis algorithm provides the possibility of calculating all the relevant dynamic characteristics of manipulator work before the device is built. The, corrections can be made in order to obtain the device with the desired characteristics. We here propose a practical design procedure based on the interactive work with the dynamic analysis algorithm. This procedure is often called the design game. For this procedure, a special communication is programmed [6].

Let us first consider the constraints imposed on a manipulator during its design and the tests included in the dynamic analysis algorithm. We mention here only the main constraints. First, the motors chosen are required to be capable of producing the manipulator work at a given speed. This constraint is taken into account through the tests of actuators: test of torque-speed diagram (P-n), test of power needs, etc. This constraint is called the actuator constraint. Then, there is the requirement that the stresses in manipulator segments do not exceed certain permitted values. This constraint is called the stress constraint and is taken into account through the test of stresses. Finally, there is also the requirement that the elastic deformations of manipulator segments do not exceed some given values. More precisely the errors in position and in orientation due to elastic deformations of segments are required to be smaller than some given values. This constraint is called the elasticity constraint and is taken into account through the tests of elastic deformations.

All the tests mentioned are described in details in chapter 2. The tests of actuators have been considered in paragraphs 2.6.1, 2.6.2; the tests of stresses in 2.5.3, 2.6.3; and the tests of elastic deformations in Para. 2.5.4 and 2.6.3. Each of these tests can be either positive or negative. A negative test means that the corresponding constraint is violated.

Let us now discuss the possibility of correcting a negative test. If some constraint is violated i.e. the corresponding test is negative, there always exist several possibilities for corrections. We can change different parameters of the manipulator or of the manipulation task in order to satisfy the constraint i.e. to make the test positive. Let us consider the main constraints mentioned above.

If the actuator constraint is violated in some joint, we can correct it by:

1) choosing a new actuator;
2) changing the reduction ratio in that joint;
3) reducing the mass of working object;
4) reducing some masses of the manipulator itself;
5) increasing the task execution time (slower manipulator work);
6) other possibilities.

Thus, in the case of actuator constraint violation there are five options for corrections. The first option is quite clear. The second should be explained. Sometimes the actuator is correct but the violation results from an unsuitable reduction ratio. It is then enough to choose a more suitable value of the reduction ratio. The options 3 and 4 are clear since reduced masses require reduced motor torques. The option 5 can easily be understood if we consult Para. 2.6. where the torque-speed motor test was explained. Finally, some other options may be possible in an actual problem.

If the stress constraint is violated i.e. if the stress in some segment exceeds a given permitted value, the violation can be corrected by:

1) increasing the cross-section dimensions of the segment;
2) choosing a new cross-sestion form;
3) reducing the mass of working object;
4) reducing some concentrated masses of manipulator device (motors, reducers, etc.);

5) other possibilities.

We assume that all these options are quite understandable.

If the elasticity constraint is violated it can be corrected by:

1) increasing the cross-section dimensions of manipulator segments (one or several segments);
2) choosing new forms of cross-sections;
3) reducing the masses of working object;
4) reducing some concentrated masses of the device (motors, actuators, etc.);
5) other possibilities.

These options are very similar to the ones mentioned for the case of stress constraint violation. But there is an important difference. The stress constraint depends upon a particular segment. This means that one segment is responsible for the violation, the one in which the permitted stress is exceeded. The corrections usually refer to that segment. In the case of elasticity constraint all segments are responsible for the violation since we consider the position or orientation error at the manipulator tip. Hence, the corrections of segments are more complicated here.

After this analysis of the constraints imposed and the possibilities for correcting the violations we can derive the interactive design procedure. The interactive algorithm starts with some chosen parameters of manipulator. The dynamic analysis is performed including the desired tests. If some test is negative (the corresponding constraint violated), the computer prints the options for corrections. We choose one of the options, then type the new values of the corresponding parameters and start the algorithm again. This is repeated until all tests are positive. Then we check whether the characteristics obtained satisfy all our requirements. For instance, it may happen that all constraints are satisfied (all tests positive) but we wish further reduction of energy consumption. In such cases we introduce some new corrections of the relevant parameters and repeat the whole procedure. This design game algorithm can be represented by the scheme given in Fig. 4.1.

Let us describe an example of the communication between the computer

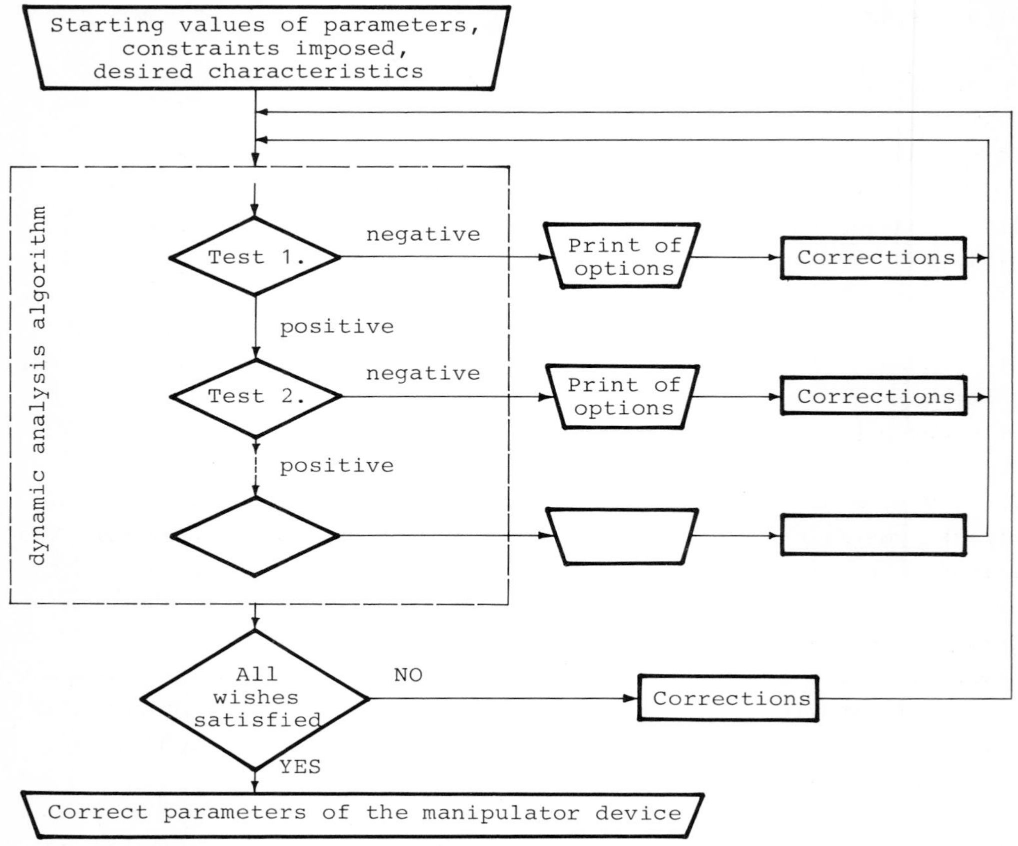

Fig. 4.1. Scheme of the interactive design procedure

algorithm and a user. Suppose that the algorithm has started with some given parameters, and the D.C. motors which are not correct for the manipulator and the task considered. We begin the description with the computer printout:

```
Computer: ***** TIME ITERATION No. 25 ********
          ACTUATOR IN JOINT 3 IS NOT CORRECT;
          TORQUE-SPEED TEST IS NEGATIVE;

          OPTIONS:

          1 - CHOOSE A NEW ACTUATOR
          2 - CHANGE THE REDUCTION RATIO
          3 - REDUCE THE WORKING OBJECT MASS
          4 - REDUCE SOME OTHER MASSES
```

5 - INCREASE THE EXECUTION TIME
6 - OTHER POSSIBILITIES

TYPE THE NUMBER OF THE OPTION CHOSEN:

User: 1

Computer: TYPE THE NEW ACTUATOR CHARACTERISTICS
(STALL TORQUE[N]; NO LOAD SPEED[r.p.m]; POWER[W])

User: 12. ; 3000. ; 2000.

With this new actuator the tests are positive i.e. the actuator suits its application.

This was only a fictitious example demonstrating the communication with the design game algorithm. Here is a detailed example which illustrates the procedure of manipulator design using the game algorithm.

Example. We consider a problem of designing a manipulator VE-2 (Fig. 4.2.a) having six degrees of freedom. It is intended for spot-welding tasks and some assembly tasks. Its kinematical scheme is arthropoid and shown in Fig. 4.2.b.

Some manipulator parameters are adopted and given in Fig. 4.2.b,c. The other parameters have to be determined. Let us first consider the question of segments geometry. Segments 2 and 3 are in the form of rectangular tubes made of light alloy AℓMg3. The cross sections (Fig. 4.3.) are defined by H_{x2}, H_{y2} and H_{x3}, H_{y3}, so these parameters have to be determined.

The actuators which drive the joints S_2, S_3 are placed in manipulator base. The torques are transported to the corresponding joints where there are Harmonic Drive reducers. The reducers in joints S_2, S_3 have the masses of about 7kg and the mechanical efficiencies $\eta = 0.8$. It is necessary to choose the actuators which will drive the joints S_2, S_3. The manipulator is equipped with spring compensators for the joint S_2.

Thus, the parameters to be determined are: cross section dimensions of segments 2 and 3, and parameters of actuators S_2 and S_3, and finally, the reduction ratio.

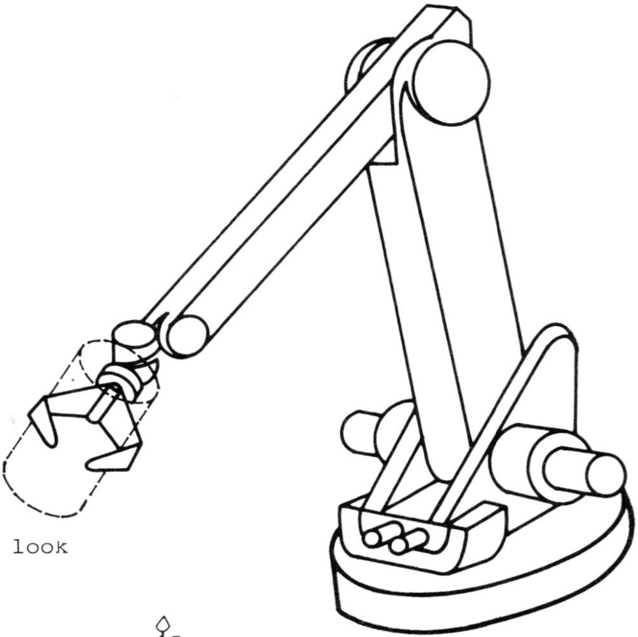

(a) External look

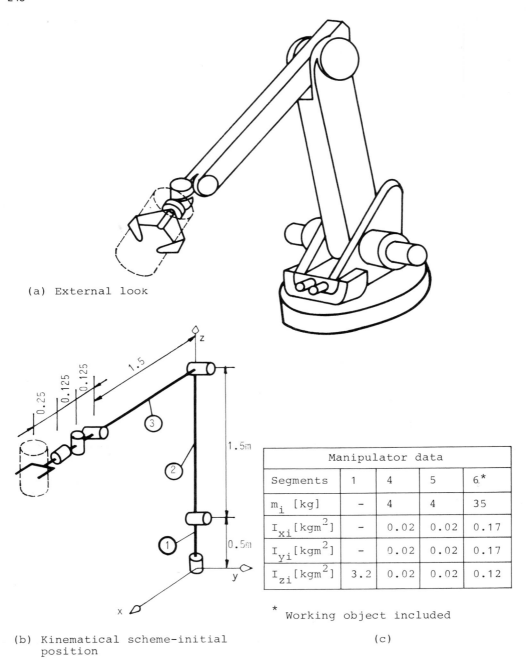

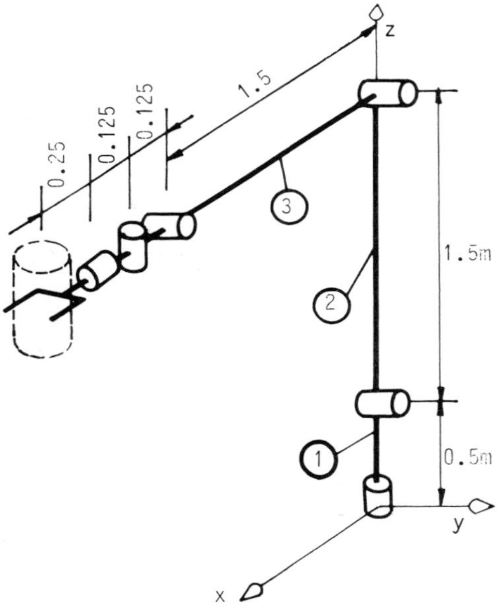

(b) Kinematical scheme-initial position

Manipulator data				
Segments	1	4	5	6*
m_i [kg]	-	4	4	35
I_{xi}[kgm^2]	-	0.02	0.02	0.17
I_{yi}[kgm^2]	-	0.02	0.02	0.17
I_{zi}[kgm^2]	3.2	0.02	0.02	0.12

* Working object included

(c)

Fig. 4.2. Arthropoid manipulator VE-2

The manipulator is tested on the assembly task shown in Fig. 4.4. The object should be moved along the trajectory $A_oA_1A_2A_3$. The velocity profile on each straight-line part of the trajectory is triangular. The changes in orientation are also shown in Fig. 4.4. The execution time is

$$T = T^1 + T^2 + T^3$$

$$T^1 = T(A_o \to A_1) = 2s$$

$$T^2 = T(A_1 \to A_2) = 1s$$

$$T^3 = T(A_2 \to A_3) = 1s$$

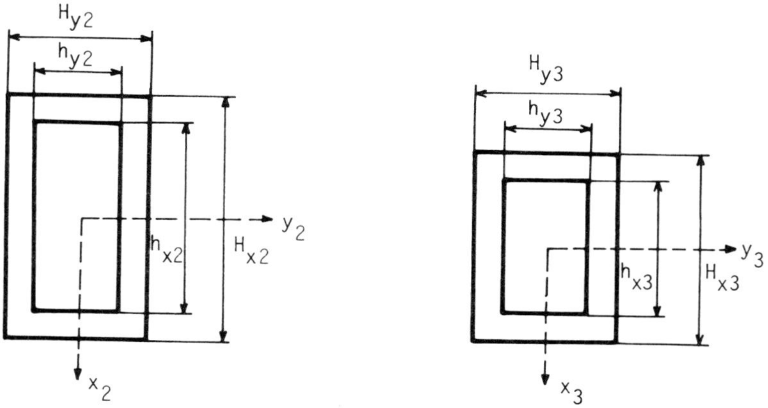

$$\frac{h_{y2}}{H_{y2}} = 0.9 \qquad \frac{h_{x2}}{H_{x2}} = 0.85 \qquad \frac{h_{y3}}{H_{y3}} = 0.9 \qquad \frac{h_{x3}}{H_{x3}} = 0.85$$

Fig. 4.3. The cross sections of segments

The manipulator starting position is given in Fig. 4.2.b.

The constraint is that the position error due to elastic deformations be smaller than $u_p = 0.002$ m.

The design (i.e. determination of parameters) is carried out by using the design game procedure. The procedure started with the initial cross section dimensions: $H_{x2} = 0.25$m, $H_{y2} = 0.13$m, $H_{x3} = 0.16$m, $H_{y3} = 0.08$m. We first choose the 2kW D.C. motors as the actuators for the joints S_2, S_3. Catalog parameters (stall torque P^m_M and no-load rotation speed n^m_M) were: $P^m_{2M} = 10$Nm, $n^m_{2M} = 2500$, $P^m_{3M} = 10$ Nm, $n^m_{3M} = 2500$. The reduction ratios were $N_2 = 150$ and $N_3 = 150$.

The procedure started with these initial values. The elastic deformation test was negative and the algorithm printed the options. We changed the cross section dimensions of the segment 2. into the values: $H_{x2} = 0.26$m, $H_{y2} = 0.14$m. The test was negative again. Then we changed the dimensions of segment 3.: $H_{x3} = 0.17$, $H_{y3} = 0.09$. Now, the elastic deformations test was positive, but the test of actuators was negative. We chose a new motor for the joint S_2 having the catalog parameters: $P^m_{2M} = 12$Nm, $n^m_{2M} = 3000$. The test of motor S_2 was negative again, but since it was near to being O.K. we decided not to change the motor again but to change only the reduction ratio N_2 in that joint; we adopted the new value $N_2 = 200$. In that way the test of S_2 was positive but the test of actuator S_3 was still negative. The results obtained allowed us to conclude that it would be enough if we changed the reduction ratio N_3. We chose the new value $N_3=200$ and the test of S_3 was positive.

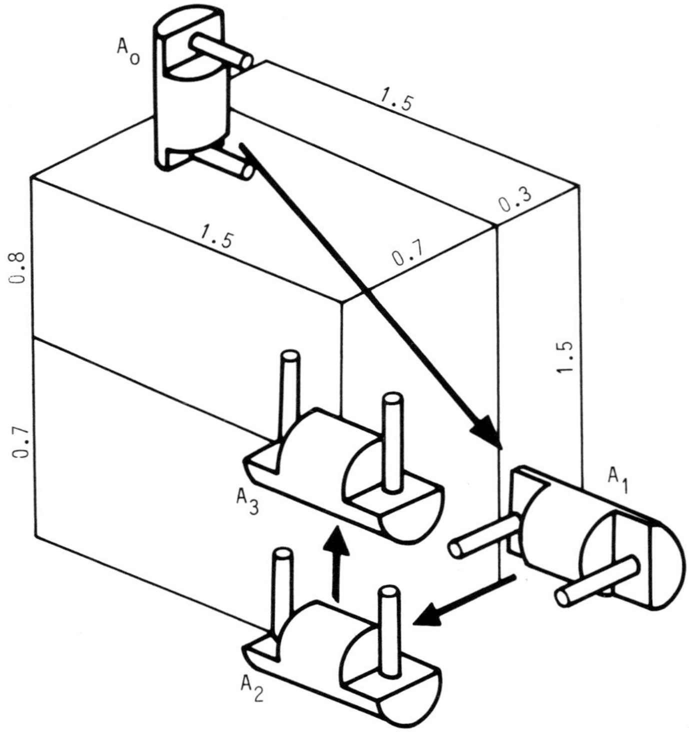

Fig. 4.4. Manipulation task

The game was finished and thus we obtained the manipulator parameters satisfying the requirements imposed. The cross section dimensions of segments are: $H_{x2} = 0.26m$, $H_{y2} = 0.14$, $H_{x3} = 0.17$, $H_{y3} = 0.09$. The actuators which drives the joint S_2 has the catalog parameters: stall torque $P^m_{2M} = 12Nm$ and no-load speed $n^m_{2M} = 3000$. The reduction ratio in the joint S_2 is $N_2 = 200$. For the joint S_3 the motor parameters and the reduction ratio are: $P^m_{3M} = 10$, $n^m_{3M} = 2500$, $N_3 = 200$.

This simplified example illustrates the proposed interactive design procedure.

4.2. Optimal Choice of Manipulator Parameters

4.2.1. Optimality criteria

It has already been said that the manipulator parameters obtained by a nonsystematic use of dynamic analysis algorithm or even by the interactive procedure cannot be considered as fully optimal, since no optimality criterion was used. In fact, this statement is not completely true. Let us explain this. In any advanced design we always keep in mind a kind of optimality criterion. For instance, it is our intention to design a device which consumes as little energy as possible. There are also some other criteria we are taking care of: operation speed etc. So, the optimality criterion is perhaps not explicitly defined but it still exists. But, there is no systematic way of defining how the parameters should be changed in order to obtain optimal ones with respect to the criterion adopted.

Let us first define the optimality criteria. From the standpoint of the efficiency of industrial manipulators in practical use, two aspects can be distinguished: operation speed and energy consumption. Hence three criteria are defined for the evaluation and optimization of manipulators:

(a) Operation speed criterion (or time criterion)
(b) Energy consumption criterion
(c) Combined criterion.

Let us explain the criteria in more detail.

(a) Speed criterion. Let T denote the execution time of the set manipulation task. Optimization with respect to the speed criterion is reduced to the determination of manipulator configuration which permits the highest work speed i.e., the shortest time T. Thus, the minimization of T is performed.

(b) Energy criterion. Let E denote the total energy the manipulator consumes in performing the task. Optimization means the minimization of E i.e. finding the configuration which ensures minimal consumption of energy.

(c) Combined criterion. It is a kind of an average of the previous two criteria, so it takes into account both the operation speed and energy consumption.

These criteria are used as the criteria for minimization in the choice of optimal parameters during design. But they can also be used for the evaluation and comparison of various manipulators offered by the manufacturers and which can be found at the market. The criteria (a) and (b) will be used in the optimization procedures, while the combined criterion (c) will be used for the evaluation and comparison of the devices offered.

The criteria mentioned serve for the comparison of both manipulators and functional movements and tasks, because each task in practice can be performed by means of mutually different movements, different velocity profiles, with different execution times.

It should be mentioned that it is also possible to evaluate or choose the manipulation systems according to supplementary criteria associated with the realization of control algorithms. Since this is out of the scope of this book, we shall discuss only some dominant influences in the choice of manipulators based on "mechanical" criteria.

4.2.2. Constraints

It has been stated that the choice of optimal parameters involves the minimization of the criterion adopted. Besides the criterion, the constraints should be defined in order to obtain the complete optimization problem. The constraints relevant for a design process have already been discussed in Para. 2.6. dedicated to the tests of dynamic characteristics. The same question was touched in 4.1. where the problem of avoiding the constraint violation was discussed. Let us summarize the discussion on constraints. The first constraint is the so-called constraint of reachability. It is the request that the manipulator can satisfy the required geometry of motion. This problem is usually considered at the very beginning of design process. This constraint usually determines some parameters (for instance, the lengths of manipulator segments, types of joints, etc.). So, we later assume that the constraint is satisfied and do not include it in the optimization procedure.

Thus, from a practical point of view the three main constraints can be defined as:

(i) Constraint of drives. This is the requirement that the driving actuators in manipulator joints can produce the driving forces and torques necessary for the manipulation task given.

(ii) Constraint of stresses. This is the requirement that the stresses in manipulator segments do not exceed certain permitted values.

(iii) Constraint of elasticity. This is the requirement that the errors in position and orientation due to elastic deformation of manipulator segments do not exceed certain permitted values.

The constraint (i) is of main importance if we are choosing the actuators. The constraints (ii) and (iii) are of special importance if some parameters of the manipulator segments (for instance the cross-section dimensions) are being chosen.

Now, the optimization consists of the minimization of the relevant criterion along with the satisfaction of the constraints imposed.

4.3. The Choice of Manipulator Segments Parameters based on the Energy Criterion

In this paragraph we describe a systematic procedure for the determination of the optimal parameters of manipulator segments based on energy consumption criterion. First, it should be said that the kinematic scheme of the manipulator mechanism is chosen on the basis of its purpose. We may try several schemes and then choose the most suitable one. This choice is not a part of the optimization procedure; it means that each scheme is optimized separately and they are then compared. Thus, in each optimization the kinematic scheme is considered to be known.

It should also be said that we do not optimize the parameters of all segments. Some segments are completely determined by the constructive solutions adopted. We shall explain this fact in more detail in the example, but let us say here that, for instance, the actuators chosen determine completely the segments which form the gripper (the last three segments if a six d.o.f. manipulator is considered). We may say that there usually exist one or two segments which should be optimized. If a cylindrical (Fig. 2.44) or spherical (Fig. 4.6) manipulator is in question, there is usually one main segment forming the manipulator arm and this is the one to be optimized. With anthropomorphic (Fig. 2.53) or arthropoid (Fig. 4.2) manipulators there are two main segments which form the manipulator arm and which should be optimized. We can make some further simplifications in order to reduce the number of parameters to be optimized. The lengths of segments can be considered as known since they directly follow from the reachability conditions imposed. The forms of segment cross-section (not the cross-section dimensions but only its general form) can be different. Several forms of cross-section can be checked, but in one optimization procedure this is considered to be known. In this way the choice of optimal parameters of manipulator segments reduces to the determination of the optimal dimensions of cross-sections. Usually there are several dimensions which define a cross-section. In the case of a circular cross-section (Fig. 4.5a) it is enough to prescribe one dimension: the radius. If the cross-section is in the form of a circular tube (Fig. 4.5b) then there are two dimensions: the inside and the outside radius. In the case of a rectangular tube (Fig. 4.5c) there are four dimensions. Let us also point out that there often is more than one segment to be optimized and these segments can have different cross-section forms. Hence, in general, a multiparameter optimization follows.

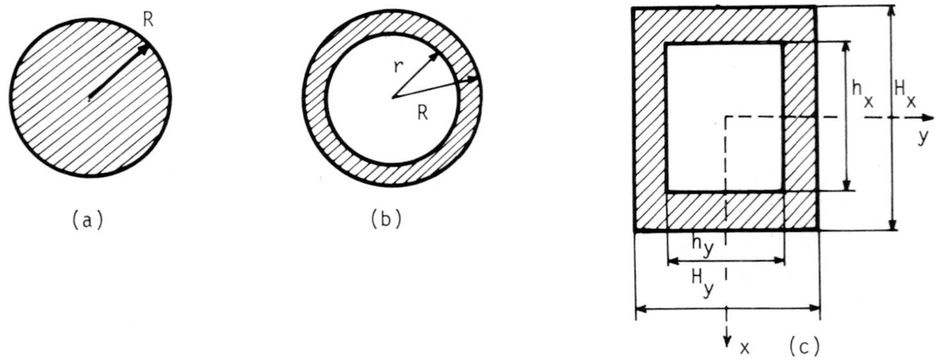

Fig. 4.5. Some possible cross-sections of manipulator segments

If the energy consumption criterion is considered it is then clear that the masses of segments should be reduced as much as possible. Hence the cross-section dimensions should be reduced too. But an answer as to the dimensions which should be adopted in order to obtain the minimal energy consumption along with the satisfaction of the constraints imposed cannot be given before the completion of the optimization procedure. It is evident that in this optimization the elasticity constraint and the constraint of stresses have to be considered only.

We now discuss the problem of multiparameter optimization. Let us consider one point in parameter space. In order to compute the value of criterion function (energy consumption) and to check the constraints it is necessary to carry out the whole dynamic analysis procedure. Since each optimization algorithm requires the calculation of criterion function and the checking of constraints in many points of parameter space, the optimization may become very time-consuming. So, our intention is to reduce the number of parameters.

We shall discuss this problem in more detail.

4.3.1. One-parameter optimization

Sometimes the number of parameters to be optimized can be reduced even to one. Let us explain this.

From the engineering point of view we can reduce the number of independent parameters by introducing a constant ratio between them. For instance, in the case of circular tube cross section we may adopt constant ratio between the inside and the outside radius $\psi = r/R$. In this way we reduce the number of parameters which should be optimized: it is enough to determine R since $r = \psi R$. If we consider a manipulator with one main segment forming the arm we then face a one-parameter-optimization problem: the determination of R. We shall illustrate this by an example. The other examples will show how some more complex configurations can be optimized via one parameter optimization.

Before giving the examples let us describe the optimization procedure. In the case of one parameter the procedure is quite simple. The parameter should be reduced as much as possible. The first possibility is to decrease the parameter successively until some constraint is violated. The last permissible value is the optimal one. In order to find this optimal value in shorter time we may use some of one-dimensional-search techniques such as the binary search or the golden ratio search [10].

Example 1.

Let us consider a spherical manipulator VE-4 (Fig. 4.6). The segments 1, 2, 4 and 5 are completely determined by the constructive solutions adopted. The third segment has the form of a cylindrical tube made of steel ($\rho = 7{,}85 \cdot 10^3$ kg/m^3, $E = 2{,}1 \cdot 10^{11}$ N/m^2, $\sigma_p = 5 \cdot 10^8$ N/m^2, safety coefficient $\nu = 3$). This tube can be pulled out of the second segment up to 0.8 m. This results in the segment being 1.2 meters long. Thus the cross-section dimensions of this segment remain to be chosen. The cylindrical tube cross-section is defined by the outside radius (R) and the inside radius (r), as shown in Fig. 4.5b. In order to reduce the number of independent parameters we adopt the constant ratio $\psi = r/R = 0.85$. Thus there remains only one parameter to be optimized. It is the outside radius R.

Let us define the manipulation task. It is shown in Fig. 4.7. The working object has to be moved along the trajectory $A_oA_1A_2$ with a triangular velocity profile on each straight-line part. We adopt: $T^2 = 2T^1$ where T^1 is the execution time $T(A_o \rightarrow A_1)$ and $T^2 = T(A_1 \rightarrow A_2)$. We perform the optimization with several different values of the total execution time ($T = T^1 + T^2$) in order to check whether the results depend on this execution time. Each optimization is performed by using the binary search procedure. The results obtained are shown in Fig. 4.8. The binary search has been carried out for eight values of T: 4.5s, 3s, 2.4s, 1.8s, 1.5s, 1.2s, 0.9s, 0.6s. We first introduced only the constraint of elasticity (iii) by the request that the position error due to elastic deformations be less than 0.0015m. In this way we obtained the curve (iii) showing how the optimal radius R^{opt} changed with the execution time. Later we used only the constraint of stresses (ii) and the curve (ii) was obtained. Which of these two constraints is relevant for an example depends on the given values of permitted stress, safety coefficient, and the permitted elastic deformation. In this example, it is clear from Fig. 4.8. that the constraint (iii) is relevant. In any case, we can see that the results for the optimal radius (R^{opt}) depend on the execution time only to a small extent. In fact the influence of the execution time is stronger for larger operation speeds and it is considerable only in the domain of very fast motions. But, such speeds

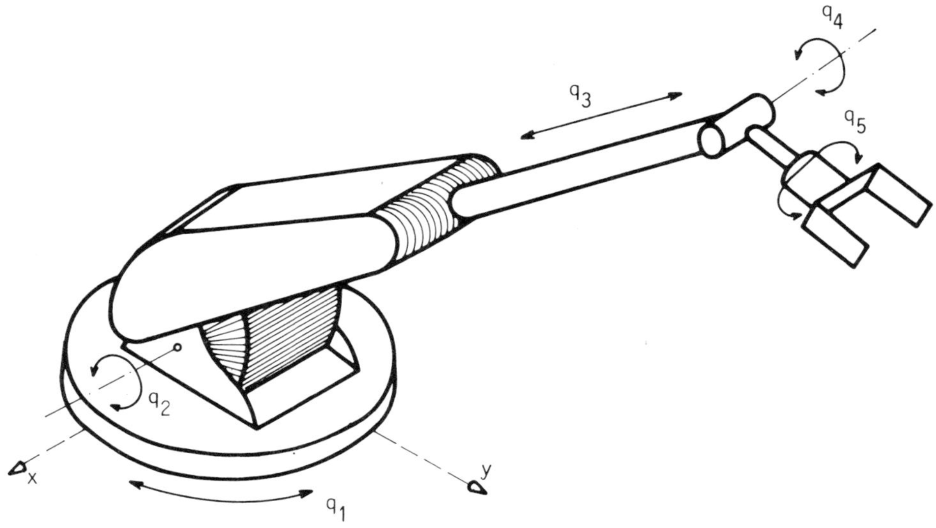

(a) External look of a spherical manipulator with four rotational and one linear degrees of freedom

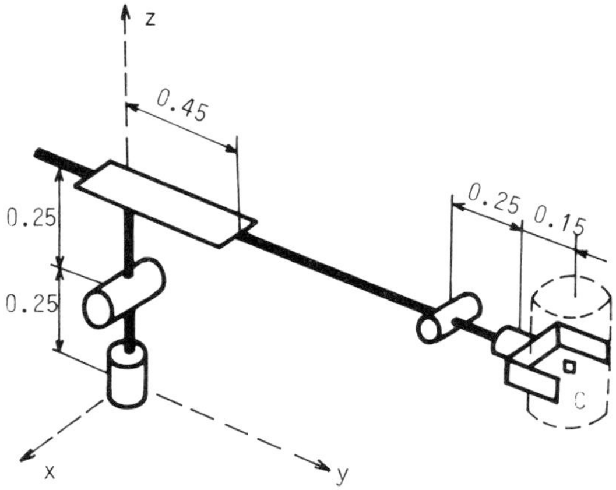

(b) kinematic scheme - initial position

Manipulator data				
SEGMENTS	1	2	4	5*
m_i [kg]	-	27.	3	15.
I_{xi}[kgm^2]	-	2.9	0.02	0.04
I_{yi}[kgm^2]	-	-	0.004	0.04
I_{zi}[kgm^2]	0.9	3.2	0.02	0.04

* working object included

(c)

Fig. 4.6. Spherical manipulator VE-4

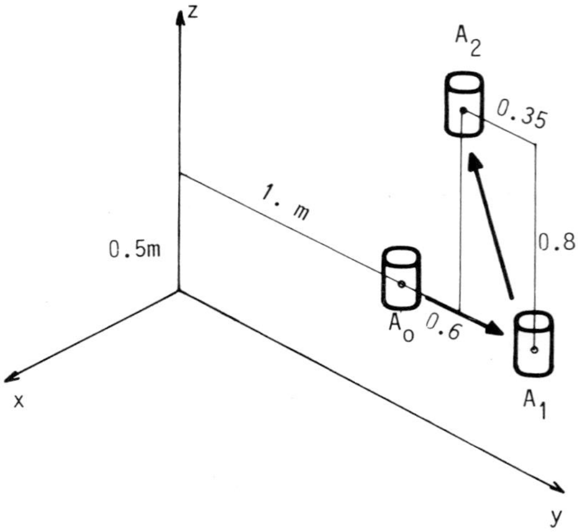

Fig. 4.7. Scheme of manipulation task

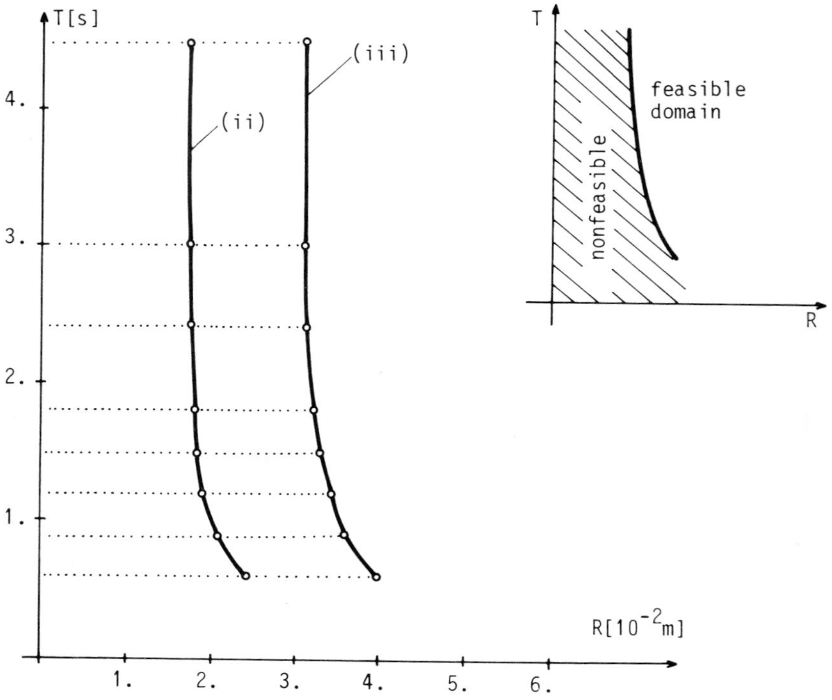

Fig. 4.8. Optimization results: R^{opt} depending on T

are usually too high for practical application.

Hence, we conclude that it is enough to perform the optimization with one value of execution time, the shortest one. That time corresponds to the largest operation speed which can be required from the device in its practical operation.

Example 2.

Let us again consider the arthropoid manipulator VE-2 (Fig. 4.2). It has been described in the example in 4.1. The problem is to choose the values of cross-section dimensions of segments 2 and 3 (Fig. 4.3). There are eight parameters defining these two cross-sections: h_{x2}, H_{x2}, h_{y2}, H_{y2}, h_{x3}, H_{x3}, h_{y3}, H_{y3}. In order to reduce the number of parameters to be optimized we first introduce the constant ratios from Fig. 4.3. i.e.

$$\frac{h_{y2}}{H_{y2}} = 0.9, \quad \frac{h_{x2}}{H_{x2}} = 0.85, \quad \frac{h_{y3}}{H_{y3}} = 0.9, \quad \frac{h_{x3}}{H_{x3}} = 0.85 \qquad (4.3.1)$$

But, there still remain four independent parameters. If we adopt

$$\frac{H_{x2}}{H_{y2}} = 1.5, \quad \frac{H_{x3}}{H_{y3}} = 1.5 \qquad (4.3.2)$$

then there are only two independent parameters: H_{x2} and H_{x3}. If our intention is to reduce the problem to one-parameter optimization, one possibility is to adopt $H_{x2} = H_{x3}$ which means that the two segments (2 and 3) are equal. We now perform the optimization for the reamining one independent parameter H_{x2}.

The manipulation task considered is described in the example in 4.1. (Fig. 4.4).

The stress constraint and the constraint of elastic deformation are imposed. We require that the position error due to elastic deformation be smaller than 0.002m.

The optimization by using the binary search gives the optimal value of the parameter H_{x2}: $H_{x2}^{opt} = 0.22$m. The other dimensions follow from (4.3.1), (4.3.2): $H_{x2} = H_{x3} = 0.22$m, $H_{y2} = H_{y3} = 0.147$, $h_{x2} = h_{x3} = 0.187$, $h_{y2} = h_{y3} = 0.132$. For these dimensions the energy consumption is E = 2509 J.

We suppose that the assumption about equal segments (2 and 3) was not quite suitable for it is probably far from the best ratio between two segments. So, let us consider another possibility. We adopt (4.3.1) and (4.3.2) and introduce the ratio: $H_{x2} = 1.4\ H_{x3}$. This ratio (1.4) follows from experience or some rough estimation. There is again the one-parameter optimization problem with the independent parameter H_{x2}. The binary search gives the optimal value $H_{x2}^{opt} = 0.235$m. For the segment 3 it is now $H_{x3} = 0.168$. The other dimensions follow from (4.3.1), (4.3.2): $H_{y2} = 0.157$, $h_{x2} = 0.2$, $h_{y2} = 0.141$, $H_{y3} = 0.112$, $h_{x3} = 0.143$, $h_{y3} = 0.101$. For these dimensions the energy consumption is E = 2217 J and it is less than in the previous case.

4.3.2. Two-parameter and multi-parameter optimization

It has already been said that from the engineering point of view the number of independent parameters can be reduced. It has been shown through examples how this number can be reduced even to one. For this one-parameter optimization we can use some of the known one-dimensional search techniques. In our optimization algorithm the binary search is used. The results obtained in this way are not exactly optimal since for the true optimum it is necessary to carry out a multi-parameter optimization. Let us discuss the possibility of multi-parameter optimization. We first consider the case with two independent parameters i.e. the two-parameter optimization. This problem is very complicated because there is no explicit function of the criterion and no explicit expressions for the constraints. Both the value of the criterion and the answer as to whether a point in parameter space is feasible or not follow from the dynamic analysis algorithm. This fact restricts to a great extent the possibility of selecting the optimization method among those given in literature. Here we use one variant of the feasible directions method [10]. This method is almost the only one that suits the problem considered.

Let us describe the optimization procedure. It is illustrated in Fig. 4.9. The procedure starts at a feasible point A. Four probes are made around the point A. We locate the improved feasible point B and accept it. We proceed in the same manner until we reach point D. At D, no probe produces an improved feasible point. We then choose a new point E by interpolating between the best feasible point H and the best non-feasible point G. Such a procedure leads us towards the optimum (quadratic point in Fig. 4.9).

This procedure can also be used for optimization when there are more than two independent parameters. One should take care of the fact that if the number of parameters increases the procedure may become very time-consuming. Hence, we suggest two or three independent parameters. We think that in almost all problems we are interested in, the number of independent parameters can be reduced to two or three. Let us see an example.

Example. We consider again the arthropoid manipulator VE-2 (Fig. 4.2). It was described in the example in 4.1. The problem is to choose the values of cross-section dimensions of segments 2 and 3. (Fig. 4.3). There are eight parameters defining these two cross-sections. In the example 2 in 4.3.1. the number of independent parameters was reduced by introducing the constant ratios (4.3.1) and (4.3.2). In that way there remain two independent parameters H_{x2} and H_{x3}. In 4.3.1. the problem was further reduced to one-parameter optimization. It was first done by considering the two segments (2 and 3) to be equal ($H_{x2} = H_{x3}$). After that, the same was done in the other way by introducing the constant ratio between the two segments ($H_{x2} = 1.4H_{x3}$). Here, we have a rather

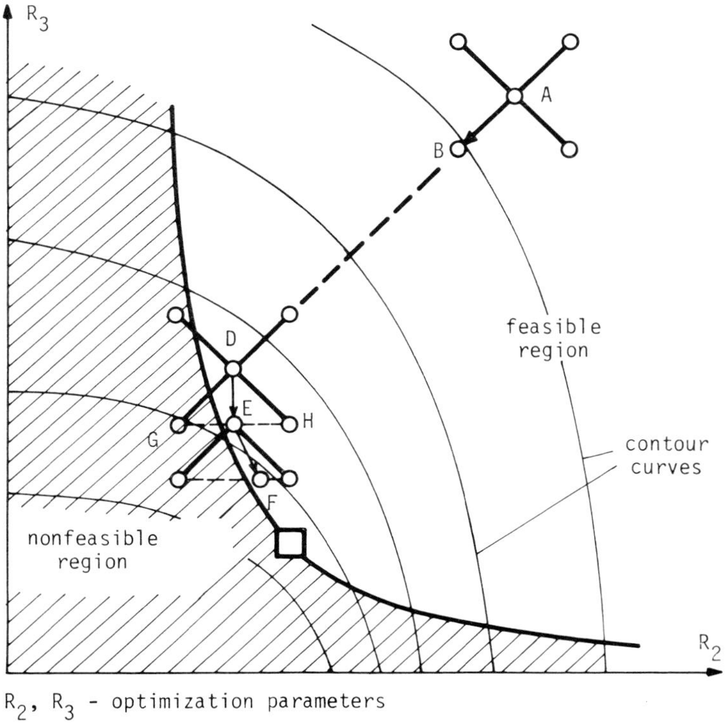

Fig. 4.9. Method of feasible directions for two-parameter optimization

different approach. We adopt (4.3.1) and (4.3.2) and consider H_{x2} and H_{x3} as two independent parameters. This two-parameter-optimization problem was solved by using the feasible directions method described above. The procedure started with initial values $H_{x2} = H_{x3} = 0.3$m and gave the optimal values: $H_{x2}^{opt} = 0.265$, $H_{x3}^{opt} = 0.15$. The other dimensions follow from (4.3.1) and (4.3.2): $H_{y2} = 0.177$, $h_{x2} = 0.255$, $h_{y2} = 0.159$, $H_{y3} = 0.1$, $h_{x3} = 0.128$, $h_{y3} = 0.09$. For these dimensions the energy consumption was E = 2171 J.

Para. 4.3. dealt with the determination of manipulator segments dimensions. The energy consumption was used as the optimality criterion. It was shown that this approach was very suitable and efficient for the determination of segments dimensions. If the number of independent parameters can be reduced to one, the binary search technique is suggested for the optimization procedure. If there are two or more independent variables then the feasible-directions method is suggested. These two optimization procedures have been programmed and incorporated in the program package. The organization of this package will be discussed later.

4.3.3. Standard form segments

In Para. 2.7. it was said that the dynamic analysis algorithm can operate with arbitrary form segments as well as with several so called standard forms. In discussions from Ch. 2 and Ch. 3 it was shown that the following data have to be known for each segment: vectors defining the position of joints with respect to segment center of gravity, mass of segment, tensor of inertia, cross-section inertial moment, cross-section moduli (for bending and torsion), etc. Thus, the fact that any change in segment dimensions produces a change in all these data can cause many problems in optimization procedure. But, we also note that some segment forms appear very often in manipulator devices. These are, for instance, circular and rectangular tubes. Hence, we call them standard forms and introduce additional subroutines which allow the algorithm to operate with such segments. If a segment is of a standard form then it is enough to prescribe its cross-section dimensions, its length and material. The subroutines mentioned compute then all the segment data needed (the data mentioned above). Thus, any change in segment dimensions produces automatically a change in the whole manipulator dynamics. This possibility is very useful in optimization procedures since some of these dimensions are often used as optimization parameters.

The first standard form is a circular tube. Its cross-section (Fig. 4.5b) is defined by two parameters: external radius R and the ratio $\psi = r/R$.

The second standard form is a rectangular tube (Fig. 4.5c). Cross section is defined by four parameters: H_x, $\psi_x = h_x/H_x$, H_y, $\psi_y = h_y/H_y$.

Finally, a user may define his own standard form.

4.4. Optimization Based on Working Speed Criterion

It the previous paragraph it was suggested that the dimensions of manipulator segments be chosen on the basis of energy consumption criterion. In fact such results are optimal only for the execution time considered. The results depend on the execution time but since they depend only a little we suggest that in practical problems higher speed should be taken and then the results can be considered to be really optimal. In such an optimization only the elastic deformation constraint and the stress constraint were considered.

In this paragraph we search for another optimum. The time (speed) criterion is used. The minimal execution time will be found and the corresponding optimal values of segments dimensions. These values represent now the optimum. In this optimization problem we have to take care of all three constraints (i), (ii), (iii) defined in 4.2.2. Since the constraint of the drives is considered too it is clear that the optimization is performed for some driving actuators chosen. Thus, the results depend on the actuators. This discussion is rather theoretical but it is useful for the good understanding of the later procedures: the choice of driving actuators and the choice of optimal reduction ratios for joint drives.

We shall now describe the procedure for time criterion minimization [2]. The procedure and the results will be illustrated by diagrams. These diagrams are given here to qualitatively demonstrate the form of some characteristics i.e. to illustrate some dependences. However, it should be said that all the diagrams from this paragraph are not imaginary. They have been taken from a real numerical example. A few detailed examples verifying this procedure and these diagrams can be found in [2].

Let us consider a manipulator and suppose that the number of independent parameters defining its segments can be reduced to one. Let R be that independent parameter (for instance it can be some cross-section radius). Suppose that the manipulator is driven by some known motors (all catalogue characteristics known) and that the reduction ratio in manipulator joints is N = 100. We now impose all three constraints: constraint of drives (i), constraint of stresses (ii), and constraint of elasticity (iii). If T is the execution time of the manipulation task given then the problem consists of finding the optimal value of R (i.e. R^{opt}) which allows the minimal execution time T along with satisfying the constraints imposed. The minimization procedure starts with some initial value of R. The dynamic analysis algorithm is carried out several times successively reducing the time T. This reduction is repeated until the drive constraint (i) is violated, i.e., until the driving motors cannot produce the working speed required. Thus, we obtain the minimal execution time for the value of R considered. Now, R is reduced and the procedure repated. In this way the curve $T_{min}(R)$ i.e. minimal execution time depending on R is obtained. This curve represents the constraint of the drives (i), and is shown in Fig. 4.10. The procedure of reducing R is repeated until the constraint (ii) or the constraint (iii) is violated i.e., until the stresses in segments exceed the permissible values or the elastic deformation becomes greater than it is permitted. Which of these two constraints will be relevant depends on the given values of permitted stress, safety coefficient, and the permitted elastic deformation. The curves in (R, T) plane representing the constraints ((ii) and (iii)) have already been discussed in 4.3.1. (Fig. 4.8). Now, if further reduction of R is required, T has to be increased. Consequently the absolute minimum of execution time (maximal speed) appears in the intersection of the constraint (i) and the constraint (ii) or (iii) Fig. 4.10. illustrates these results, i.e., the curve $T_{min}(R)$ and the constraints (ii) and (iii). The minimum appears in the point M designated by a square. The coordinates of this point are the optimal value of R and the absolute minimum of T, i.e., (R^{opt}, T^{abs}_{min}).

We have already seen (Para. 4.3. and Fig. 4.8) that the energy optima for different values of T lie on the curves (ii) or (iii). Since these curves depend on R only a little we can conclude from Fig. 4.10. that

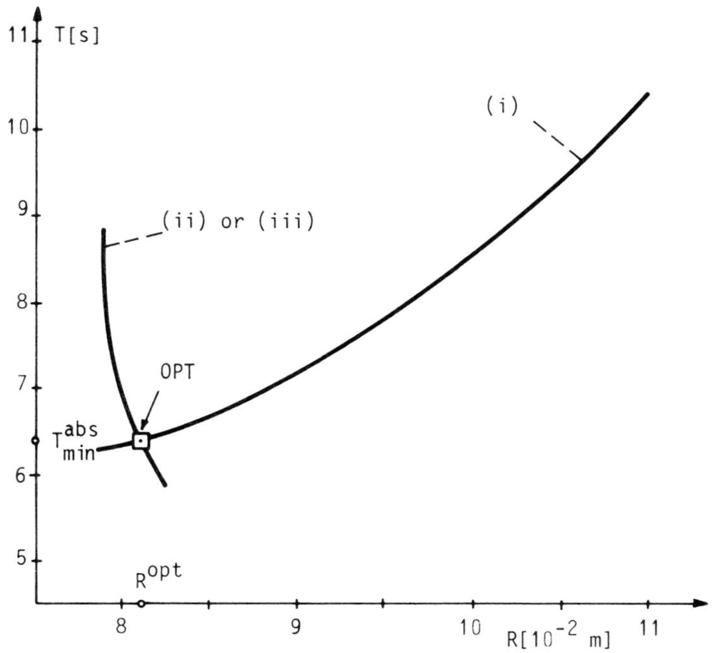

Fig. 4.10. Dependence $T_{min}(R)$

the energy optima and the speed optimum appear for the values of R which are very close to each other. This fact supports our suggestion that the segments dimensions should be obtained by energy criterion minimization.

Let us now remember that the minimization was performed for the reduction ratio N = 100 which need not be the best one. We shall now show that this value really is not the best one. The influence of the reduction ratio will be discussed and the optimal value determined which allows the shortest execution time i.e. the greatest operation speed. In Fig. 4.11. there are presented the curves $T_{min}(R)$ for different values of reduction ratio N. When N increases, the curve moves down then turns and moves up. Thus, a minimum appears. This turning happens earlier for smaller values of R, which means that the optimum of N depends on R. On the basis of the curves in Fig. 4.11. we can construct the curves representing the minimal execution time depending on the ratio N. It is a family of curves each of them corresponding to some value of the parameter R. Fig. 4.12. shows two of these curves corresponding to two

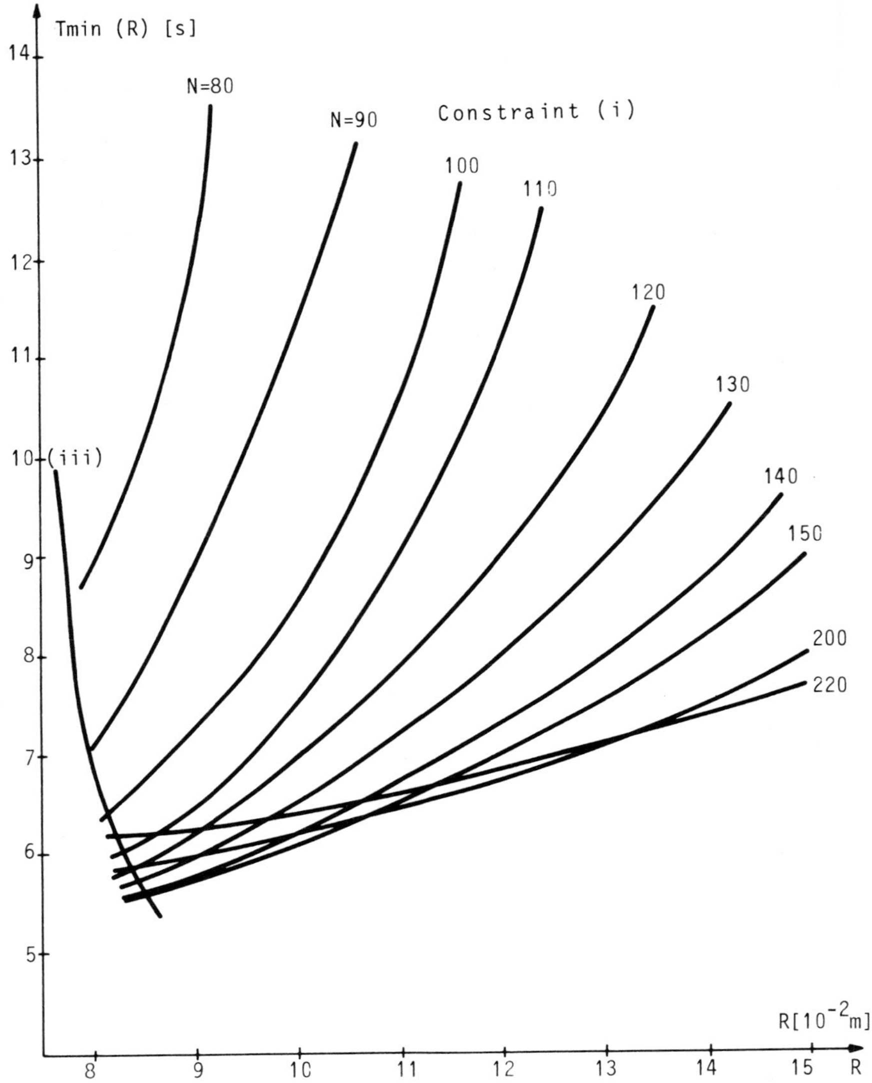

The constraint (i) is recorded for several values of reduction ratio N. Each curve (i) produces its corresponding optimum at the intersection with the curve (iii).

Fig. 4.11. Dependence $T_{min}(R)$ for different values of N

different values of R. Let us remember that R was said to be some parameter of manipulator segments, for instance a cross-section radius of a segment. Now, if the value of the parameter R is known then the corresponding optimal value of the reduction ratio can be found from the curves in Fig. 4.12. This family of curves depends only on the actuator chosen and so it is irrelevant how we obrain R: on the basis of energy criterion (as in 4.3), from experience, or in any other way. This is the basis for the procedure of choosing the reduction ratio. Optimal reduction ratio corresponds to the minimum of execution time. From Figs. 4.11. and 4.12. one can conclude that with the increase of the radius R the minimum of time with respect to ratio N moves towards greater values of N. This conclusion seams quite logical since it is expected that heavier segments (greater radius R) require greater value of reduction ratio.

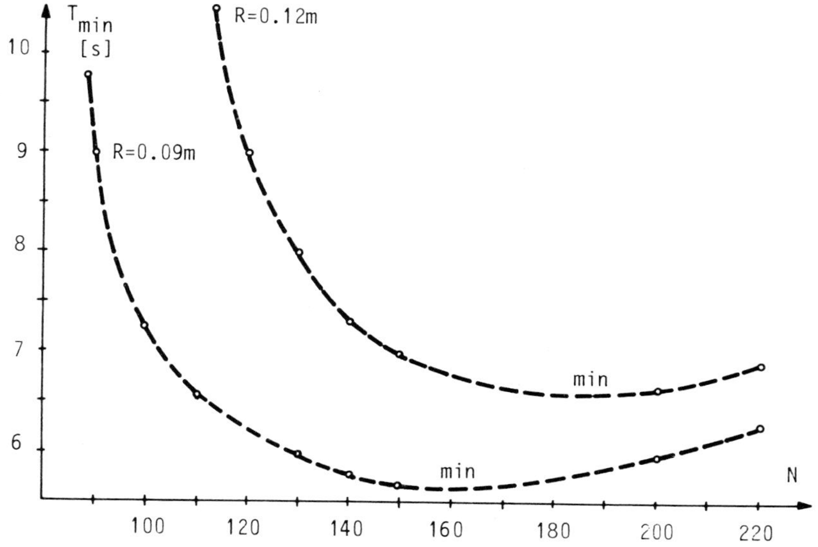

Fig. 4.12. Dependence of $T_{min}(N)$ for two values of R

4.5. Choice of Actuators and Reducers

This paragraph deals with the methodology of choosing the actuator units which will drive the manipulator being designed. We assume here that all the parameters of the manipulator segments have already been determined. That problem was discussed in 4.3. Thus the manipulator mechanism is completely known: the kinematical scheme, the form and the dimensions of segments, masses, etc. The next step is to select the actuators which will drive this mechanism. It should be pointed out that we shall not search for some completely optimal parameters of actuators but we shall only select the most suitable actuator from those offered by manufacturers. Together with the selection of actuators we determine the reduction ratios in manipulator joints. The reduction ratio is optimized, i.e. it is determined in such a way as to allow the greatest operation speed.

4.5.1. Selection of D.C. motors

Checking method. Let us consider a manipulator mechanism which is to be driven by permanent magnet D.C. motors. We face the problem of selecting suitable motors from the ones offered by manufacturers. In Ch. 2 we have described the dynamic analysis algorithm which includes the subroutine for testing the actuators (Para. 2.6.1). Thus, for any given actuator we can obtain the answer as to whether it suits our manipulator. So, the idea for selecting the motor is to check several of them and find the one satisfying all the tests imposed. However this simple idea appears to be more complex since the results of tests depend strongly on the reduction ratio. For instance, a motor which is quite suitable if used with some appropriate reducer can become unsuitable (negative tests) if the reducer is not correctly chosen. In order to clarify the influence of reduction ratio let us consider the main test of D.C. motors: the test of torque-speed characteristics. These characteristics can be expressed in the plane (n, P), where P is the necessary driving torque in the joint and $n = \frac{\dot{q}}{2\pi}\,60$ is the rotation speed in the joint (r.p.m. = revolutions per minute). Since P and $\dot{q}$ are obtained directly from the dynamic analysis of manipulation task given, the P-n characteristic does not depend on the reduction ratio. The torque-speed characteristics can also be expressed in the plane (n^m, P^m), where P^m is the necessary motor torque and n^m is the motor rotation speed. Since the motor acts through the reducer, it is

$$n^m = Nn, \qquad P^m = \frac{P}{\eta N} \tag{4.5.1}$$

where N is the reduction ratio and η is the mechanical efficiency of the reducer. The constraint is represented by the maximal characteristic of the motor $P^m_{max} - n^m$ i.e. maximal motor torque depending on rotation speed. It has already been said (Para. 2.6.1) that this maximal characteristic is a nearly straight line. Fig. 4.13a shows the necessary characteristic P-n and the maximal characteristic $P_{max} - n$ (both in (n, P) plane) for two values of the reduction ratio N. The necessary characteristic does not depend on N but the maximal characteristic does. Fig. 4.13b shows the same relations but in (n^m, P^m) plane. Now the necessary characteristic does depend on N and the maximal one does not. In both figures we see that for the ratio N_1 the test is positive, i.e., the motor is suitable, and for N_2 the constraint is violated and the test is negative. So, we conclude that motors and reducers should be determined together. But, while the choice of motors is simply a matter of selection from the ones offered, the reduction ratio is chosen in an optimal way enabling the maximal operation speed. In 4.4. we have explained the theoretical basis for the optimal choice of reduction ra-

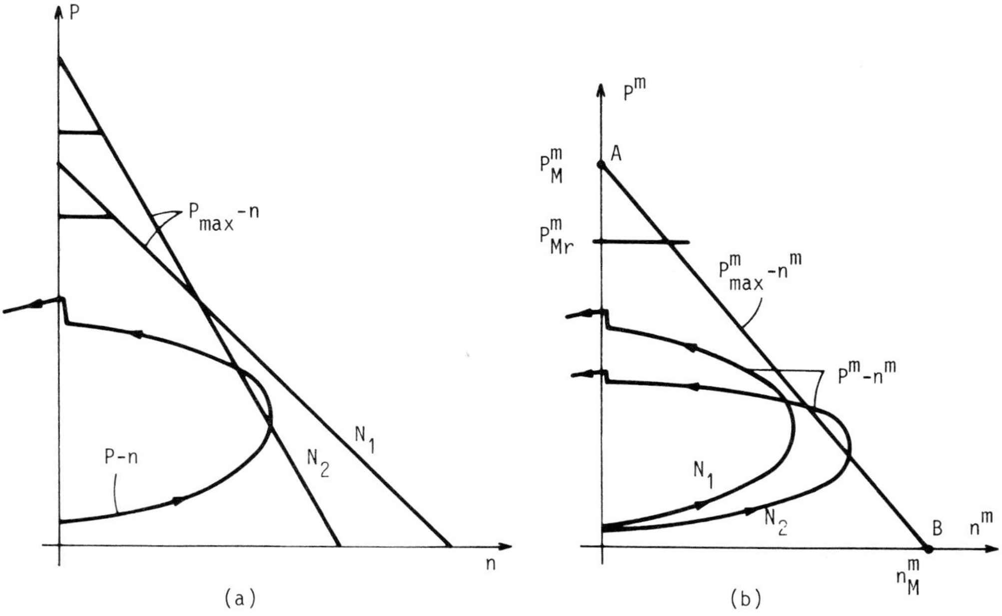

Fig. 4.13. Torque-speed characteristics depending on reduction ratio

tio. In Fig. 4.12. there are curves representing the minimal execution time depending on ratio N. Since we have designed the robotic mechanism before choosing the motors and reducers, the external radius R is known and we can find the optimal reduction ratio N^{opt} which provides the minimal execution time by using the corresponding curve from Fig. 4.12.

It should be pointed out that such a value of N^{opt} is optimal only for the motor considered since the curves 4.12. depend on the motor.

Let us now summarize the procedure of choosing D.C. motors and reducers. We first determine a set of motors which should be considered. The data about these motors are given to the computer. For each motor the computer finds the corresponding optimal reduction ratio, by checking several values of N until it finds the minimum of execution time. In that way the optimal ratio N^{opt} and minimal execution time T_{min} are obtained for each motor considered. We now choose the motor having the least T_{min}, i.e., allowing the greatest operation speed.

Let us now discuss the data about the motors which should be given to the computer. This depends on the tests imposed. If the power requirements are tested then the data about motor power are necessary. If the torque-speed characteristic is tested then it is necessary to give the maximal characteristic $P^m_{max} - n^m$. Since this characteristic is almost a straight line it is enough to give two values: P^m_M (point A in Fig. 4.13b) and n^m_M (point B). P^m_M is the maximal motor torque corresponding to $n^m \to 0$ and n^m_M is the maximal rotation speed corresponding to $P^m \to 0$. P^m_M is often called stall torque and n^m_M is called no-load speed. If expressed in terms of rad/s no-load speed is marked by ω^m_M. When the real value of maximal torque (P^m_{Mr}) is less than P^m_M it is then necessary to prescribe P^m_{Mr} (Fig. 4.13b). If power-dynamic power characteristic is tested then the computer needs P^m_M, rotor resistance and inertia moment (R_r and J_r), and the constants of torque and electromotor force (C_M, C_E). Finally, if overheating test is to be made then nominal motor torque P^m_{nom} is needed.

Systematic selection procedure. Although the checking method previously desribed can sometimes be very suitable, it does not allow a systematic choice of the most appropriate motor. Hence, besides this checking method we propose a systematic selection method based on power analysis. In 2.6.1. a test dealing with power and dynamic power was briefly presented. We now give a theoretical background [11] and further elaboration of this methodology.

Let us consider one joint of a manipulator having rotation speed $\dot{q}$ and torque P. The index indicating the number of joint is omitted for simplicity. The dynamic model of a D.C. motor driving the joint has the form (2.9.12) - (2.9.15). If the latter two equations of the matrix model (2.9.15) are written separately then:

$$J_r\ddot{q} = -B_c\dot{q} + C_M i_r - P, \tag{4.5.2}$$

and

$$L_r \frac{di_r}{dt} = -C_E\dot{q} - R_r i_r + u, \tag{4.5.3}$$

where J_r is rotor moment of inertia; L_r and R_r are its inductivity and resistance; i_r is rotor current; C_M and C_E are constants of torque and electromotor force; B_c is viscous friction coefficient; and u is control voltage. In these equations there is made no difference between the values which refer to motor shaft (before reducer) and the ones referring to joint shaft (after reducer). If the corresponding reducer has reduction ratio N and mechanical efficiency η then (4.5.2) and (4.5.3) become

$$J_r\ddot{q}^m = -B_c\dot{q}^m + C_M i_r - P^m \tag{4.5.4}$$

and

$$L_r \frac{di_r}{dt} = -C_E\dot{q}^m - R_r i_r + u \tag{4.5.5}$$

where the variables having the upper index m refer to motor shaft, and without this index the same variables refer to joint shaft. This means that

$$\dot{q} = \dot{q}^m/N \tag{4.5.6}$$

$$P = P^m N\eta \tag{4.5.7}$$

P is the torque on joint shaft which is necessary to produce the desired motion. P^m is the corresponding torque on the motor shaft. Thus P^m is the output torque of the motor. The term $C_M i_r$ appearing in (4.5.4) will be called the electric torque in order to distinguish it from output motor torque P^m. Thus, we introduce the electric torque as

$$P^e = C_M i_r \tag{4.5.8}$$

Now the equation (4.5.5) can be written in the form

$$u = C_E \dot{q}^m + \frac{R_r}{C_M} P^e + \frac{L_r}{C_M} \frac{dP^e}{dt} \tag{4.5.9}$$

If we use the maximal control voltage u_{max} then (4.5.9) can be transformed into

$$\frac{C_E}{u_{max}} \dot{q}^m + \frac{R_r}{C_M u_{max}} P^e + \frac{L_r}{C_M u_{max}} \dot{P}^e = 1$$

or

$$\frac{\dot{q}^m}{\omega_M^m} + \frac{P^e}{P_M^m} + T_e \frac{\dot{P}^e}{P_M^m} = 1 \tag{4.5.10}$$

where

$$\omega_M^m = \frac{u_{max}}{C_E} \quad \text{and} \quad P_M^m = \frac{u_{max} C_M}{R_r}$$

are no-load speed and stall torque, and $T_e = \frac{L_r}{R_r}$ is electromagnetic constant.

The equation (4.5.10) represents the extremal plane of motor capabilities (Fig. 4.14).

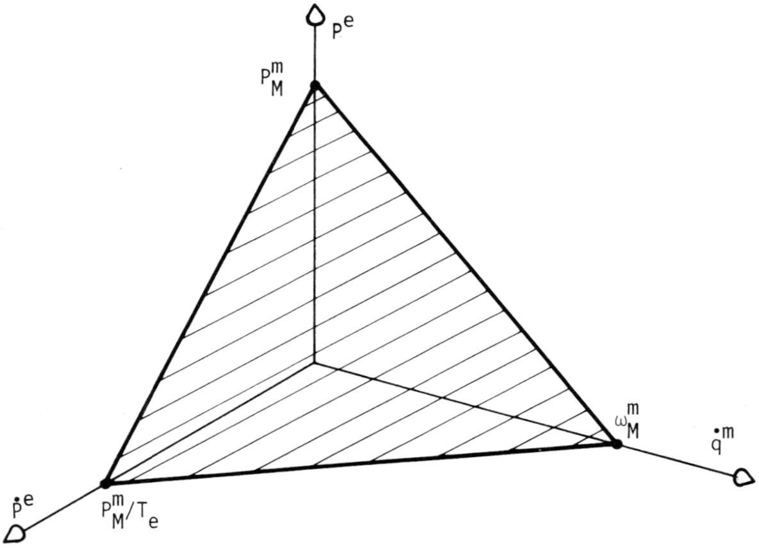

Fig. 4.14. Plane of maximal motor capabilities

If we neglect the rotor inductivity i.e. if T_e is small enough then (4.5.10) becomes

$$\frac{\dot{q}^m}{\omega_M^m} + \frac{P^e}{P_M^m} = 1 \tag{4.5.11}$$

or

$$\frac{N\dot{q}}{\omega_M^m} + \frac{P^e}{P_M^m} = 1 \tag{4.5.12}$$

and we can represent it by a two dimensional Fig. 4.15. By neglecting the rotor inductivity we have adopted the second order model of D.C. motor instead of the third order model (see Para 2.9.2).

The equation (4.5.11) represents the maximal torque characteristic. It is a straight line connecting ω_M^m and P_M^m. This characteristic was used in 2.6.1. for testing motor torque capabilities. Let us note that in such a test we neglected friction and rotor acceleration terms ($B_c\dot{q}^m$ and $J_r\ddot{q}^m$) since we used P_{max}^e instead of P_{max}^m. But, the power analysis we are now dealing with takes care of this acceleration effect.

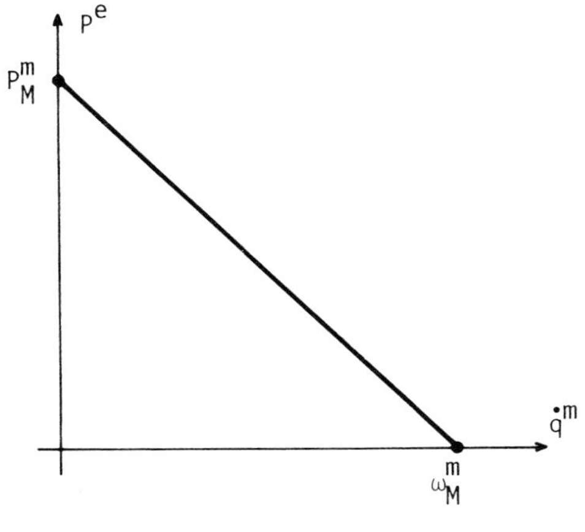

Fig. 4.15. Line of maximal motor capabilities

Let us now transform the equation (4.5.4). We first neglect friction B_c and then introduce (4.5.6) - (4.5.8) thus obtaining:

$$P^e N = J_r N^2 \ddot{q} + \frac{P}{\eta} \tag{4.5.13}$$

In the equations (4.5.12) and (4.5.13) some of the variables refer to joint shaft and some of them to motor shaft, depending on their nature. By elimination of P^e the equations (4.5.12) and (4.5.13) give the form

$$\frac{N\dot{q}}{\omega_M^m} + \frac{J_r N^2 \ddot{q} + P/\eta}{P_M^m N} = 1$$

i.e. a quadratic equation with respect to reduction ratio N:

$$\left(\frac{\dot{q}}{\omega_M^m} + \frac{J_r \ddot{q}}{P_M^m}\right) N^2 - N + \frac{P}{\eta P_M^m} = 0 \qquad (4.5.14)$$

By solving this equation for the unknown N we obtain the feasible interval for reduction ratio:

$$N_{1/2} = \left| \frac{1 \mp \sqrt{1 - 4\left(\frac{\dot{q}}{\omega_M^m} + \frac{J_r \ddot{q}}{P_M^m}\right) \frac{P}{\eta P_M^m}}}{2\left(\frac{\dot{q}}{\omega_M^m} + \frac{J_r \ddot{q}}{P_M^m}\right)} \right| \qquad (4.5.15)$$

The absolute value is introduced since u_{max} can be negative.

In order that there exists a reduction ratio N which allows the motor to produce the required motion with required joint torque, the value under the square root must be nonnegative i.e.

$$1 - 4\left(\frac{\dot{q}}{\omega_M^m} + \frac{J_r \ddot{q}}{P_M^m}\right) \frac{P}{\eta P_M^m} \geq 0 \qquad (4.5.16)$$

i.e.

$$\frac{P\dot{q}/\eta}{\omega_M^m P_M^m / 4} + \frac{P\ddot{q}/\eta}{(P_M^m)^2 / 4J_r} \leq 1 \qquad (4.5.17)$$

Let us introduce power $Q = P\dot{q}$ and dynamic power (or acceleration power) $DQ = P\ddot{q}$ with the corresponding values referring to motor shaft

$$Q^m = P\dot{q}/\eta \qquad \text{and} \qquad DQ^m = P\ddot{q}/\eta \qquad (4.5.18)$$

Let us also introduce maximal motor power

$$Q_M^m = P_M^m \, \omega_M^m / 4 \qquad (4.5.19)$$

and maximal dynamic power

$$DQ_M^m = (P_M^m)^2/4J_r = Q_M^m/T_{em} \tag{4.5.20}$$

where the electromechanical constant T_{em} is defined by

$$T_{em} = \frac{J_r R_r}{C_M C_E} \tag{4.5.21}$$

Thus (4.5.17) obtains the form:

$$\frac{Q^m}{Q_M^m} + \frac{DQ^m}{DQ_M^m} \leq 1 \tag{4.5.22}$$

If the friction is not neglected then the same consideration and the same expessions hold but whith the no-load speed which becomes

$$\omega_M^m = \frac{u_{max}}{C_E + \frac{R_r B_c}{C_M}}$$

and the electromechanical constant which becomes

$$T_{em} = \frac{J_r R_r}{C_M C_E + B_c R_r}$$

where B_c is viscous friction coefficient.

Let us note that the real value of maximal torque P_{Mr}^m is usually less than the theoretical value P_M^m and thus the real value of maximal dynamic power DQ_{Mr}^m is less then DQ_M^m.

In Para. 2.6.1. we introduced the power-dynamic power diagram (Figs. 2.45. and 4.16). Such a Q^m - DQ^m diagram represents a necessary characteristic holding for the manipulation task given. The characteristic is obtained by means of dynamic analysis algorithm. The equation (4.5.22) is a maximal characteristic since it has been derived for maximal control voltage. Thus, a straight line (4.5.22) defines the feasible domain i.e. it represents a constraint (Fig. 4.16). In this way we have proved the straight line constraint introduced in 2.6.1. The difference between DQ_M^m and DQ_{Mr}^m contributes to the restriction of feasible domain (Fig. 4.16)

The test is performed by comparing the necessary characteristic against the constraint. If the whole Q^m - DQ^m diagram lies within the feasible domain it means that the motor is chosen correctly. If the constraint is violated the motor is "incorrect". We note that this test does not depend on reduction ratio. Hence it is used to derive the procedure for motor selection. Reduction ratio is chosen after selecting the motor. But, one characteristic of reducer has to be known in advance, it is mechanical efficiency η since it is needed for calculation of Q^m, DQ^m by using (4.5.18). Thus, we adopt some value of η before reduction ratio (i.e. reducer) is chosen.

Let us now explain the motor selection procedure. The intention is to find the necessary value of maximal motor power (i.e. motor power capacity) Q^m_M and then select an appropriate motor. We want this capacity to be as small as possible. First, the Q^m - DQ^m diagram is computed and absolute values taken. In that way the whole diagram is placed in the first quadrant of the coordinate plane. Then, we analyse this diagram in order to find the so called "critical points". If there is only one critical point (point K in Fig. 4.17a) then the minimal feasible value of motor power capacity is

$$\min Q^m_M = Q^m_K + DQ^m_K \cdot T_{em} \tag{4.5.23}$$

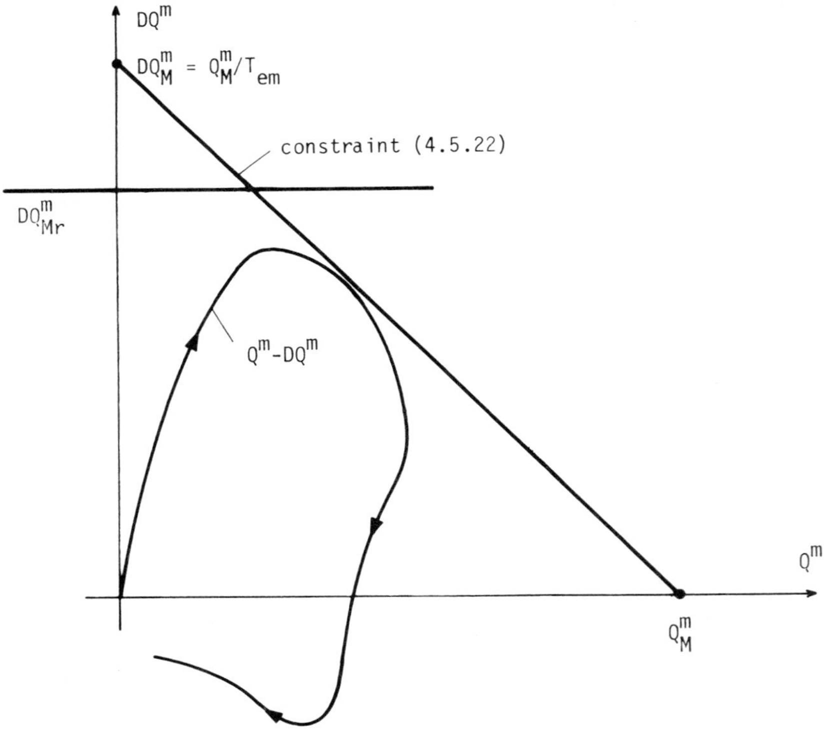

Fig. 4.16. Q^m - DQ^m diagram and a corresponding constraint

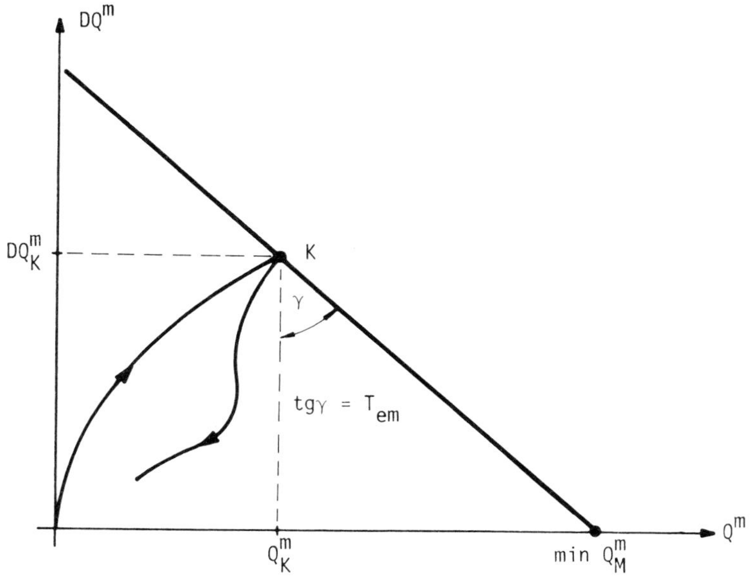

Fig. 4.17a. Motor selection - one critical point

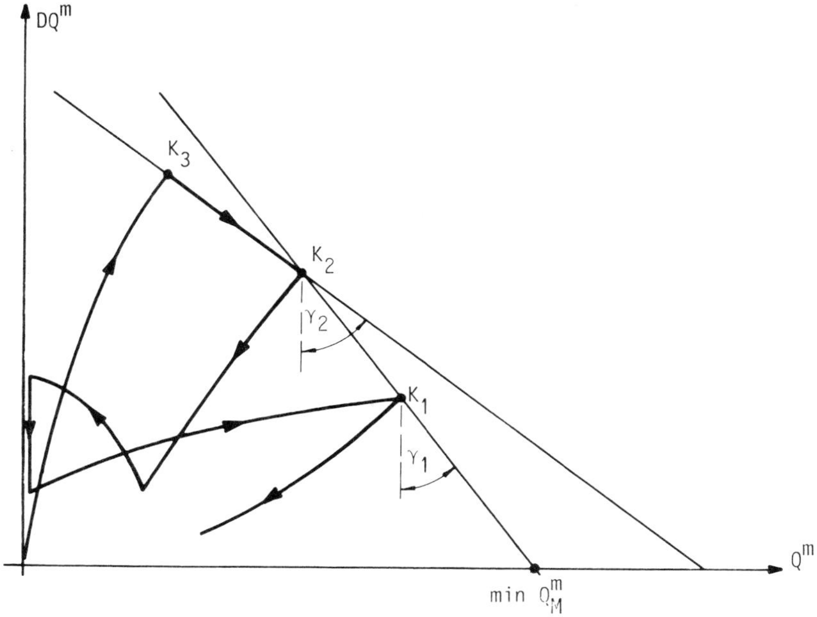

Fig. 4.17b. Motor selection - several critical points

It is clear that the optimal selection needs the minimal value of electromechanical constant T_{em}. The formula (4.5.23) is used for selecting the motor from catalogue. We notice that Q_K^m is the dominant term. Hence, we take into consideration the motors having the power capacity a bit greater than Q_K^m. Among these motors we search for the one having minimal T_{em}. Finally, we check whether the power capacity of the motor chosen Q_M^m and its constant T_{em} satisfy

$$Q_M^m \geq \min Q_M^m = Q_K^m + DQ_K^m T_{em}.$$

The case of several critical points is slightly more complicated. Let us explain it through the example in Fig. 4.17b. At the beginning K_1 is considered as a critical point. Minimal power capacity is then:

$$\min Q_M^m = Q_{K_1}^m + DQ_{K_1}^m \cdot T_{em} \tag{4.5.24}$$

But, T_{em} is now restricted by

$$T_{em} \leq tg\gamma_1 = \frac{Q_{K_1}^m - Q_{K_2}^m}{DQ_{K_2}^m - DQ_{K_1}^m}$$

If an appropriate motor cannot be found in catalog (e.g. γ_1 is so small that T_{em} is restricted too much, and a suitable value cannot be found) then power capacity should be increased. Now K_2 is a critical point. Minimal capacity is

$$\min Q_M^m = Q_{K_2}^m + DQ_{K_2}^m \cdot T_{em}$$

and T_{em} is restricted by

$$tg\gamma_1 \leq T_{em} \leq tg\gamma_2.$$

Such a procedure can be repeated several times until an appropriate motor is found. The subroutine for finding the critical points is derived and incorporated in the motor choice program. During this selection care should be taken about the difference between the real and the theoretical value of motor power capacity. For this reason, after motor selection the algorithm makes a Q^m - DQ^m test in order to check whether the motor is chosen correctly.

After selecting the motor, the next stage is the choice of reducer. The feasible interval $[N_1, N_2]$ for reduction ratio has been defined by (4.5.15). This interval depends on P, $\dot{q}$, $\ddot{q}$ and thus holds only for the time instant considered. Since P, $\dot{q}$ and $\ddot{q}$ are changing the interval is changing too. Let the interval corresponding to a time instant t_k be marked by $[N_1, N_2]^k$. We now make the intersection of all intervals $[N_1, N_2]^k$:

$$[N_1, N_2] = \bigcap_k [N_1, N_2]^k \tag{4.5.25}$$

and at the end of manipulation task we obtain an interval $[N_1, N_2]$ valid for all time instants (i.e. for the whole task).

The relation (4.5.15) did not take care of the difference between the real and theoretical value of maximal torque (P^m_{Mr} and P^m_M) and accordingly of the difference between DQ^m_{Mr} and DQ^m_M also. Taking these differences into consideration we obtain an additional relation restricting the feasible interval of reduction ratio:

$$P^m_{Mr} N = J_r N^2 \ddot{q} + \frac{P}{\eta} \tag{4.5.26}$$

or

$$N_{1/2} = \left| \frac{P^m_{Mr} \mp \sqrt{\left(P^m_{Mr}\right)^2 - 4J_r \ddot{q} P/\eta}}{2J_r \ddot{q}} \right| \tag{4.5.27}$$

This relation is used together with (4.5.15).

Now, the reduction ratio is chosen from the feasible interval $[N_1, N_2]$ defined by (4.5.25), (4.5.15), (4.5.27). The answer as to the value from the interval which should be adopted is not unique. One approach is the following. All the results we are dealing with hold for a given manipulation task or a set of test-tasks. Thus, the reduction ratio interval is obtained for some given execution time (i.e. given operation speed) and some given manipulator payload. If our manipulator is oriented to fast motion then we adopt some smaller value of reduction ratio still belonging to the feasible interval $[N_1, N_2]$. If we expect some greater payload then we adopt a larger value.

Another approach is to find the optimal value of reduction ratio as it was explained in checking selection method. This value allows the greatest operation speed but the procedure for reduction ratio selection is very extensive.

At the end it should be stressed that this systematic procedure for motor and reducer selection is completely automatized.

Let us note that the whole selection method is based on nominal dynamics. In this way we find a motor which is optimal for nonperturbed (nominal) motion. In order to be sure that it will operate correctly in perturbed motion we may choose it making a certain power reserve. After the motor is choosen and the control strategy defined we make a simulation of perturbed motion and check whether the motor can satisfy the tracking requirements. The same remark holds for the choice of hydraulic actuators.

4.5.2. Selection of hydraulic actuators

Hydraulic servo systems have found wide spread use in systems of industrial robotics, because they have some advantages over other control systems.

Servohydraulic drives can perform quickly and precisely with high repeatability when precise positional control is required. With hydraulic actuators translatory or rotary motions are easy to obtain without intermediate gear boxes and expensive ball screws. The high usable force density (system pressures up to 210 bar) allows large force for small component volume. The hydraulic fluid helps to cool the drive, whereas electric motors need artificial cooling or must be oversized for reasonable averhaul time intervals. The high acceleration available from electrohydraulic drives is combined with smooth positioning and high traverse speed (up to 1800 mm/s or 3 rad/s) which reduce to a minimum the idle time required to move from rest point to working regions.

An example of a hydraulic control scheme is given in Fig. 4.18.

In this paragraph we consider the problem of optimal choice of hydraulic actuators [9]. To be precise we have to choose a hydraulic cylinder, servovalve and reducer. It is the assembly marked by (*) in Fig. 4.18. It is the fact that there is usually no reducer combined with hydraulic cylinder but we still make a discussion including a reducer in order to obtain the general methodology which could easily be changed to suit the rotary actuators including high speed types. Let us also mention that the optimization understands the selection of optimal units from the ones offered by manyfacturers.

In Para. 2.6.2. we discussed testing of a hydraulic actuator. The static characteristics of actuator were used (Figs. 2.42a, 2.42b). Here, we propose a dynamic approach based on dynamic model of cylinder and servovalve. The dynamic model of servovalve-cylinder assembly was shortly explained in 2.9.2.

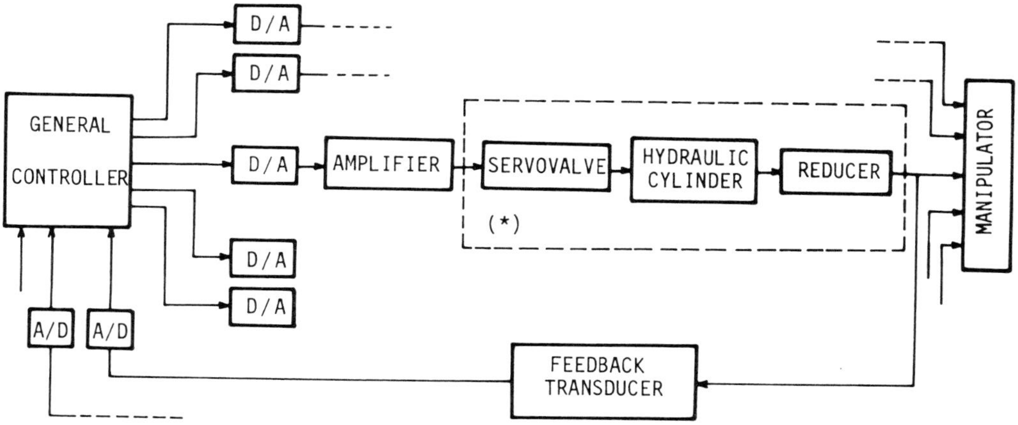

Fig. 4.18. Electrohydraulic control system for manipulation robots

Let us consider an electro hydraulic actuator shown in Fig. 4.19. In this figure as well as in the following expressions the joint index "i" is omitted.

Dynamic behaviour of such an actuator can be described in linear form by equations [12, 13]

$$V_s = A\dot{\ell}^m + c_\ell \Delta p + \frac{V}{4\beta}\Delta\dot{p} \qquad (4.5.28)$$

$$F^p = A\Delta p = m\ddot{\ell}^m + B_c\dot{\ell}^m + F^m \qquad (4.5.29)$$

$$V_s = V_s' - k_c\Delta p \qquad (4.5.30)$$

$$k_q i = c_1 V_s' \qquad (4.5.31)$$

V_s' and V_s are the theoretical and the real flow (m^3/s); A is the piston area (m^2); $c_\ell = (c_{i\ell}+c_{e\ell})/2$, where $c_{i\ell}$, $c_{e\ell}$ are the coefficients of internal and external leakage $(m^3/s/N/m^2)$; Δp is the difference in pressure between the two piston sides (N/m^2); V is the cylinder working capacity-volume; β is the compressibility coefficient of the fluid (N/m^2); m is the piston mass (kg); ℓ^m is the piston displacement (m);

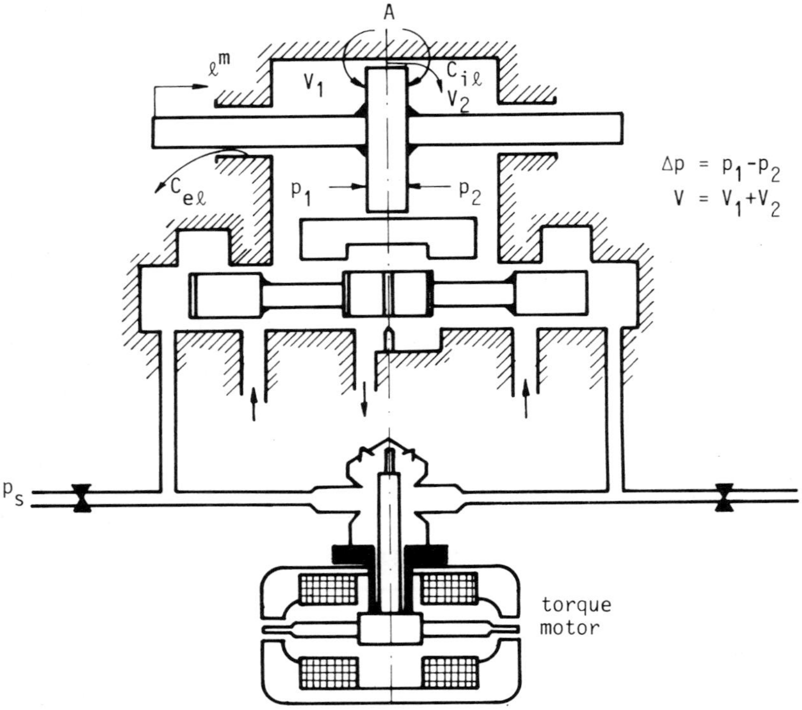

Fig. 4.19. Electrohydraulic actuator

B_c is the viscous friction coefficient (Ns/m); F^m is the force of external load (N); k_c is the slope of servovalve characteristic in the working point ($m^3/s/N/m^2$); k_q is the coefficient of servovalve ($m^3/s^3/mA$); c_1 is the characteristic of servovalve ($1/s^2$); i is the input current; F^p is the pressure force.

The system of equations (4.5.28) - (4.5.31) can be transformed into the form

$$F^p = A\Delta p = m\ddot{\ell}^m + B_c\dot{\ell}^m + F^m \tag{4.5.32}$$

$$k_q i = c_1 A\dot{\ell}^m + c_1(c_\ell + k_c)\Delta p + c_1 \frac{V}{4\beta} \Delta\dot{p} \tag{4.5.33}$$

If we apply the maximal input current i_{max}, then from (4.5.33) it follows that

$$\frac{\dot{\ell}^m}{v_M^m} + \frac{F^p}{F_M^m} + T_m \frac{\dot{F}^p}{F_M^m} = 1 \tag{4.5.34}$$

where

$$v_M^m = \frac{k_q i_{max}}{c_1 A} \tag{4.5.35}$$

is no-load speed,

$$F_M^m = \frac{A k_q i_{max}}{c_1 (c_\ell + k_c)} \tag{4.5.36}$$

is stall force, and

$$T_m = \frac{V}{4\beta(c_\ell + k_c)} \tag{4.5.37}$$

is a time constant.

Equation (4.5.34) defines the plane representing the constraint of maximal dynamic capabilities of actuator (Fig. 4.20)

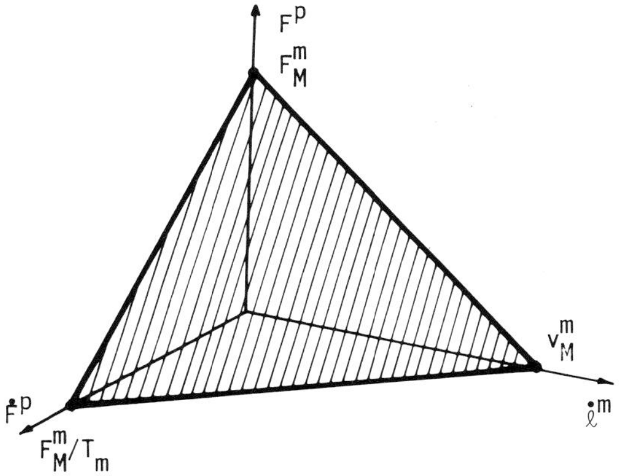

Fig. 4.20. Plane of maximal actuator capabilities

At the beginning we adopted the third order mathematical model of the electro hydraulic actuator (eqs. (4.5.28) - (4.5.31)) which led us to equation (4.5.34). Let us remember that something analogous was obtained for the third order model of the electrical D.C. motor (eq. (4.5.10) and Fig. 4.14). In the case of D.C. motor we immediately reduced the order of model (obtaining the second order model) by neglecting the time constant. In that way we obtained eq. (4.5.11) and the plane constraint (Fig. 4.15). In the case of electrohydraulic actuator and equation (4.5.34) we could do the same. But, we shall keep working with the third order model in order to locate the difficulties that will arise.

If we consider actuator-reducer assembly, then

$$\dot{\ell}^m = N\dot{\ell} \tag{4.5.38}$$

$$F^m = \frac{F}{N\eta} \tag{4.5.39}$$

where N is the reduction ratio and η is the mechanical efficiency of reducer. We also note that the velocity $\dot{\ell}^m$ holds on the motor side of the the reducer (i.e. before the reducer) and $\dot{\ell}$ holds outside of the reducer. The same consideration holds for F^m and F.

It should be said that no reducer is combined with hydraulic cylinders in real devices, but our discussion still includes a reducer in order to obtain a general methodology which could easily be changed to suit the rotary actuators including high speed types. If we wish our actuator to work without the reducer, then the feasible interval of reduction ratios must contain N=1.

When a cylinder actuator (translational motion) drives a translational joint, then F=P, where P is the drive in the joint. If the same actuator drives a rotational joint, there can exist a nonlinear dependence between the force F and torque P. In the same manner, for the first case it holds $\ell=q$, $\dot{\ell}=\dot{q}$, and in the second case there can exist a nonlinear dependence between the linear displacement ℓ and angle q, as well as between the linear velocity $\dot{\ell}$ and angular velocity $\dot{q}$. The calculation of these nonlinear dependences is incorporated in the algorithm for dynamic analysis.

Substituting (4.5.38), (4.5.39) into (4.5.32) one obtains

$$F^p = mN\ddot{\ell} + B_c N\dot{\ell} + \frac{F}{N\eta} \tag{4.5.40}$$

and, after substitution of (4.5.38) and (4.5.39) into (4.5.34) it follows that

$$\frac{N\dot{\ell}}{v_M^m} + \frac{mN\ddot{\ell} + B_c N\dot{\ell} + \frac{F}{N\eta}}{F_M^m} + \frac{T_m\left(mN\dddot{\ell} + B_c N\ddot{\ell} + \frac{\dot{F}}{N\eta}\right)}{F_M^m} = 1 \tag{4.5.41}$$

If we solve equation (4.5.41) for the unknown N, we obtain the feasible interval for reduction ratio:

$$N_{1/2} = \frac{1 \pm \sqrt{1-4\left[\frac{\dot{\ell}}{v_M^m}\left(1+\frac{v_M^m B_c}{F_M^m}\right) + \frac{m\ddot{\ell}}{F_M^m}\left(1+\frac{B_c T_m}{m}+\frac{T_m\dddot{\ell}}{\ddot{\ell}}\right)\right]\frac{F/\eta}{F_M^m}\left(1+\frac{T_m\dot{F}/\eta}{F/\eta}\right)}}{2\left[\frac{\dot{\ell}}{v_M^m}\left(1+\frac{v_M^m B_c}{F_M^m}\right) + \frac{m\ddot{\ell}}{F_M^m}\left(1+\frac{B_c T_m}{m}+\frac{T_m\dddot{\ell}}{\ddot{\ell}}\right)\right]} \tag{4.5.42}$$

In order that there exists a reduction ratio N which allows the actuator to produce the required motion with the required joint force, the value under the square root must be nonnegative i.e.:

$$4\left[\frac{\dot{\ell}}{v_M^m}\left(1+\frac{v_M^m B_c}{F_M^m}\right)+\frac{m\ddot{\ell}}{F_M^m}\left(1+\frac{B_c T_m}{m}+\frac{T_m \dddot{\ell}}{\ddot{\ell}}\right)\right]\frac{F/\eta}{F_M^m}\left(1+\frac{T_m \dot{F}/\eta}{F/\eta}\right) \leqslant 1 \qquad (4.5.43)$$

where from

$$\frac{(F/\eta+T_m\dot{F}/\eta)\dot{\ell}}{\dfrac{(F_M^m)^2}{4}\dfrac{v_M^m}{B_c v_M^m+F_M^m}}+\frac{(F/\eta+T_m\dot{F}/\eta)\ddot{\ell}}{\dfrac{(F_M^m)^2}{4}\dfrac{1}{m+T_m B_c}}+\frac{(F/\eta+T_m\dot{F}/\eta)\dddot{\ell}}{\dfrac{(F_M^m)^2}{4}\dfrac{1}{mT_m}} \leqslant 1 \qquad (4.5.44)$$

Let us introduce power

$$Q = (F + T_m\dot{F})\dot{\ell}, \qquad (4.5.45)$$

acceleration power (dynamic power)

$$DQ = (F+T_m\dot{F})\ddot{\ell}, \qquad (4.5.46)$$

and jerk power

$$JQ = (F+T_m\dot{F})\dddot{\ell}. \qquad (4.5.47)$$

These variables hold for the mechanism i.e. outside the reducer (after it). The corresponding values holding for the actuator (before the reducer) are

$$Q^m = (F/\eta + T_m\dot{F}/\eta)\dot{\ell} \qquad (4.5.48)$$

$$DQ^m = (F/\eta + T_m\dot{F}/\eta)\ddot{\ell} \qquad (4.5.49)$$

$$JQ^m = (F/\eta + T_m\dot{F}/\eta)\dddot{\ell}. \qquad (4.5.50)$$

Maximal values for these variables can be obtained from (4.5.44).

$$Q_M^m = \frac{(F_M^m)^2}{4}\frac{v_M^m}{B_c v_M^m + F_M^m} \qquad (4.5.51)$$

$$DQ_M^m = \frac{(F_M^m)^2}{4}\frac{1}{m+T_m B_c} \qquad (4.5.52)$$

$$JQ^m_M = \frac{(F^m_M)^2}{4} \frac{1}{mT_m} \tag{4.5.53}$$

Let us note that the value $(F+T_m\dot{F})\dot{\ell}$ can be called power only conditionally since the exact notion of power understands only $F\dot{\ell}$.

Now, equation (4.5.44) can be written in the form

$$\frac{Q^m}{Q^m_M} + \frac{DQ^m}{DQ^m_M} + \frac{JQ^m}{JQ^m_M} = 1 \tag{4.5.54}$$

and it defines a constraint plane as shown in Fig. 4.21. The three dimensional diagram Q^m-DQ^m-JQ^m obtained by means of dynamic analysis algorithm must be within the feasible domain constrained by plane (4.5.54) (Fig. 4.21). If this is the case, then the actuator is correct and if there is a violation of the constraint then the actuator is not correct.

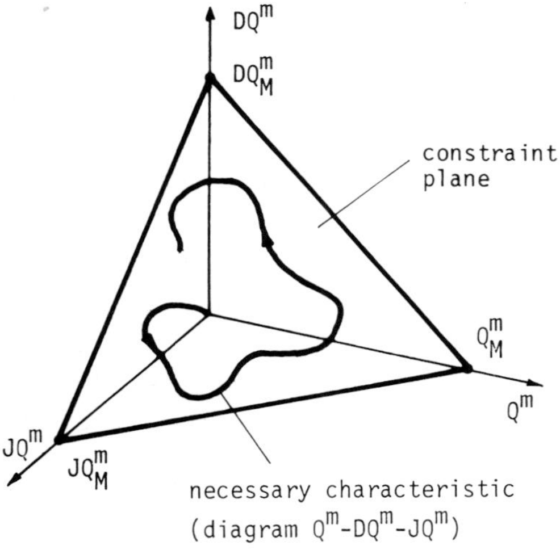

Fig. 4.21. Q^m-DQ^m-JQ^m diagram and the corresponding constraint

This whole consideration is based on the third order mathematical model of electrohydraulic actuator. The problem we face is the fact that the Q^m-DQ^m-JQ^m diagram depends on the actuator due to the time constant T_m. Hence, it is not possible to calculate the diagram first and then choose the appropriate actuator. Another disadvantage lies in three-dimensionality of the problem. For this reasons we reduce the order of mathematical model by adopting the assumption that the compressibility coefficient is large enough that the time constant T_m can be neglected. In this way we adopt the second order model.

Now (4.5.48) - (4.5.53) becomes

$$Q^m = F\dot{\ell}/\eta \tag{4.5.55}$$

$$DQ^m = F\ddot{\ell}/\eta \tag{4.5.56}$$

$$JQ^m = F\dddot{\ell}/\eta \tag{4.5.57}$$

and

$$Q_M^m = \frac{(F_M^m)^2}{4} \frac{v_M^m}{B_C v_M^m + F_M^m} \tag{4.5.58}$$

$$DQ_M^m = \frac{(F_M^m)^2}{4m} \tag{4.5.59}$$

$$JQ_M^m \to \infty \tag{4.5.60}$$

constraint equation (4.5.54) obtains the form

$$\frac{Q^m}{Q_M^m} + \frac{DQ^m}{DQ_M^m} = 1 \tag{4.5.61}$$

and this two dimensional constraint can be represented by a line in Fig. 4.22.

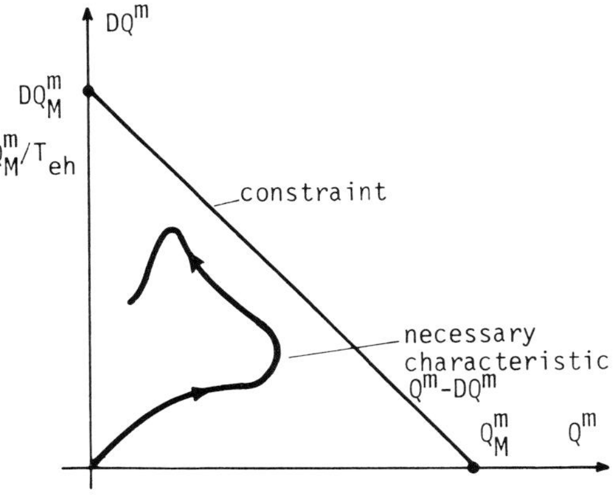

Fig. 4.22. Q^m-DQ^m diagram and the corresponding constraint

The real Q^m-DQ^m diagram obtained by means of dynamic analysis algorithm now does not depend on the actuator, but it is still necessary to know the mechanical efficiency η of reducer. Hence we adopt some value of η before the reducer is chosen.

Now we can first calculate the Q^m-DQ^m diagram and then search for an appropriate actuator and reducer.

For the second order model, the feasible interval for reduction ratio (eq. (4.5.42)) becomes:

$$N_{1/2} = \frac{1 \pm \sqrt{1 - 4\left[\frac{\dot{\ell}}{v_M^m}\left(1 + \frac{v_M^m B_C}{F_M^m}\right) + \frac{m\ddot{\ell}}{F_M^m}\right]\frac{F/\eta}{F_M^m}}}{2\left[\frac{\dot{\ell}}{v_M^m}\left(1 + \frac{v_M^m B_C}{F_M^m}\right) + \frac{m\ddot{\ell}}{F_M^m}\right]} \tag{4.5.62}$$

We introduce the hydraulic time constant

$$T_{eh} = \frac{Q_M^m}{DQ_M^m} = \frac{mv_M^m}{F_M^m + v_M^m B_c} = \frac{m(c_\ell + k_c)}{A^2 + B_c(c_\ell + k_c)} \qquad (4.5.63)$$

and transform (4.5.61) into the form:

$$\frac{Q^m}{Q_M^m} + \frac{DQ^m}{DQ_M^m} = \frac{Q^m}{Q_M^m} + \frac{DQ^m}{Q_M^m/T_{eh}} = 1 \qquad (4.5.64)$$

Now, the choice of hydraulic actuator (cylinder - servovalve assembly) can be made in the same manner as it was done for D.C. electromotors. We can use the method of "critical points" and obtain the minimal power capacity

$$\min Q_M^m = Q_K^m + DQ_K^m T_{eh} \qquad (4.5.65)$$

where Q_K^m, DQ_K^m correspond to critical points. Also there is a constraint on the value of T_{eh}. A more detailed explanation of critical points method is given in 4.5.1.

The problem of hydraulic actuator choice is more complex than the choice of D.C. motor since we have to choose two components: a cylinder and a servovalve. The problem is that neither the power capacity of such an assembly nor its hydraulic time constant T_{eh} can be found in the catalog. Thus, in each step of the selection procedure we have to compute these values on the basis of catalog data of the two components (cylinder and servovalve). For this calculation relations (4.5.65), (4.5.63), (4.5.51), (4.5.35), and (4.5.36) are used. This procedure although quite possible and sometimes powerful, requires an extensive calculation to be done by the user, especially because it is a two--component optimization. We can reduce the problem by adopting one component and conducting the procedure for the best choice of the other.

Afer choosing the complete actuator we calculate the feasible interval for reduction ratio (4.5.62).

We now suggest a practical methodology for the choice of cylinder-servovalve assembly. The cylinder is chosen first. The choice is performed on the basis of static characteristics of cylinder. Some ideas for such a choice are explained in Para. 2.6.2. Let them be repeated. The dynamic analysis algorithm computes the time histories of force F^m, speed $\dot{\ell}^m$, and acceleration $\ddot{\ell}^m$. The maximal value (critical value) of force F_c^m is found. This value is tested against the catalog values for different actuators:

$$F_c^m \leq F_n^p = A\Delta p_n \tag{4.5.66}$$

where the index "n" indicates the nominal values of force and pressure. The piston area can be found from (4.5.67):

$$A \geq \frac{F_c^m}{\Delta p_n} \tag{4.5.67}$$

The nominal value of pressure is sometimes given in catalogues (see MOOG catalogues). If not, then we adopt

$$\Delta p_n = \frac{2}{3} p_s \tag{4.5.68}$$

where p_s is pressure supply, i order to obtain maximal output power [19, 20].

Relation (4.5.67) does not take care of cylinder dynamics. If we can estimate the mass of piston in advance (cylinder has not yet been chosen), then we can include its dynamics and obtain

$$A \geq \frac{(F^m + m\ddot{\ell}^m + B_c\dot{\ell}^m)_c}{\Delta p^n} \tag{4.5.69}$$

where the index "c" indicates critical value (maximal value).

We have defined the critical force as the maximal value. But if maximal values appear for a short time interval, then a critical force can be smaller than the maximal one in order that the cylinder operates for a longer time period near its nominal state.

Now, the cylinder is chosen on the basis of (4.5.67) or (4.5.69).

The next stage is the choice of servovalve which is optimal for the cylinder adopted. We start from relation (4.5.65) i.e.

$$Q_M^m \geq Q_K^m + DQ_K^m T_{eh} \tag{4.5.70}$$

Substituting from (4.5.51), (4.5.35), (4.5.36) and from (4.5.63), relation (4.5.70) becomes:

$$\frac{A^2 k_q^2 i_{max}^2}{4c_1^2(c_\ell+k_c)[B_c(c_\ell+k_c)+A^2]} \geqslant Q_K^m + DQ_K^m \frac{m(c_\ell+k_c)}{A^2+B_c(c_\ell+k_c)} \tag{4.5.71}$$

The catalogs give the values of k_q and k_c in terms of nominal flow V_{sn}, maximal current i_{max}, nominal pressure Δp_n, and pressure supply p_s.

$$\frac{k_q}{c_1} = \frac{V_{sn}}{i_{max}}, \qquad k_c = \frac{V_{sn}}{2(p_s-\Delta p_n)} \tag{4.5.72}$$

Introducing (4.5.72) into (4.5.71) the quadratic inequality with respect to V_{sn} is obtained

$$\begin{aligned} &V_{sn}^2\left[A^2 4\left(p_s-\Delta p_n\right)^2-4\left(mDQ_K^m+B_cQ_K^m\right)\right] - \\ &\quad - V_{sn}\left[8\left(p_s-\Delta p_n\right)\left(2mDQ_K^m c_\ell+2B_cQ_K^m c_\ell-Q_K^m A^2\right)\right] - \\ &\quad - 16c_\ell\left(p_s-\Delta p_n\right)^2\left(mDQ_K^m c_\ell+B_cQ_K^m c_\ell+A^2Q_K^m\right)\geqslant 0 \end{aligned} \tag{4.5.73}$$

We now choose a servovalve having minimal flow V_{sn}. It can be done by solving inequality (4.5.73). This procedure can be simplified if we neglect the leakage (c_ℓ). Then (4.5.73) reduces to

$$V_{sn} \geqslant \frac{2A^2(p_s-\Delta p_n)Q_K^m}{A^2(p_s-\Delta p_n)^2-(mDQ_K^m+B_cQ_K^m)} \tag{4.5.74}$$

and we choose the servovalve having V_{sn} minimal but still satisfying (4.5.74).

It has been said that this choice methodology can easily be adapted to suit rotary motors. For such motors relation (4.5.74) has the form

$$V_{sn} \geqslant \frac{2\left(\frac{V'}{2\pi}\right)^2(p_s-\Delta p_n)Q_K^m}{\left(\frac{V'}{2\pi}\right)^2(p_s-\Delta p_n)^2-(IDQ_K^m+B_cQ_K^m)} \tag{4.5.75}$$

where V' is unit volume (per one full revolution) and I is the moment of inertia.

4.5.3. Some remarks on actuators selection procedure

In this paragraph we have already explained the choice of driving actuators based on knowledge of the complete manipulator dynamics (4.5.2, 4.5.3). But, let us note that the calculation of manipulator dynamics requires all masses to be known. On the other hand masses of actuators and reducers are not known since the selection is performed after calculating the dynamics. Thus, a circle is closed. In order to solve this problem we first note that torque in some joint S_i does not depend on the mass concentrated in that joint (corresponding motor or reducer) but on masses placed from S_i to the gripper only (Fig. 4.23). Thus, we can, make a recursive procedure for motor selection. It starts with the last joint (S_n). The torque in this joint does not depend on any motor mass and so we calculate the exact dynamics of that joint and select the motor and reducer.

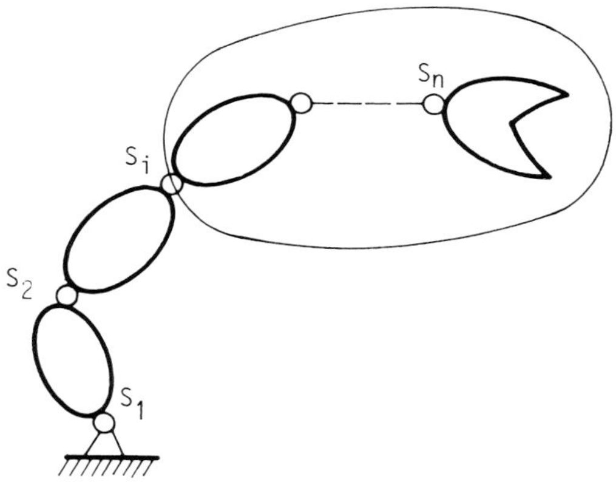

Fig. 4.23. Motor selection

Now, the mass concentrated in joint S_n is known (motor and/or reducer) and we can calculate the exact torque in joint S_{n-1} which enables us to select the motor for that joint. Then, we calculate the exact dynamic of joint S_{n-2}, select the motor, and repeat this procedure until we reach S_1. This recursive procedure is shown in Fig. 4.24. There is something which should be keept in mind when making such procedure. With manipulator devices motors may but need not be placed in manipulator joints. They can be placed in manipulator column and their torques transported to corresponding joints. The same holds for reducers.

Let as make an important remark. The choice procedures proposed are exact i.e. they are based on the complete dynamic models of both mechanism and actuators. Such a procedure is expecially suitable for high speed manipulation where dynamic effects of robot are important. But, the methodology can also be used for low speed manipulation since it is general.

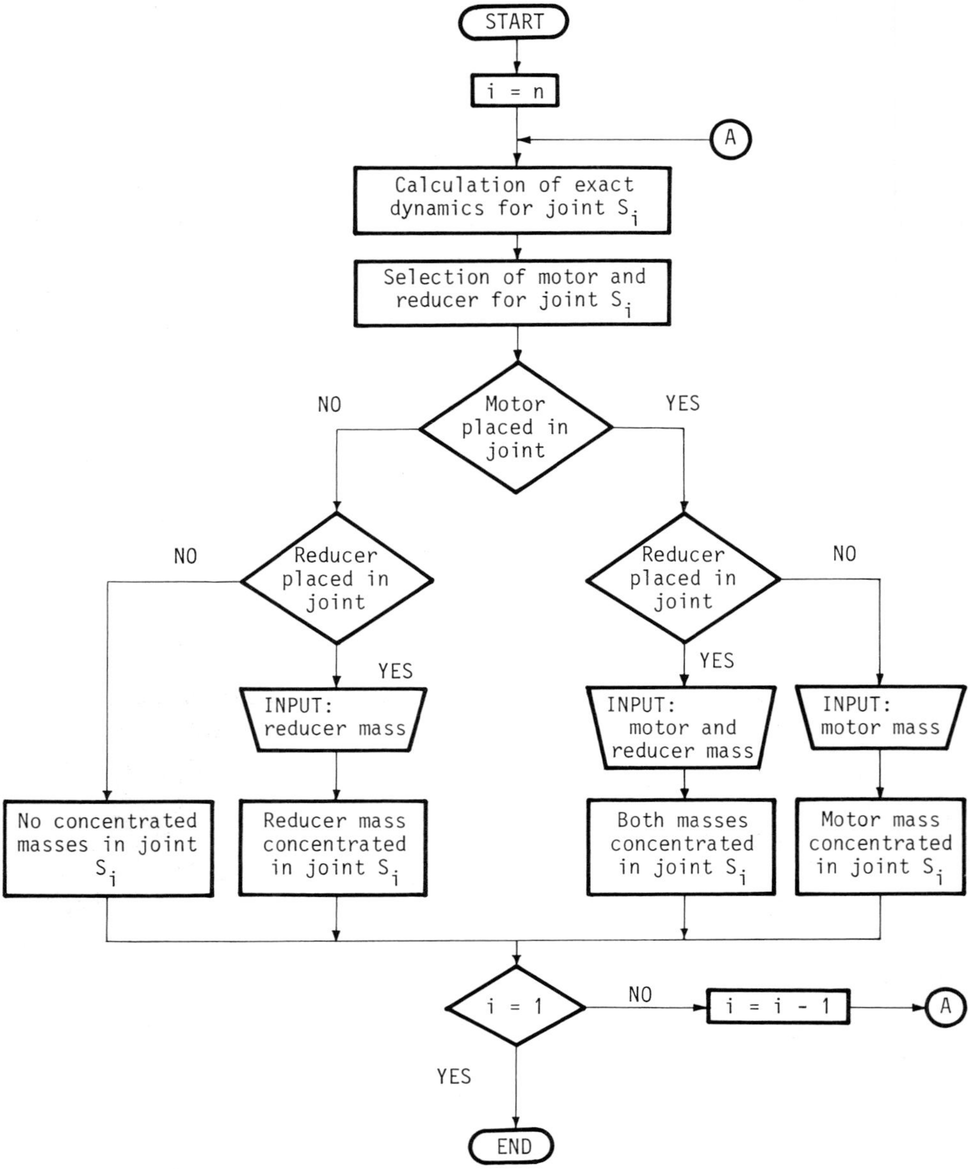

Fig. 4.24. Recursive procedure for selection of manipulator motors

4.5.4. Examples

Example 1

We consider the arthropoid manipulator UMS-E (Fig. 4.25) designed to carry loads up to 60 kg. It is a redundant manipulator having nine degrees of freedom. It is originally designed for hydraulic drives. Here we are going to choose electromotors and thus the example is rather hypothetic.

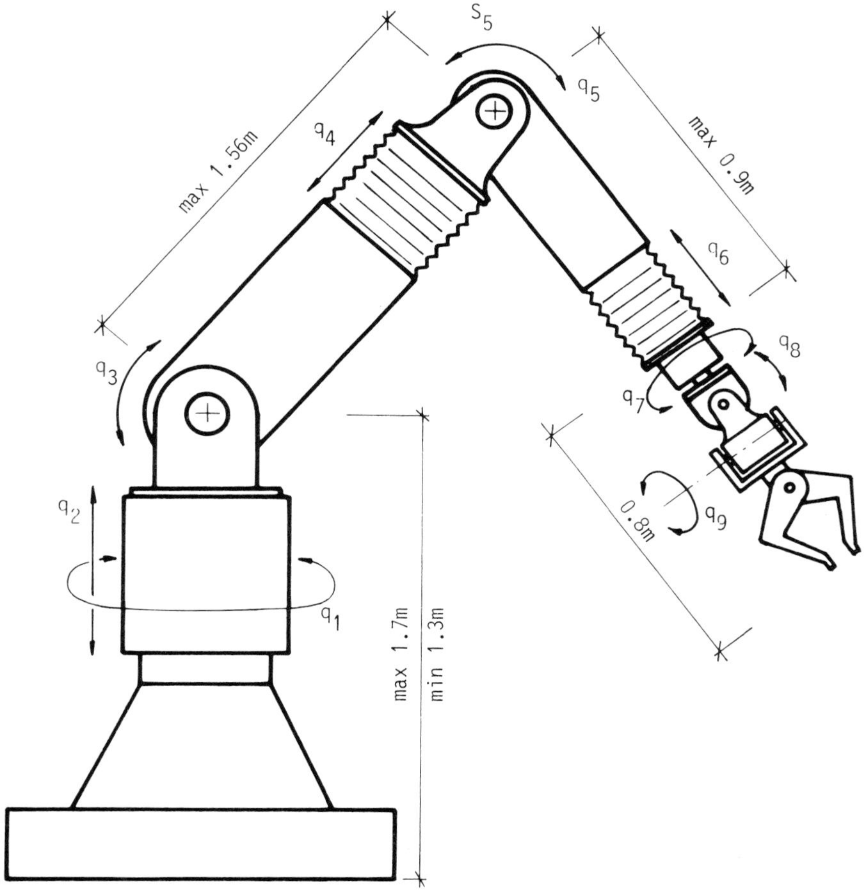

Fig. 4.25. A 9 d.o.f. manipulator UMS-E

The manipulation task consists in carrying the payload of 60 kg mass along the trajectory $A_oA_1A_2$ (Fig. 4.26) keeping all the time its initial orientation. On each straight-line part of this trajectory the velocity profile is trapezoidal with the acceleration time being 20% of transport time (t_a=0,2T). Let the mean velocity be $v_m(A_o \to A_1) = v_m(A_1 \to A_2)$= = 1 m/s. This request gives $T(A_o \to A_1)$ = 1.65 s, t_a = 0.23 s and $T(A_1 \to A_2)$= = 2.2 s, t_a=0.44 s. We assume that, in this test task, the whole motion is produced by the rotational joints i.e. the translational coordinates are kept constant.

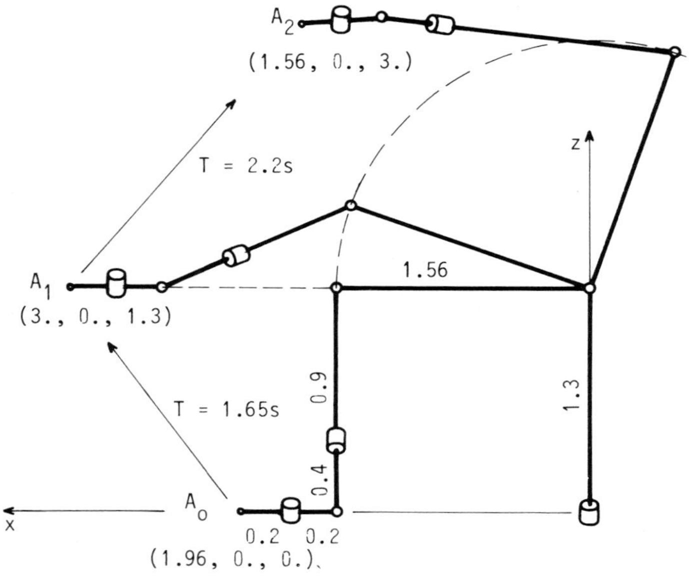

Fig. 4.26. Manipulation task

We now demonstrate the choice of D.C. electromotor which have to drive the joint S_5. Let us note that in this example there is no compensation for this joint.

In Fig. 4.27 there is a copy of the computer list showing the dialog between the designer and the computer in the proces of motor choice.

The motor data have been taken from the catalog of AXEM D.C. motors produced by CEM-PARVEX (type of motor is MC 24 P).

We seen that there is a rather narrow interval of feasible values for the reduction ratio. It is a result of the optimal choice having almost no reserve. For more powerful motors this interval is wider. For instance the 3.3 kW motor (type MC 23 S) produces the feasible interval [256.4, 335.7].

```
=== SELECTION OF MOTOR FOR JOINT No. 5 ===

YOU CAN TAKE CHECKING METHOD (C)
 OR SYSTEMATIC METHOD (S)
 Choose one and type [C/S]:S

-------------------------
SYSTEMATIC CHOICE BEGINS
-------------------------

You should give the value of reducer mechanical
efficiency (estimated value). Type the value [F10.5]:0.8

* 10 CRITICAL POINTS

--OPTION No. 1
MOTOR POWER CAPACITY SHOULD BE AT LEAST:
      A + B * Tem
where A=2823.340    ,    B=3265.555
and Tem is el.mech. time const. which is constrained by:
      Tem <    0.18869

Can you find an appropriate motor [Y/N]:Y

You should give the motor data. Type the following:
  maximal input voltage {SI} [F10.5]:140.
  const. Cm and Ce of torque and el.mot. force {SI}
  [2F10.5]:0.435,0.435
  rotor resistance and moment of inertia {SI}
  [2F10.5]:0.285,0.0032
  stall torque (real value) {SI} [F10.5]:11.
  power capacity {SI} [F10.5]:3000.

Q-DQ TEST IS POSITIVE AND YOUR MOTOR IS O.K.
Is the motor placed in joint [Y/N]:N

----------------------------------------
REDUCER SELECTION BEGINS (Joint No. 5)
----------------------------------------

FEASIBLE INTERVAL FOR REDUCTION RATIO IS:
     [  288.2  ,   324.5  ]

Select reducer and type the reduction ratio [F10.5]:300.

P-n TEST IS POSITIVE AND MOTOR REDUCER ASSEMBLY IS O.K.
Is reducer placed in joint [Y/N]:Y
Type the reducer mass {SI} [F10.5]:9.
```

Fig. 4.27. Example of choice comunication

Example 2

This example illustrates the procedure for the choice of electro hydraulic actuators. We consider the cylindrical six d.o.f. manipulator UMS-3B (Fig. 4.28)

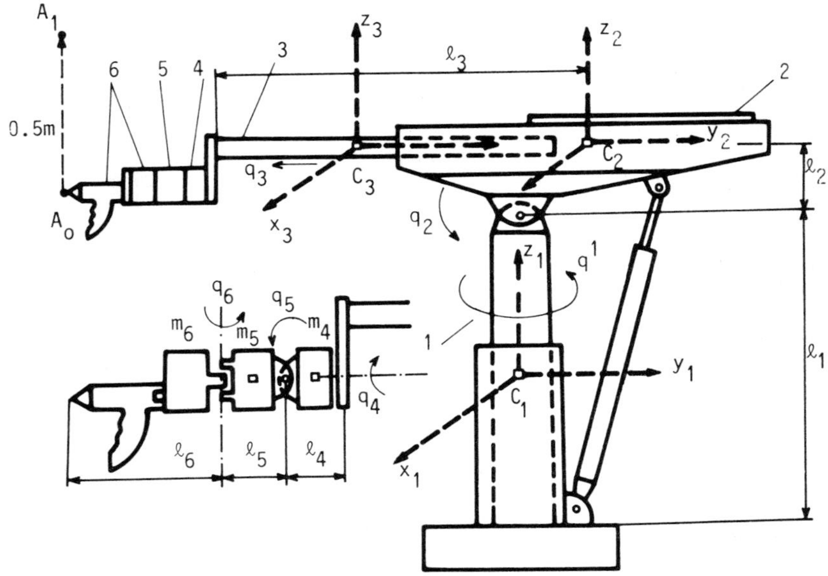

Members	1	2	3	4,5	6
Mass m_i (kg)	61.47	27.8	22.34	2.33	30.
Length ℓ_i (m)	1.2.	0.142	1.14	0.14	0.26
J_{ix} (kgm^2)	0	2.98	1.21	0.004	0.007
J_{iy} (kgm^2)	0	-	-	0.004	0.009
J_{iz} (kgm^2)	0.322	3.701	1.21	0.004	0.009

Fig. 4.28. Manipulation system UMS-3B

Let us discuss the choice of electrohydraulic actuator for the second joint (degree of freedom q_2). It is a rotational joint driven by a dylinder with translatory piston motion. Nonlinear dependences between the linear and angular variables are solved in the dynamic analysis algorithm.

The manipulation task consists in moving the spraygun from the point A_o to the point A_1 (Fig. 4.28) keeping all the time its initial orientation in space. The motion is performed with trapezoidal velocity profile with the transport time T = 0.625 s and the acceleration time t_a = 0.2 T = 0.125 s. It means that the mean velocity is v_m = 0.8 m/s and the maximal velocity is v_{max} = 1 m/s.

The dynamic analysis algorithm gave the F^m-$\dot{\ell}^m$ diagram and the Q^m-DQ^m diagram as shown in Figs. 4.29 and 4.30.

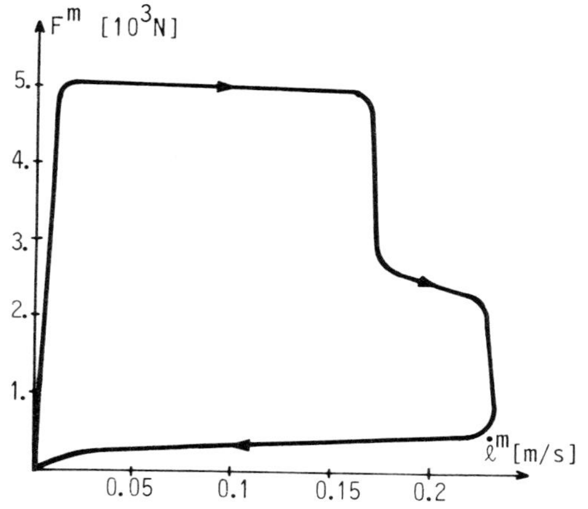

Fig. 4.29. F^m-$\dot{\ell}^m$ diagram for joint S_2

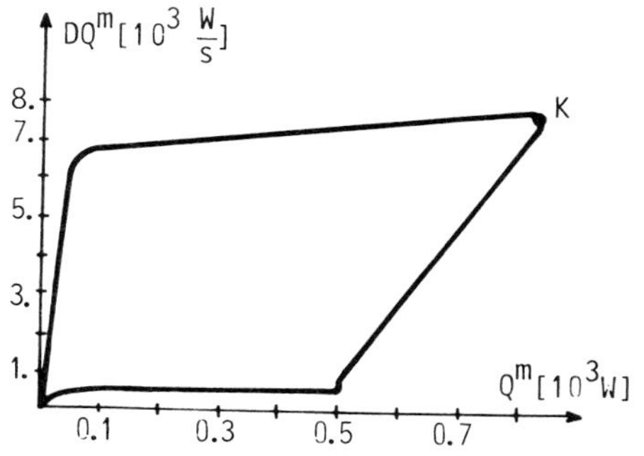

Fig. 4.30. Q^m-DQ^m diagram for joint S_2

From the F^m-$\dot{\ell}^m$ diagram the maximal value of force is F_c^m = 5163.3 N. We decided to use MOOG servovalve having Δp_n = 70 bar. Now, from (4.5.67) it follows that $A \geq 7.38 \cdot 10^{-4}$ m^2. We adopt hydraulic cylinder KNAPP Z.9.32/20 having

$$A = 8.04 \cdot 10^{-4}\ m^2, \qquad V = 0.4824 \cdot 10^{-3}\ m^3$$

$$m = 1.85\ kg, \qquad B_c = 30\ \frac{Ns}{m}$$

Let us now choose an appropriate servovalve. From the Q^m-DQ^m diagram the critical point is: Q_K^m = 833.03 W, DQ_K^m = 7622.6 W/s. From (4.5.74) it follows that $V_{sn} > 22.3$ litre/min. We adopt MOOG servovalve 76-125 having

$$V_{sn} = 28.5 \text{ litre/min}$$

$$i_{max} = 15 \text{ mA (parallel coils)}$$

$$\frac{k_q}{c_1} = 3.17 \cdot 10^{-5} \frac{m^3/s}{mA}, \qquad k_c = 6.786 \cdot 10^{-11} \frac{m^5}{Ns}$$

From (4.5.35) and (4.5.36) one obtains no-load speed and stall force: $v_M^m = 0.59$ m/s, $F_M^m = 5633.7$ N.

Now, the feasible interval for reduction ratio is $[N_1, N_2] = [1.63, 1.93]$ which means that the cylinder-valve assembly adopted cannot perform the given task without the reducer. Since we want our actuator to operate without the reducer we choose a stronger assembly: hydraulic cylinder KNAPP Z 9.40/25 having $A = 12.6 \cdot 10^{-4}$ m^2, m = 2.65 kg, and MOOG servovalve 76-103 having $k_q/c_1 = 4.22 \cdot 10^{-5}$ m^3/s/mA, $k_c = 9.04 \cdot 10^{-11}$ m^5/Ns. For this assembly the interval for reduction ratio is $[N_1, N_2] = [0.76, 1.88]$ and it contains N=1. Thus this assembly can operate without the reducer.

Example 3

This example illustrates the choice of rotary hydraulic actuator. Let us consider again the manipulator UMS-E (Fig. 4.25) and the manipulation task from Example 1. We choose again the actuator for the joint S_5 but this time it is a rotary hydraulic actuator.

The maximal torque is P_{5c} = 2507.4 Nm. By adopting Δp_n = 70 bar we obtain

$$\frac{V'}{2\pi} = \frac{P_{5c}}{\Delta p_n} = 376 \cdot 10^{-6} \text{ m}^3$$

We now adopt the motor PRVA PETOLETKA*) 145-7000 having $V'/2\pi$ = $= 433.12 \cdot 10^{-6}$ m^3.

*) The largest Yugoslav industry for hydraulics and pneumatics.

The critical point is $Q_K^m = 2258.7$ W, $DQ_K^m = 2612.5$ W/s. Relation (4.5.75) gives $V_{sn} \geq 79.79$ litre/min. We adopt MOOG servovalve 72-154 having

$$V_{sn} = 95 \text{ litre/min}$$

$$i_{max} = 200 \text{ mA}$$

$$\frac{k_q}{c_1} = 7.9 \cdot 10^{-6} \frac{m^3/s}{mA}, \qquad k_c = 2.257 \cdot 10^{-10} \frac{m^5}{Ns}$$

For the reduction ratio we obtain $[N_1, N_2] = [0.8, 3.22]$ which means that the assembly adopted can operate without the reducer. The maximal output power, according to (4.5.51), is $Q_M^m = 2670$ W which is close to the value obtained for D.C. motor in Example 1.

Remark. The choice methodology proposed is intended for fast dynamics, but the examples are taken from up-to-date applications with the velocities which we can meet in practice.

4.6. Organization of the CAD Program Package

This paragraph presents the general organization of the program package for computer-aided design of manipulation robots. It consists of four parallel branches each of them representing one procedure which can be solved. This general organization is shown in Fig. 4.31. At the beginning we choose one of procedures offered. These are:

Dynamic analysis. This algorithm calculates all relevant dynamic characteristics, and also performs tests of these characteristics. It includes selection of print-out where we define the characteristics we are interested in. We also make selection of tests from those offered. This dynamic analysis branch is shown in Fig. 4.32.

Interactive design. This branch is shown in Fig. 4.33. A detailed scheme is given in Paragraph 4.1. This part of program package contains some communication subroutine allowing a "comfortable" interactive work with the CAD algorithm. It is very suitable to use the possibility of operation with standard form segments (see 4.3.3).

Determination of optimal segments dimensions. This algorithm represents

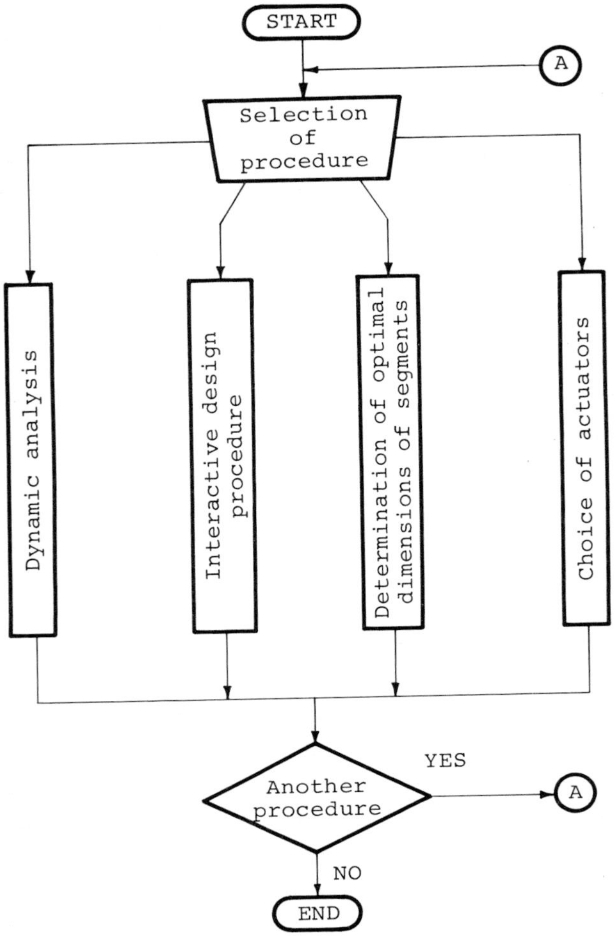

Fig. 4.31. General scheme of CAD program package

the design of manipulator mechanical part. It has been discussed in Paragraph 4.3. The program realization of such an optimal design procedure can be described by a general scheme given in Fig. 4.34. As has already been said, some manipulator parameters follow directly from constructive solutions adopted and some of them do not. Hence, we find and indicate free parameters. Further, it is necessary to reduce the number of independent parameters by adopting some values (on the basis of experience) or introducing some constant radii among them. After definition of constraints optimization can be performed.

Using standard form segments is strongly recommended. The corresponding

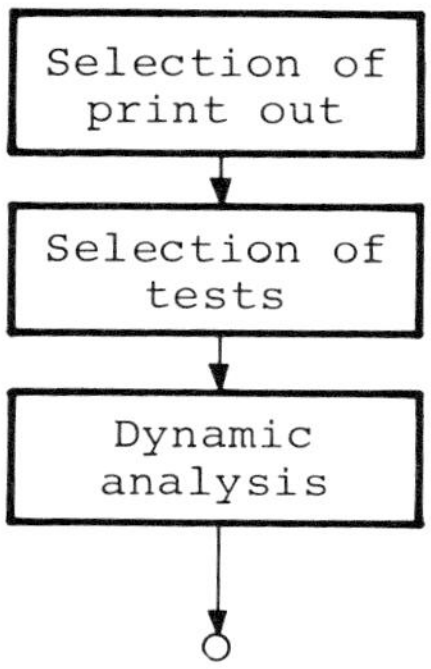

Fig. 4.32. Dynamic analysis branch

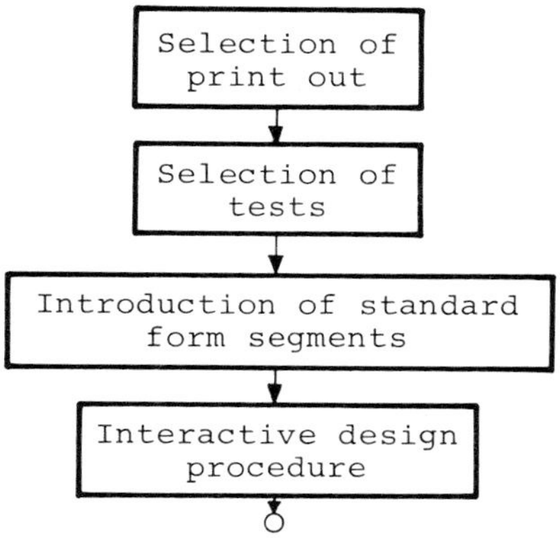

Fig. 4.33. Interactive design branch

subroutines are included in the program package.

Let us note a few things. A systematic design of the whole manipulator device (mechanical part and actuators) requires first the choice of segments dimensions and after that selection of appropriate driving actuators. It is clear that the masses of actuators and reducers influence the manipulator dynamics and hence they have to be taken into consideration when choosing segments dimensions. But, the exact masses are not known since actuators choice follows after segments design. Thus, in segments optimization procedure we adopt certain values of motors and reducers masses (for safety reasons we usually adopt some larger values).

Choice of actuators and reducers. This problem has been discussed in Paragraph 4.5. In general, the procedure can be presented in the form shown in Fig. 4.35. It includes the definition of joints for which we want to choose actuators.

After each procedure a user is asked if he wants another procedure (e.g. after segments design we may proceed to actuators choice).

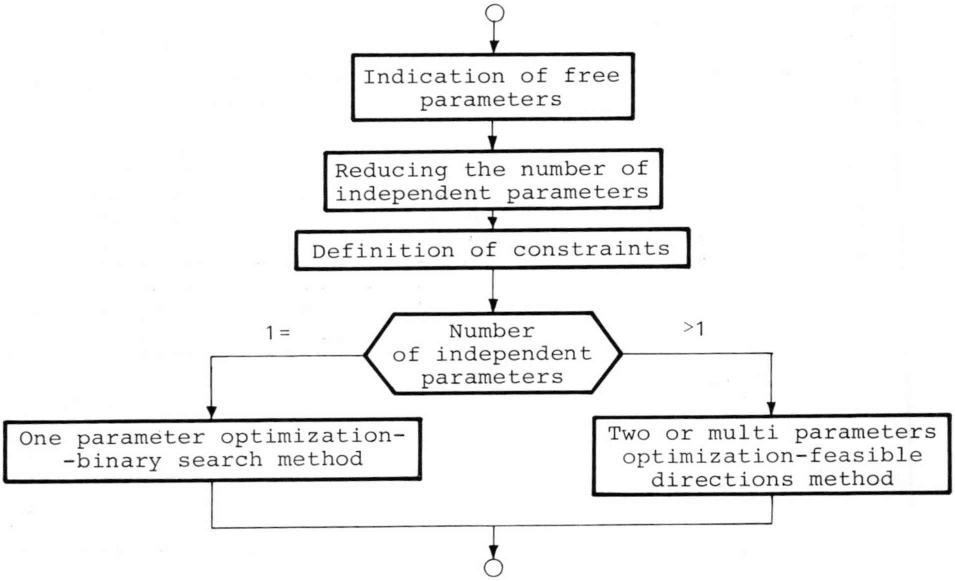

Fig. 4.34. Optimization of segments dimensions

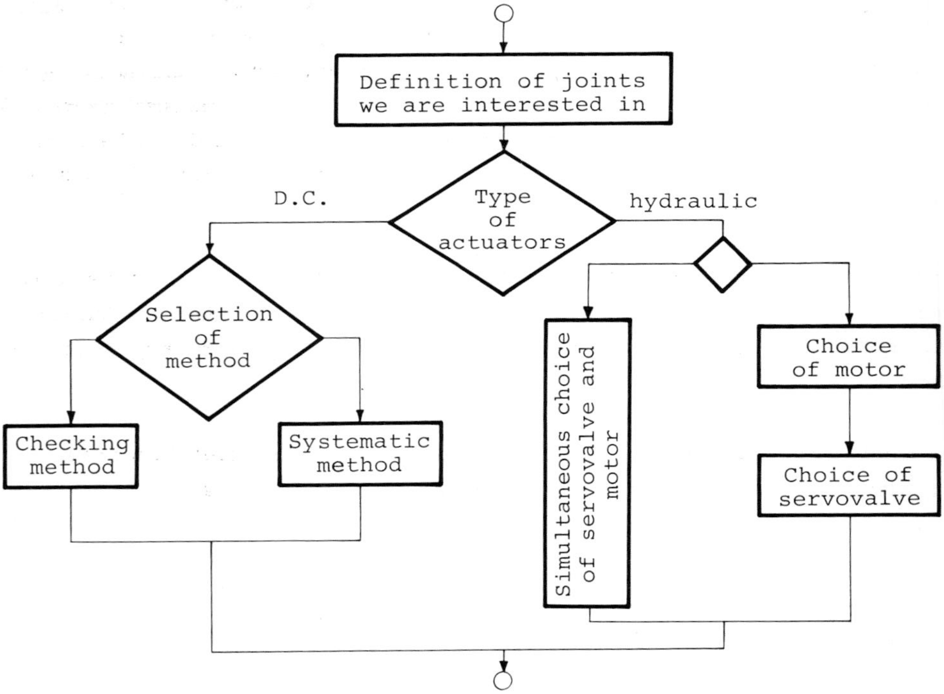

Fig. 4.35. Scheme of program branch for the choice of actuators

References

[1] Vukobratović M., Legged Locomotion Robots and Anthropomorphic Mechanisms, Monograph, 1975, Institute "M.Pupin", Beograd, Yugoslavia.

[2] Vukobratović M., Potkonjak V., Dynamics of Manipulation Robots: Theory and Application, Monograph, Springer-Verlag, 1982.

[3] Vukobratović M., Potkonjak V., "Contribution to Computer-Aided Design of Industrial Manipulators Using Their Dynamic Properties", Journal of IFToMM, Mechanisms and Machine Theory, Vol. 16, No. 2, 1982.

[4] Truckenbrodt A., "Dynamics and Control Methods for Moving Flexible Structures and Their Application to Industrial Robots", Proc. of 5th World Congres of Theory of Machines and Mechanisms, publ. ASME, 1979.

[5] Sunada H.W., Dynamic Analysis of Flexible Spatial Mechanisms and Robotic Manipulators", Ph. D. Thesis, University of California, Los Angelos, 1981.

[6] Potkonjak V., Vukobratović M., "Interactive Procedure for Computer-Aided Design of Industrial Robot Mechanism", Proc. 13th International Symposium on Industrial Robots, Chicago, 1983.

[7] Vukobratović M., Potkonjak V., Hristić D., "Contribution to Computer-Aided Design of Industrial Manipulators", Proc. 11th ISIR, Tokyo, 1981.

[8] Vukobratović M., Potkonjak V., Hristić D., "Computer Procedure for the Design of Industrial Manipulators", Proc. Fourth RO.MAN.SY-81, Warsaw, Poland, 1981.

[9] Vukobratović M., Potkonjak V., Katić D., "Computer-Aided Choice of Electrohydraulic Actuators for Manipulation Robots", Proc. 14th ISIR, Gothenburg, 1984.

[10] Gottfried B.S., Weisman J., Introduction to Optimization Theory, Prentice-Hall, 1973.

[11] Petrov B.E., Polkovnikov A.A., Dynamic Possibilities of Servo Systems, (in Russian), Energy, Moscow, 1976.

[12] Merrit E.H., Hydraulic Control Systems, John Wiley & Sons, 1967.

[13] Guillon M., Hydraulic Servo Systems, Butterworths, 1969.

Subject Index